Green Chemistry

This book is a complete guide to the science and practice of sustainable chemistry. It explains how chemicals can be designed, produced, and used in ways that reduce harm to humans and the environment. The book combines fundamental concepts, practical applications, and industrial aspects, making it a valuable resource for students, researchers, and professionals worldwide.

Key Features and Unique Selling Points:

- **Foundations of Green Chemistry:** Explains the main principles and importance of sustainable chemical practices.
- **Practical Processes and Techniques:** Shows how to design, synthesize, and analyze chemicals efficiently and safely.
- **Industrial Applications:** Provides examples of recycling, re-refining, and circular economy strategies in industry.
- **Bio-based and Renewable Methods:** Covers production of chemicals and biofuels using microbes and waste resources.
- **Waste Management and Disposal:** Offers guidance on safe handling, disposal, and remediation of chemical wastes.
- **Economic and Technological Aspects:** Explains financial and technological factors that support green chemistry adoption.

This book not only provides a clear understanding of green chemistry but also demonstrates how to apply these principles to real-world challenges. Readers will gain the knowledge and tools needed to design chemical processes that are efficient, cost-effective, and environmentally responsible. Whether you are a student learning about sustainable chemistry, a researcher exploring new technologies, or a professional implementing green practices in the chemical industry, this book is an essential resource for advancing a cleaner, safer, and greener chemical future.

Green Chemistry

Principles, Processes, and Pathways

Edited by
Avneet Kaur, Sukhvinder Singh Purewal,
Marcelo Franco, Ashutosh Kumar Rai,
and Muhammad Irfan

CRC Press
Taylor & Francis Group
Boca Raton London New York

CRC Press is an imprint of the
Taylor & Francis Group, an informa business

Designed cover image: Shutterstock

First edition published 2027
by CRC Press
2385 NW Executive Center Drive, Suite 320, Boca Raton FL 33431

and by CRC Press
4 Park Square, Milton Park, Abingdon, Oxon, OX14 4RN

CRC Press is an imprint of Taylor & Francis Group, LLC

ISBN: 978-1-041-07857-9 (hbk)
ISBN: 978-1-041-07858-6 (pbk)
ISBN: 978-1-003-64258-9 (ebk)

DOI: 10.1201/9781003642589

Typeset in Times
by KnowledgeWorks Global Ltd.

Contents

About the Editors

Dr. Avneet Kaur is currently working as Assistant Professor, Department of Chemistry, University Institute of Sciences (UIS)-University Centre for Research and Development (UCRD), Chandigarh University, Mohali, Punjab, India. During her PhD tenure, she has published five research papers in journals of high repute. Her research interest in PhD was primarily focused on two different areas. First, the thermophysical and theoretical investigation of binary mixtures of choline-based ionic liquids with hydrogen bond donor solvents, e.g., water, straight chain alcohols, and alkane diols, which have the efficacy to be used at industrial level. Second is the fabrication of choline-based ionic liquid-assisted nanostructures for electrochemical sensing of pollutants which ultimately projects them as smart design for various engineering and industrial processes. Having authored more than 35 research papers/book chapters with publishers of international repute, Dr. Kaur has demonstrated her expertise by publishing two edited books: 1) *Nanotechnology in the Food Industry: Applications, Recent Trends, and Future Perspectives* and 2) *Edible Insects: Handling Methods, Nutritional Value, and Commercial Uses* with CRC Press, Taylor & Francis Group. Over the past three years, Dr. Kaur has been actively engaged in the field of food chemistry. In 2023, she was honored with the Best Poster Award at a national conference in recognition of her research excellence. She is a life member of the Indian Science Congress Association (ISCA) and is currently serving as a Guest Editor for a special issue on Pigmented Cereals in the *Journal of Food Measurement and Characterization*, a prestigious Springer journal. Dr. Kaur is an active reviewer of various journals of international repute.

Sukhvinder Singh Purewal is currently serving as Assistant Professor at the University Centre for Research and Development (UCRD), Chandigarh University, Mohali, Punjab, India. He was awarded a Major Research Project (2018–2022) by the SEED Division, Department of Science and Technology (DST), New Delhi. His research interests include solid-state fermentation, extraction of bioactive compounds from natural sources, antioxidant research, and fruit processing. He has authored over 50 research papers in peer-reviewed international journals and has received multiple best poster awards at both national and international conferences. He is an active member of several professional organizations such as the Association of Microbiologists of India (AMI), Mycological Society of India (MSI), Association of Food Scientists and Technologists (India) (AFSTI), and the Indian Science Congress Association (ISCA). In 2023, Dr. Purewal was honored with the prestigious "Young Scientist Award" by the renowned Association of Microbiologists of India (AMI) in recognition of his outstanding contributions to the field of Food Microbiology. Furthermore, Dr. Sukhvinder Singh Purewal has demonstrated his expertise by publishing an authored book, *Millets: Properties, Processing, and Health Benefits*, along with ten edited books with CRC Press, Taylor & Francis Group, and Springer: 1) *Maize Nutritional Composition, Processing, and Industrial Uses*, 2) *Chickpea and Cowpea: Nutritional Profile, Processing, Health Prospects, and Commercial Uses*, 3) *Nanotechnology in the Food Industry: Applications, Recent Trends, and Future Perspectives*, 4) *Colored Cereals: Properties, Processing, Health Benefits, and Industrial Uses*, 5) *Compendium of Clove: Navigating Agriculture, Chemistry, Processing, and Health Benefits*, 6) *Liquorice Properties, Processing, and Health Benefits*, 7) *Recent Advances in Citrus Fruits*, 8) *Rye: Processing, Nutritional Profile and Commercial Uses*, 9) *Edible Insects: Handling Methods, Nutritional Value, and Commercial Uses*, and 10) *Agri-Nanotechnology: Innovations for Sustainable Agriculture and Environmental Restoration*. Dr. Purewal also serves as an Associate Editor for the highly reputed *Journal of Food Measurement and Characterization* of Springer. In 2025, his outstanding contributions to antioxidant research have earned him a place among the top 0.5% of researchers worldwide in this field, as reported by ScholarGPS.

Marcelo Franco PhD in Chemistry from the Federal University of Rio de Janeiro. Currently, he serves as a full professor at the State University of Santa Cruz (UESC), focusing on the research of agro-industrial by-product reuse and the production of high-value-added compounds. Dr. Marcelo Franco actively participates in postgraduate programs in Chemistry (UESC), Food Engineering and Science (UESB), and the Northeast Biotechnology Network (UFBA). He is the Vice-Coordinator of the Innovation and Technological Nucleus at UESC (2024–2026) and a member of the Evaluation and Assessment Chamber of FAPESB (2024–2026). Dr. Marcelo Franco collaborates with international researchers from various countries and leads the Research Group on Biotransformation and Organic Biocatalysis at UESC. His academic output includes patents, over 100 scientific articles, and book chapters, reflecting an H-index of 28 (Scopus), 26 (Web of Science), and 31 (Google Scholar – i10 = 75).

Ashutosh Kumar Rai has been serving as an assistant professor in the Department of Biochemistry, College of Medicine, Imam Abdulrahman Bin Faisal University (IAU), Dammam, Saudi Arabia, since 2017. In this role, he contributes extensively to teaching, research, laboratory supervision, and academic administration. Dr. Rai earned his PhD in Applied Biochemistry in 2012 from the School of Biotechnology, Banaras Hindu University (BHU), India, with a specialization in microbial biotechnology and molecular biology. His doctoral research focused on cyanotoxins (microcystins) and plant growth-promoting bacteria. He holds a Master of Science degree in Applied Biochemistry from the Department of Biochemistry, VBS Purvanchal University, India, and a Bachelor of Science degree in life sciences. Following his PhD, Dr. Rai worked at the International Centre for Genetic Engineering and Biotechnology (ICGEB), New Delhi, on a project aimed at enhancing biofuel production in algae using genetic engineering strategies. He later contributed to a metaproteomics-based research project at the National Bureau of Agriculturally Important Microorganisms (NBAIM-ICAR). Dr. Rai also served at the School of Biotechnology, Jawaharlal Nehru University (JNU), New Delhi, as an SERB-DST Young Scientist and Principal Investigator, where he led a project centered on host-pathogen interactions. With a cumulative 17 years of academic and research experience including over 11 years post-PhD. Dr. Rai has gained expertise across a wide range of biochemical, molecular biology, and biotechnology techniques. He has authored more than 50 peer-reviewed research articles and book chapters published in reputable international journals and scientific volumes. His current research interests include microbial biotechnology, molecular biology and proteomics, biomass and biofuel production, and environmental microbiology. In addition to his research activities, Dr. Rai serves as an editorial board member and peer reviewer for multiple international scientific journals.

Muhammad Irfan, PhD, is working as Chairman in the Department of Biotechnology, University of Sargodha, Pakistan. He received his doctorate degree from Shenyang Agricultural University, P.R. China. Previously, he served as researcher at Food & Biotechnology Research Center, PCSIR Laboratories Complex, Ferozepur Road, Lahore, Pakistan. He is an editorial/advisory board member of various international reputed journals. He has published more than 230 publications in national and international reputable journals with citations of more than 3700, H-index of 30, and i-10 index of 92. He has supervised more than 40 MPhil scholars and 3 PhDs and 3 candidates under supervision. Dr. Muhammad Irfan's current research focuses on microbial enzymes from fermentation, bioactive compound production, lignocellulose biotechnology, biofuels, single-cell protein production, and genomics (molecular markers in genetic diversity analysis, QTL mapping in plants).

Preface

Green chemistry has become an essential part of modern science and industry because of the need to protect the environment while producing chemicals efficiently. The book edited by Avneet Kaur, Sukhvinder Singh Purewal, Marcelo Franco, Ashutosh Kumar Rai, and Muhammad Irfan brings together the latest research, technologies, and strategies to guide scientists, engineers, and students in creating sustainable chemical processes.

Chapter 1, Green Chemistry, Efficiency, and Sustainability: A Framework for a Sustainable Future, introduces the idea of sustainability in chemistry and sets the stage for understanding how efficiency and environmental responsibility can go hand in hand. Principles that guide green chemistry are explained in **Chapter 2**, which helps to understand designing safer and more sustainable chemical processes. The design, synthesis, and analysis of such processes are explored in **Chapter 3**, giving practical examples of applying green chemistry in the laboratory and industry. Real-world industrial applications are the focus of **Chapters 4 and 5**. **Chapter 4**, Re-refining Technologies for Circularity and Sustainability in the Oil and Gas Supply Chain, explains how industries can recycle and reuse materials efficiently. **Chapter 5**, From Formation to Remediation: Maneuvering the Regulatory Framework for Waste Chemicals, guides readers through regulations and strategies for managing chemical waste safely. The next set of chapters (**Chapters 6–10**) highlights biological and renewable methods for producing chemicals. **Chapter 6**, Production of Chemicals Using Microbial Consortia, introduces the use of microorganisms for sustainable chemical production. **Chapter 7**, Sustainable Biohydrogen Production from Waste via Dark Fermentation and Its Efficient Disposal, and **Chapter 8**, Bioethanol and Biobutanol: Production, Optimization, and Disposal Guidelines, focus on biofuels. **Chapter 9**, Biogas Process Optimization for Green Methane and Biohydrogen Generation and Sustainable Disposal Strategies, and **Chapter 10**, Biogas (Methane, Biohydrogen): Production, Optimization, and Disposal Guidelines, provide practical approaches for optimizing bioenergy processes and handling waste safely. Finally, **Chapters 11** and **12** discuss the economic aspects of green chemistry. **Chapter 11**, Financial and Techno-economic Aspects of Green Chemistry, explains how financial planning and cost analysis support sustainable practices. **Chapter 12**, Economic and Technological Drivers of Green Chemistry, highlights the market and technological forces that encourage the adoption of green chemistry worldwide.

Together, these chapters provide a complete guide to green chemistry, combining science, technology, and practical applications. This book is designed to help researchers, students, and industry professionals understand, develop, and implement chemical processes that are efficient, safe, and environmentally friendly.

Contributors

Sadia Abdullah
Department of Botany, Rawalpindi Women University, Satellite Town, Rawalpindi, Pakistan

Mushtaq Ahmad
Department of Plant Sciences, Quaid-i-Azam University, Islamabad, Pakistan
Department of Physiology and Microbiology, Samarkand State University named after Sharof Rashidov, University Blv. 15, Samarkand, Uzbekistan

Camila de Moura Strieder
Laboratory of Materials and Scientific Computing (LabMAC), Department of Chemical and Food Engineering, Federal University of Santa Catarina, Florianópolis, Santa Catarina, Brazil

Abdolvahhab Fetanat
Department of Electrical Engineering, Faculty of Energy and Data Science, Behbahan Khatam Alanbia University of Technology, Behbahan, Iran

Abeera Gul
Sino-Pak Biodiversity Research Centre for Agro-Biological Resources, Quaid-i-Azam University, Islamabad, Pakistan

Namdeo R. Jadhav
Krishna Institute of Pharmacy, Krishna Vishwa Vidyapeeth (Deemed to be University), Karad, India

Bisma Javed
Department of Botany, Rawalpindi Women University, Satellite Town, Rawalpindi, Pakistan

Domingos Lusitâneo Pier Macuvele
Laboratory of Materials and Scientific Computing (LabMAC), Department of Chemical and Food Engineering, Federal University of Santa Catarina, Florianópolis, Santa Catarina, Brazil
Center for Studies in Science and Technology (NECET), Department of Science, Engineering, Technology and Mathematics, University of Rovuma, Lichinga, Niassa, Mozambique

Mamoona Munir
Department of Botany, Rawalpindi Women University, Satellite Town, Rawalpindi, Pakistan

Mumna Munir
Student/Research Scholar, Sino-Pak Biodiversity Research Centre for Agro-biological Resources, QAU, Islamabad, Pakistan

Nigar Mursal
Institute of Botany, Ministry of Science and Education of the Republic of Azerbaijan, Baku, Azerbaijan

Natan Padoin
Laboratory of Materials and Scientific Computing (LabMAC), Department of Chemical and Food Engineering, Federal University of Santa Catarina, Florianópolis, Santa Catarina, Brazil

M. Rajasimman
Department of Chemical Engineering, Annamalai University, Annamalai Nagar, Tamil Nadu, India

R. Rajeshkannan
Department of Chemical Engineering, Annamalai University, Annamalai Nagar, Tamil Nadu, India

Humberto Gracher Riella
Laboratory of Materials and Scientific Computing (LabMAC), Department of Chemical and Food Engineering, Federal University of Santa Catarina, Florianópolis, Santa Catarina, Brazil

Valentin Romanovski
Department of Materials Science and Engineering, University of Virginia, Charlottesville, VA, USA

Panchamoorthy Saravanan
Department of Petrochemical Technology, University College of Engineering-BIT Campus, Anna University, Tiruchirappalli, Tamil Nadu, India

Omer Faruk Sarioglu
Department of Molecular Biology and Genetics, Istanbul Medeniyet University, Istanbul, Turkey
Science and Advanced Technologies Research Center (BILTAM), Istanbul Medeniyet University, Istanbul, Turkey

Anand D. Sawant
School of Physical and Chemical Sciences, Sanjay Ghodawat University, Kolhapur, India

Diwakar Z. Shende
Advanced Separation and Analytical Laboratory (ASAL), Department of Chemical Engineering, Visvesvaraya National Institute of Technology (VNIT), Nagpur, India

Cíntia Soares
Laboratory of Materials and Scientific Computing (LabMAC), Department of Chemical and Food Engineering, Federal University of Santa Catarina, Florianópolis, Santa Catarina, Brazil

Shazia Sultana
Department of Plant Sciences, Quaid-i-Azam University, Islamabad, Pakistan
Department of Physiology and Microbiology, Samarkand State University named after Sharof Rashidov, University Blv. 15, Samarkand, Uzbekistan

Mohsen Tayebi
Research and Technology Group, Bidboland Gas Refining Company, Omidiyeh, Iran

Ashwini S. Thakre
Advanced Separation and Analytical Laboratory (ASAL), Department of Chemical Engineering, Visvesvaraya National Institute of Technology (VNIT), Nagpur, India

Gülbahar Bilgiç Tüzemen
Nevşehir Haci Bektas Veli University, Faculty of Engineering-Architecture, Department of Metallurgy and Material Engineering, Nevşehir, Turkey

Kailas L. Wasewar
Advanced Separation and Analytical Laboratory (ASAL), Department of Chemical Engineering, Visvesvaraya National Institute of Technology (VNIT), Nagpur, India

Abbreviations

3D	Three-Dimensional
ACS GCPR	American Chemical Society Green Chemistry Institute Pharmaceutical Roundtable
ACS	American Chemical Society
AD	Anaerobic Digestion
AE	Atom economy
AGREE	Analytical GREEnness Metric Approach
AI	Artificial Intelligence
Green ELN	Green Electronic Laboratory Notebook
AMGS	Analytical Method Greenness Score
APEA	Aspen Process Economic Analyzer
API	Application Programming Interface
BASF	BadischeAnilin- and Soda-Fabrik
BMP	Biochemical Methane Potential
BOD	Biochemical Oxygen Demand
BPA	Bisphenol A
C/N	Carbon-to-Nitrogen Ratio
CCUS	Carbon Capture, Utilization, and Storage
CDP	Carbon Disclosure Project
CECs	Contaminants of Emerging Concern
CFCs	Chlorofluorocarbons
CFD	Computational Fluid Dynamics
CHAs	Chemical Hazard Assessments
CHP	Combined Heat and Power
CNG	Compressed Natural Gas
COD	Chemical Oxygen Demand
CP	Cyclophosphamide
CSTRs	Continuous Stirred-Tank Reactors
DDT	Dichloro Diphenyl Trichloroethane
DEHP	Di-2-Ethylhexyl Phthalate
DES	Deep Eutectic Solvents
DJSI	Dow Jones Sustainability Index
EBRD	European Bank for Reconstruction and Development
E-Factor	Environmental Factor
EGSB	Expanded Granular Sludge Bed
EPA	Environmental Protection Agency
EPI	Environmental Performance Index
ES	Environment/Environmental Systems
ESG	Environmental, Social, Governance
EU	European Union
GAC	Granular Activated Carbon
GC-MS	Gas Chromatography Coupled with Mass Spectrometry

GHG	Greenhouse Gas
GRI	Global Reporting Initiative
GSK	Glaxo Smith Kline
GWP	Global Warming Potential
H_2S	Hydrogen Sulfide
HRT	Hydraulic Retention Time
IF	Ifosfamide
IKEA	Ingvar Kamprad Elmtaryd Agunnaryd
INN	Glufosfamide
IRR	Internal Rate of Return
ISATs	Inherent Safety Assessment Tools
ISD	Inherent Safer Design
ISO	International Organization for Standardization
ISPCWP	Intergovernmental Science-Policy Panel on Chemicals, Waste and Pollution
IUPAC	International Union of Pure and Applied Chemistry
LCA	Life Cycle Assessment
LCC	Life Cycle Costing
LCCA	Life Cycle Cost Analysis
LC-QTOF-MS	Liquid Chromatography coupled with tandem Time-Of-Flight Mass
LWT	Low-Waste Technologies
MCC	Multiple-Column Chromatography
MECs	Microbial Electrolysis Cells
ML	Machine Learning
MOFs	Metal-Organic Frameworks
MS	Membrane Separation
MSW	Municipal Solid Waste
NEPA	National Environmental Policy Act
NPs	Nanoparticles
NPV	Net Present Value
NZVI	Nanoscale Zero-Valent Iron
OFETs	Organic Field-Effect Transistors
OLEDs	Organic Light-Emitting Diodes
OLR	Organic Loading Rate
OPEX	Operating Expenditures
PEMs	Proton Exchange Membranes
PHAs	Polyhydroxyalkanoates
PI	Profitability Index
PLA	Polylactic Acid
PMI	Process Mass Intensity
PVC	Polyvinyl Chloride
REACH	Registration, Evaluation, Authorization,and Restriction of Chemicals
RME	Reaction Mass Efficiency
ROI	Return on Investment

scCO_2	Supercritical Carbon Dioxide
SCFs	Supercritical Fluids
SMR	Steam Methane Reforming
SRB	Sulfate-Reducing Bacteria
ST	Sustainable Technology
TAN	Total Ammonia Nitrogen
TEA	Techno-Economic Assessment
TEM	Transmission Electron Microscopy
TPAD	Temperature-Phased Anaerobic Digestion
TRL	Technology Readiness Level
tRNA	Transfer Ribonucleic Acid
TSCA	Toxic Substances Control Act
UNEP	United Nations Environment Programme
UNIDO	United Nations Industrial Development Organization
UNSDGs	United Nations Sustainable Development Goals
US	United States
USA	United States of America
UV	Ultraviolet
VMSs	Volatile Methylsiloxanes
VOC	Volatile Organic Compounds
VS	Volatile Solids
ZWT	Zero-Waste Technologies

1 Green Chemistry, Efficiency, and Sustainability
A Framework for Sustainable Future

Anand D. Sawant and Namdeo R. Jadhav

1.1 INTRODUCTION

1.1.1 Origin and Historical Perspective

Chemistry has been an essential part of human life in modern society. But if one looks at the timeline down in history of evolution of chemistry, one may find discoveries, ideas, inventions, and numerous successful and failed experiments that have paved the foundation of modern chemistry. Prior to scientific methods get acceptance, the great thinkers were proposing four-part theory of root of matter viz. air, water, earth, fire, and idea of atom as an invisible particle of matter, elements, and substances. Many of the concepts of alchemy were later found valid and contributed to the development of modern chemistry concepts (Allen, 2009). With progression of periodic table and discovery of elements, the scientific impressions in chemistry have been taken its shape. One can found a linkage in chemical industry revolution, which had brought innovation and transformations, if looks at history of evolution of modern chemistry. In world the countless revolution hold place alongside industrial revolution. Throughout 18th century, the flourishing of the chemical industry started with requirement of rapid bleaching technology for cotton industry. In the middle of 18th century, they resulted into progress of process for mass creation of sulfuric acid in lead chambers. It's been used for preparing efficient bleaching agents in cotton industry (Russell, 1987). Afterward the development of Leblanc soda process like manufacturing of sodium carbonate which was providing alkali to industries for instance soap, glass, and other related manufacturing processes. The Solvay process achieved widespread adoption owing to its cost-effectiveness (Liu, 2014). The shifting to organic chemistry from heavy chemical processes was held within 19th century through the chemical industry innovation. Some high explosives like nitrocellulose, nitro glycerin, and dynamite came in industries as foremost artificial aniline-based dye. Few plastics like celluloid and artificial silk rayon came into progression through the cellulose

DOI: 10.1201/9781003642589-1

processing. They formed a brick in the development of large-scale industries. New medicinal and pharmaceutical products manufacturing were initiated to meet the crucial outcome for expanding chemical industry. In between the same period, the major component of chemical industry, that is, polymer industry was introduced through polymer chemistry of polyethylene and polystyrene. Nowadays the size of market of polymer industry is more than 350 million metric ton (Din et al., 2020). The twentieth century witnessed real growth in the chemical industry. As the world population increased the demand for food production putting pressure on agricultural produce increased. To help grow crop production, fertilizer industry came into being and brought in the green revolution. Great foundation was paved by the path-breaking discovery of the process of ammonia synthesis by Haber, who developed and patented a catalytic process where nitrogen and hydrogen were reacted under heat and pressure to generate ammonia. This laboratory experiment was transformed to industrial process by BASF engineer Carl Bosch who later awarded with Nobel Prize for the work done by him. The technique is popularly known as Haber-Bosch process (Tamaru, 1991). The huge manufacturing facility was first commissioned at BASF in first half of 20th century (Figure 1.1). This process further brought in a revolution in nitrogen-based fertilizer industry which was producing around 150 million metric ton of ammonia until 2021. According to data of US Department of Agriculture, there was 1500 kg per hectare of corn produced by US farmers till 1930, and today the production average is 10,000 kg per hectare. Despite such benefits of Haber-Bosch process, it is now understood that the process has caused serious environmental impact contributing to around 1% of human-made carbon dioxide. The fertilizers are polluting water bodies and threatening existence of some aquatic species (Boyce et al., 2024).

The first image below is title Oppau plant 1912 and second image has title Installation of a high-pressure reactor in 1935.

FIGURE 1.1 Ammonia manufacturing sites of BASF during early 20th century.

Source: BASF Corporate History, Ludwigshafen a. Rh./Germany.

With many such inventions there was other side of story and filling chemistry legacy with both triumphs and unintended consequences. *Silent Spring* is a landmark environmental science book written by Rachel Carson and published in 1962. It exposed the shocking ecological and health belongings of indiscriminate pesticide use particularly DDT and challenged the unchecked control of the chemical industry. Carson conscientiously documented how pesticides weren't just vanishing pests but also harming fish, birds, and even humans. She found that the identified chemicals should be called "biocides" due to their broad-spectrum toxicity. The title of the book suggests a haunting image: a spring season with no birdsong and silenced by chemical poisoning.

Some of the notable examples where well-designed discoveries led to serious health even though environmental fallout:

1. **DDT (Dichlorodiphenyltrichloroethane):** It initially came up as a mystery pesticide all through World War II and onward used broadly in agriculture. After that DDT was introduced as cause particularly thinning bird eggshells, ecological harm, and accumulation in the food chain. Its dangers were spotlighted famously by Rachel Carson's *Silent Spring* (Turusov et al., 2002).
2. **CFCs (Chlorofluorocarbons):** Once known for their stability even used in refrigeration and aerosols, CFCs were reported by Rowland and Molina to reduce the ozone layer and responsible to enhanced UV radiation and global health risks for instance skin cancer (Molina & Rowland, 1974).
3. **Asbestos:** Valued for its fire resistance and insulation properties, asbestos was used extensively in construction. Decades later, it was linked to severe respiratory diseases, including mesothelioma (Mossman & Gee, 1989).
4. **Thalidomide:** Introduced in 1950s as a sedative and treatment for morning sickness, thalidomide triggered severe birth defects such as sealed limbs, in thousands of children before it was withdrawn from the market for same use (Vargesson, 2015).
5. **Antibiotics:** Antibiotics were popular after World War II. Later it was found that development of bacterial resistance to antibiotics has led to serious concern in healthcare (Ventola, 2015).
6. **Lead in gasoline and paint:** Lead additives improved engine performance and paint durability, but chronic exposure led to widespread neurotoxicity, especially in children (Needleman, 2000).
7. **BPA (Bisphenol A):** Used in plastics and food containers, BPA mimics estrogen and has been linked to endocrine disruption, prompting regulatory scrutiny and reformulations (Rochester, 2013).

Despite fierce opposition from chemical companies, Silent Spring galvanized public awareness and led to a nationwide ban on DDT for agricultural use in the United States (Chiarappa, 2011). It also helped spark the modern environmental movement and contributed into the formation of the US Environmental Protection Agency (EPA). Carson's work is often credited with changing the way we thought our relationship with nature shifting from domination to stewardship. *Silent Spring* had a profound ripple effect on environmental policy, initially in the United States and then globally. Mostly significant outcomes contain policy changes, for example:

1. **Inception of the USEPA:** It's been eight years of publication in 1970 the EPA was established to consolidate federal monitoring, research, and enforcement of environmental protection. It was the response to increasing public concern Carson helped ignite (Bernard, 1983).
2. **Ban on DDT:** EPA banned the use of DDT in agricultural applications in 1972 because of its ecological and health impacts. Later it was banned almost everywhere in the world (Turusov et al., 2002).
3. **Landmark Environmental Laws:**
   a. **National Environmental Policy Act** (1970)**(NEPA)**: It is an essential program for environmental effect assessments toward major federal projects.
   b. **Clean Water Act** (1972) and **Clean Air Act Amendments** (1970): Strong regulations on pollutants which set the national standards for water and air quality.
   c. **Endangered Species Act** (1973): Nearing extinction species endangered by habitat destruction and pollution.
4. **Global Influence:** It enthused environmental activities worldwide while influencing pesticide regulations and environmental governance in countries nearby Europe and beyond (Ivanova, 2007).
5. **Intergovernmental Science PolicyPanel on Chemicals, Waste and Pollution (ISPCWP):** After discussion for years, in June 2025, the 107 nations of the world have been decided to establish ISPCWP. This panel will give, countries with independent, policy-related scientific advice on chemicals, waste, and pollution prevention (Ivanova, 2007).

As a proactive reaction to the health and environmental problems brought on by conventional chemical practices, green chemistry first appeared in the late 20th century. The United States' Pollution Prevention Act of 1990, which signaled a change from controlling pollution after it happens to preventing it at the source, is closely linked to its formal origin (Anastas & Kirchhoff, 2002) (*History of Green Chemistry | Center for Green Chemistry & Green Engineering at Yale*, n.d.) (American Chemical Society, "History of Green Chemistry.").

In an attempt to motivate scientists to redesign chemical products and reduce their hazardous impacts, the US EPA established a green chemistry research program in 1991. The 1996 launch by the Presidential Green Chemistry Challenge Awards, which recognized innovations in line with these goals, gave this movement a boost (Anastas & Kirchhoff, 2002). By publishing the 12 principles of green chemistry in 1998, John Warner and Paul Anastas clarified the field and offered an exact structure for creating safer, more environmentally friendly chemical processes (Anastas & Warner, 2000). With networks and publications like the Green Chemistry journal (founded by the Royal Society of Chemistry in 1999) helping to formalize the field, the term "green chemistry" quickly gained international recognition (Clark, 1999;Zuin et al., 2021). The significance of green chemistry has been demonstrated by these cases. It serves as a potent reminder that creativity needs to be combined with vision and moral accountability. Green chemistry has a close connection with "design of chemical products that especially eliminate the use or generation of substances harmful to humans, animals, plants, and the environment in the process," according to the findings of the International Union of Pure and Applied Chemistry (Tundo et al., 2000). Green chemistry is now acknowledged as an essential strategy for tackling some of the most urgent global issues, such as pollution, resource depletion, and climate change. A more ecological and sustainable conscious future is being made possible by the growing integration of its principles into industrial processes, research, and education. As the chemical industry works to reduce its environmental impact and contribute to a healthier planet, the ongoing advancement of green chemistry highlights the significance of cooperation, education, and a sustained commitment to sustainability (Poliakoff et al., 2002).

1.1.2 Twelve Principles and Beyond

Green chemistry, sometimes called *sustainable chemistry*, which is the science of designing chemical products which eliminates the use and generation of hazardous substances during the reaction (Anastas & Kirchhoff, 2002). Apart from traditional approaches which often focus on cleaning pollution after it occurs, green chemistry aims to prevent it at the source. Though green chemistry emerged in 20thcentury, its origin can be traced back to great discoveries such as Antoine Lavoisier's Law of Conservation of Mass which paved the way for concepts like atom economy (18th century), Baron Charles Cagniard de la Tour's discovery of Super Critical Fluids (1822), Jons Jacob Berzelius the concept of Catalysis (1830). Further, Alfred Nobel discovered ways to make explosives safe (1860s), Paul Walden discovered Ionic Liquids (1904), Trevor Kletz introduced the concept of inherent process safety for accident prevention (1960s) (Li et al., 2020).

Twelve **Principles of Green Chemistry**, formulated by John Warner and Paul Anastasin in 1998, serve as a blueprint for designing safer, more sustainable chemical processes and products (Warner et al., 2004). The 12 principles are:

1. **Prevention:** It's better to prevent waste than to treat or clean it up after it's formed.
2. **Atom Economy:** Design synthetic methods to maximize the union of all materials used in the final product.
3. **Less Hazardous Chemical Syntheses:** Generate and use components with little or non-toxicity to the environment and human health.
4. **Designing Safer Chemicals:** Chemical products should be effective yet have minimal toxicity.
5. **Safer Solvents and Auxiliaries:** Reduce or avoid use of supplementary substances for instance solvents and separation agents.
6. **Design for Energy Efficiency:** Reduce requirement of energy to conduct reaction at ATM and pressure whenever possible.
7. **Use of Renewable Feedstock:** Favor renewable ingredients over depleting ones when feasible.

8. **Reduce Derivatives:** Tolerate unnecessary derivatization (for instance use of protecting or blocking groups) need generated waste and additional reagents.
9. **Catalysis:** Habit catalytic substances which are most effective and selective than stoichiometric ones.
10. **Design for Degradation:** Build products to sever into harmless materials after use for preventing environmental persistence.
11. **Realtime Analysis for Pollution Prevention:** Generate analytical methods in process monitoring to avoid hazardous substance production.
12. **Inherently Safer Chemistry for Accident Prevention:** Select substances and forms that diminish the effect for chemical accidents such as explosions or releases.

The given guidelines challenge engineers, chemists, and policymakers to design processes which remove hazardous substances, minimalize waste, and use resources most efficiently. In an era defined by climate change, social demands, and dwindling resources for safe goods, green chemistry has been shifted from narrow scope research to a global imperative (Anastas & Eghbali, 2010). Modern green technology is evolving over these principles by integrating:

- **Life Cycle and Systems Thinking** Transfer from isolated reaction-level optimizations to cradle-to-grave or cradle-to-cradle assessments which include raw materials, transport, use, and end of life.
- **Circular Economy Design** Chemistries are designed to be repairable, reusable, and recyclable for creating goods and materials that smoothly reintegrate into the supply chain.
- **Digitalization and Predictive AI** Machine learning (ML) models predict reactivity and toxicity even environmental fate prior to the start of bench work, cutting down on unnecessary experiments.
- **Regulatory Integration and Safe-by-Design** incorporating regulatory standards for instance REACH, TSCA, etc. and hazard assessments directly into Research and Development processes rather than as an afterthought.
- **Social Metrics and Equity** assessing the impact of chemical novelties on communities and making sure that underprivileged groups aren't burdened by pollution.
- **Co-Creation of Policy and Science Adoption** accelerating incentive structures (like carbon credits, green taxes, and sustainable procurement) are created by scientists, engineers, legislators, and economists.
- **New Metrics Beyond Mass** Incorporating use of water, impacts on biodiversity, human health indices and social cost of carbon into the decision matrix.
- **Bio-Inspired and Synthetic Biology Approaches** Engineering cellular "factories" and enzyme hybrids that operate under ambient conditions, using CO_2 or waste streams as feedstock (Ganesh et al., 2021).

With 12 principles and advancements, chemical products and processes could play a key role in enabling transition to more sustainable production modes across industrial sectors. These advancements are helping us to meet UN Sustainable Development. According to Ganesh et al.(2021), green chemistry can be seen as framework for advancement of sustainable future. But there are still challenges such as "more than 98% of all organic chemicals are still derived from petroleum" (The History of Green Chemistry – American Chemical Society).

Despite encouraging progress, some significant hurdles remain and need to be addressed are:

1. **Economic Pressures**
   - High upfront costs for process redesign or new equipment.
   - Market reluctance to absorb premiums for greener materials (Anastas & Eghbali, 2010).
2. **Technical Complexity**
   - Scaling lab-scale green reactions to multi-ton industrial volumes often reveals unforeseen issues (mass transfer, heat control).
   - Compatibility with legacy infrastructure can be challenging (Clark & Macquarrie, n.d.).
3. **Data and Metrics**
   - Lack of universally accepted green metrics leads to inconsistent reporting.
   - Lifecycle assessments (LCAs) are complex and resource-intensive (Kümmerer et al., 2020).
4. **Knowledge Gaps**
   - Many chemists lack training in green design principles.
   - Fragmented communication between academia, industry, and regulators hampers technology transfer (Sheldon, 2016).
5. **Regulatory Fragmentation**
   - Divergent standards across regions can deter companies from investing globally.
   - Uncertainty over long-term policy commitment reduces appetite for large capital outlays (Johnson et al., 2025).

Overcoming these barriers will require coordinated action among stakeholders from governments, research institutions, and funding bodies to multinational corporations and startups. Despite the challenges, green chemistry is poised to accelerate under several converging trends in future. This includes:

Bio-enabled Manufacturing: Advances in synthetic biology and metabolic engineering could allow microbes to churn out complex chemicals – flavorings, polymers, pharmaceuticals– using sugar, CO_2, or waste biomass as feedstock (Matlin et al., 2020). There had been a recent report on the use of a common bacterium to covert plastic waste to paracetamol (Figure 1.2), a pain killer and antipyretic.

Carbon Circularity: Breakthroughs in catalytic CO_2 conversion may turn greenhouse gas into commodity chemicals at gigaton scale. If powered by renewables, this creates a carbon-neutral or even carbon-negative chemical industry.

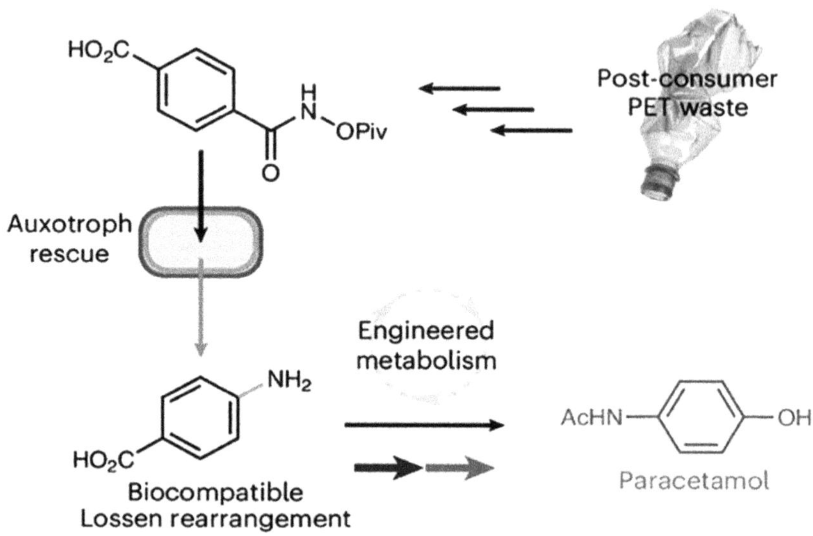

FIGURE 1.2 A biocompatible Lossen rearrangement in *Escherichia coli* to prepare paracetamol (Johnson et al., 2025).

Digitalization and AI: ML and high-throughput experimentation promise to:

- Predict greener reaction pathways earlier to lab trials.
- Optimize reaction conditions in silico by design of experiments, slashing development time and resource used (Anastas & Eghbali, 2010).

Next-Gen Metrics and Transparency: Realtime sensors, blockchain tracking, and open access platforms will deliver product footprints of consumers and regulators can be verified thereby pushing firms in the direction of continuous green improvements (Matlin et al., 2020).

Broader Societal Impact: Green chemistry's evolution is not just technical but it is cultural. As consumers wants safer products and investors prioritize Environmental, Social, and Governance (ESG) factors and companies that embed green principles achieve competitive edge, brand loyalty, and resilience counter to supply chain shocks.

Curriculum Design: It's been started with curriculum design for UGs (undergraduates) where firms like Beyond Benign are working globally to transform chemistry education by weaving sustainability straight into teaching and learning.

Here's some standout case studies of companies which have been successfully embraced **green chemistry** to innovate sustainably and decrease environmental impact:

1. **IKEA – Biobased Adhesives in Furniture:** Nowadays traditional formaldehyde-based resins have been replaced by IKEA in particleboard with **plant-derived adhesives**. This shift:

- Decreased **volatile organic compound (VOC)** emissions
- Improved indoor air quality
- Aligned with green chemistry principles such as using **renewable feedstocks** and scheming for degradation

2. **DuPont – Bio-PDO Production** developed DuPont method to produce **1,3-propanediol (Bio-PDO)** from **corn sugar** using engineered microbes. This innovation:
   - Replaced petroleum-based processes
   - Cut **greenhouse gas emissions by over 40%**
   - Reduced hazardous byproducts and energy use
3. **Pfizer – Greener Drug Synthesis** Pfizer applied green chemistry to synthesis the **sildenafil citrate** (Viagra), resulting in:
   - Fewer synthetic steps
   - Less solvent use and waste
   - Safer reagents and improved yield
4. **Dow Chemical – BPA-Free Thermal Paper:** Dow created a **heat-sensitive coating** for thermal paper which removes **bisphenol A (BPA)** which is a known endocrine disruptor. This safest alternative:
   - Protects retail workers and consumers
   - Maintains performance without toxic chemicals
5. **UNIDO Toolkit Examples – Global Impact:** The UNIDO Green Chemistry Toolkit underlines global case studies, for instance:
   - **Serbia**: Using ionic liquids to purify wastewater
   - **Colombia**: Replacing petroleum solvents with ***d*-limonene** for degreasing
   - **Egypt**: Producing **ethylene from bio-ethanol** instead of fossil fuels

These given examples shown that the green chemistry is not just better for the planet but it is also a driver of innovation, efficiency, and competitive benefit.

The 12 principles of green chemistry show us how to get from wasteful, dangerous chemical practices to a future that is sustainable and circular. Today, industries from pharmaceuticals to plastics are testing out more environmentally friendly methods. However, there are still problems with cost, scale, metrics, and policy. But green chemistry is on its way to changing the chemical business as renewable energy, biotechnology, digital tools, and smart consumers come together. To make this journey happen, chemists, engineers, business leaders, regulators, and civil society must all work together to rethink everything from the first feedstock to the last environmental fate of chemicals.

1.1.3 Toolkits and Metrics

Green chemistry is not just a philosophy but it is quantifiable. In green chemistry, the metrics perform an important role by providing the techniques which measures and compares the impact of chemical products and processes on environment. With no quantifiable metrics, it is difficult to track developments as to assess a process and products having true sustainability. Metrics provides a quick transition to objective assessment about environmental quality related to subjective claims. They help engineers and chemists to identify the areas in which energy efficiency with waste reduction and resource utilization in the generation of safer chemicals for the improvement. Effective communication in green chemistry to different stakeholders related to quantifiable date which is important for decision making while comparing different approaches and driving innovation (Matlin et al., 2020). One can take metrics (like E-factor and atom economy) as an example which helps to detect waste generation while LCA gives opinion of a product environment or process depending on its lifecycle. Various metrics have both advantages and limitations, yet they play a crucial role in guiding the development and implementation of sustainable chemical practices in both academia and industry.

Scientists use various tools and metrics to assess whether chemical processes are environmentally friendly and sustainable. Even industries are following the same path. Here the more commonly used breakdowns:

Key Green Chemistry Metrics: They help to quantify the safety, efficiency, and environmental influence of chemical processes:

- **Atom Economy:** Measures how much of the reactants end up in the final product. Higher is better!
- **E-Factor (Environmental Factor):** Ratio of waste produced to product made. Lower values indicate greener processes.
- **Process Mass Intensity (PMI):** Total mass of all materials used per mass of product. A lower PMI means less resource use.
- **Carbon Efficiency:** Percentage of carbon from reactants that ends up in the product.
- **Energy Intensity:** The total energy consumed per unit of product.
- **LCA:** Assesses environmental influences from raw material extraction to product disposal.

Popular Resources and Tools help chemists to make greener choices in real time:

- DOZN™ 2.0: An evaluation tool which quantifies green chemistry offered by Merck was known to be DOZN™ 2.0. Twelve principles of green chemistry accelerate the web-based premediated system to evaluate the environmental effect of chemical processes and chemicals. DOZN™ 2.0 delivers a way to compare and score the "greenness" of various options, enabling researchers to create most sustainable choices in their research.
- Chem21 Solvent Selection Guide: Rates solvents based on health, safety, and environmental criteria. It is developed by joint industry university partnerships.
- Analytical Method Greenness Score (AMGS) Calculator: Draws a comparison in between chromatography methods based on greenness.
- AI4Green ELN: It is an open-source electronic lab notebook which fit in sustainability metrics.
- PMI Prediction Calculator: Evaluates the efficiency of synthetic routes prior to lab work starts.

- Reagent Guides: Help in selecting greener reagents for common transformations.
- Biocatalysis Guide: An instant reference for enzyme-based transformations.
- Acid-Base Selection Tool: Helps to choose safest acids and bases based on several parameters.

These kinds of tools have been developed by organizations such as **ACS Green Chemistry Institute, Merck (then Milipore Sigma),** Chemistry **for Sustainability** and **university-industry partnerships** and are widely used in industry and academia alike.

To meet the sustainability aspects of chemistry, the nonprofit organizations like Center for Green Chemistry and Green Engineering at Yale University, Beyond Benign and Chem FORWARD are working.

The **Center for Green Chemistry and Engineering at Yale University** focuses on sustainability challenges by serving innovation and practical explanations which is covered under academic community. They made commitment for the improvement of worlds' present as well as future generations by implementing education and research of sustainable practices. The Paul Anastas was served as the director of the center.

Beyond Benign plays an important part in education and practice which focuses in making green chemistry. The mission is to foster a green chemistry community which empowers educators to transform chemistry educators for sustainable future cofounded in 2007 by John Warner and Amy Cannon.

The platform which is used in assessment of hazards and safer alternative identification advanced with safer chemicals is focused by **Chem FORWARD.**

It offers tools and resources, including a database of chemical hazard assessments (CHAs) and a program for verifying safer trade name materials, to help companies make informed decisions about chemical selection.

Globally renowned companies like Pfizer, Merck, BASF, Dow Chemical, DuPont, Novozyme, etc. and startups like Imperagen, Sudoc, DUDE Chem, Ayas Renewables, Impact Nano, Brisil, etc. are investing into development of Green Chemistry tools. Here are some real-world examples that show how companies and researchers are using them to make chemistry cleaner and smarter:

1. **Atom Economy – Ibuprofen Synthesis:** The original six-step synthesis of **ibuprofen** by Boots Company had a low atom economy due to multiple wasteful steps.
    - **Green upgrade:** BHC Company (now BASF) developed a **three-step catalytic process** with over **77% atom economy**, reducing waste and improving yield dramatically.
2. **E-Factor – Pharmaceutical Manufacturing:** The **E-Factor** (waste-to-product ratio) in pharma can be as high as 100–200.
    - **Example:** Pfizer applied green chemistry to reduce the E-Factor in the production of **sildenafil citrate (Viagra)** by optimizing solvents and reagents, cutting waste significantly.

3. **Process Mass Intensity (PMI) – GSK's Molar Efficiency:** GlaxoSmithKline (GSK) uses **PMI** and **Molar Efficiency** to evaluate early-stage drug synthesis.
   - **Case:** GSK's collaboration with the University of Strathclyde led to a new metric– **Molar Efficiency** – to identify more sustainable synthetic routes in drug discovery.
4. **LCA– Bio-PDO by DuPont:** DuPont's production of **Bio-PDO** (used in Sorona® fibers) from corn sugar underwent a full **LCA**.
   - Result: **Reduction** over **40% in greenhouse gas emissions** compared to petroleum-based PDO plus lesser energy and water use.
5. **Reaction Mass Efficiency – Green Oxidation Reactions:** In academic laboratories the **reaction mass efficiency** is useful to compare traditional oxidation reactions with greenly synthesized alternatives.
   - **Example: Hydrogen peroxide is used** instead of chromium (VI) reagents to advance efficiency and lessen toxic waste.

The given metrics are not just about academic even they are powerful tools for transforming how chemistry is done.

The development of green chemistry metrics and toolkits and their applications are important for driving the transition in the direction of a more sustainable chemical industry. By delivering a framework for sustainable chemistry, quantifying environmental effect and encouraging the use of safe and more effective processes. These tools empower engineers and chemists to design an upcoming where chemical innovation is inherently environmentally conscious.

1.1.4 Lifecycle and Systems Thinking: An Integrated Decision-Making Approach

Green chemistry principles systems thinking and their principles are complementary approaches which integrated and lead to more sustainable chemical processes and products. Life cycle thinking, a key component of systems thinking, considers the environmental and social impacts of a product throughout its entire lifespan, from raw material extraction to final disposal. This holistic approach helps chemists identify potential trade-offs and design for optimal sustainability.

LCA and systems thinking are crucial components of sustainable chemistry framework. Systems thinking considers the interconnectedness of various factors within the production of chemicals and use. By integrating these approaches, green chemistry can be more effectively applied to achieve truly sustainable chemical products and processes.

Reyes et al. (2023) underlines that most of the green chemistry metrics are mass-based that do not encompass environmental and socio-economic challenges. They present comparative account of two quantitative tools such as DOZEN 2.0 and LCA integrated with systems thinking. In this article of Journal of Chemical Education, student survey indicated that with the complimentary tools students were able to take more informed and insightful decisions. The tools enabled make sustainable decisions and increased consciousness about the role of chemists in environmental and social responsibility and justice (Figure 1.3).

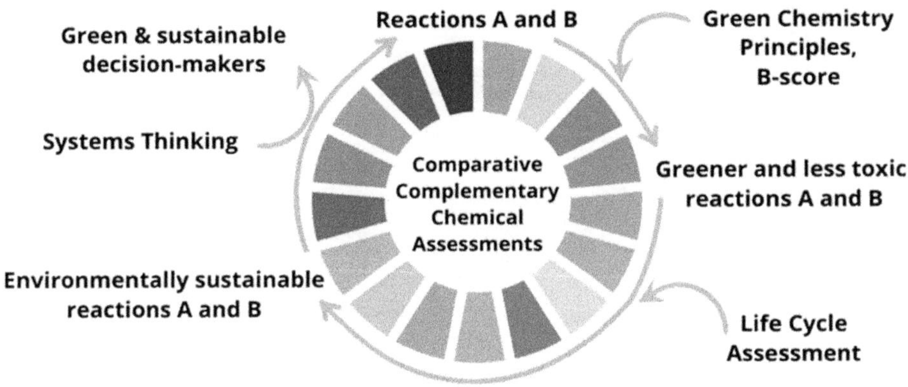

FIGURE 1.3 Life cycle assessment and systems thinking for comprehensive decision making (Reyes et al., 2023) (CCC LN 6081661420616).

D. J. C. Constable (Constable, 2021) in an article has underlined the importance of LCA and systems thinking to transform green chemistry to sustainable chemistry. According to the author, the practice of green chemistry must move beyond the environmental and human health-related roots of green chemistry toward a more systems-based, life cycle-informed, and interdisciplinary practice of chemistry (Figure 1.4).

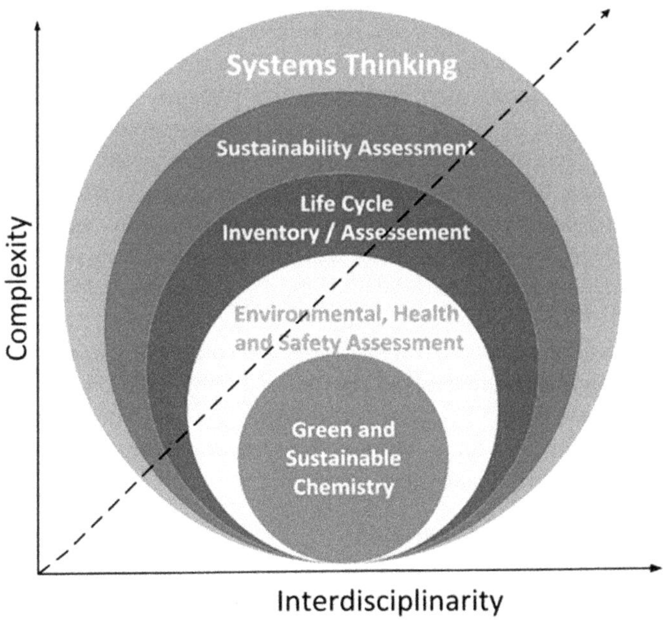

FIGURE 1.4 Interdisciplinary of LCA and ST (Constable, 2021)(CCC LN 6081681400082).

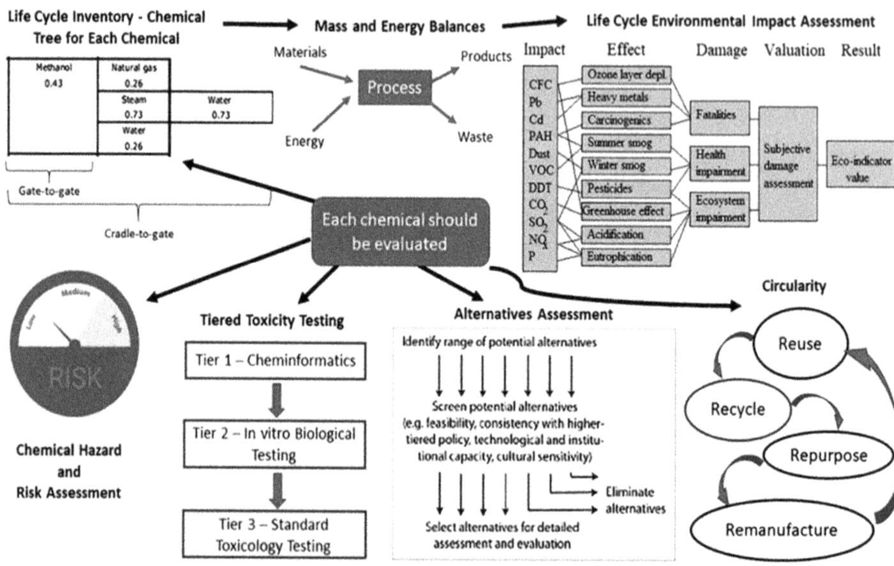

FIGURE 1.5 System level evaluation of chemical process (Constable, 2021)(CCC LN 6081681400082).

Systems thinking helps to manage complexity in sustainability. Figure 1.5 indicates systems view of a chemical reaction which further highlights the importance of systems thinking integrated with LCA.

Considering interdisciplinary and diverse inputs for systems thinking, that involves inter-system interactions, one needs to be able to define system boundaries. To understand the impact of chemistry it goes beyond chemistry to human social systems, economic systems, and environmental systems. Contributors to systems thinking are presented in Figure 1.6. To arrive at sustainable solutions to chemical problems, chemist has to acquire skills outside chemistry and work routinely with scientists, engineers, businesses, and other non-science professions.

End-of-life management is crucial for a complete green chemistry approach. This stage involves designing products that can be safely recycled, composted, or disposed of with minimal environmental impact. LCA can help identify the most sustainable end-of-life options and guide the development of effective recycling and waste management strategies. By considering the entire life cycle, green chemistry aims to create a more sustainable and circular economy for chemical products.

Circular economy design in chemistry reshapes how we think about materials not as disposable commodities but as **resources to be continuously cycled** through the economy. Here's some of the key implications:

1. **Redefining Waste as Feedstock**
   Chemistry turns into the engine for transforming waste streams such as agricultural residues, CO_2, or plastic waste into appreciable inputs for new entities or products. They need innovative catalysis, separation science, and process intensification (Caicedo et al., 2016).

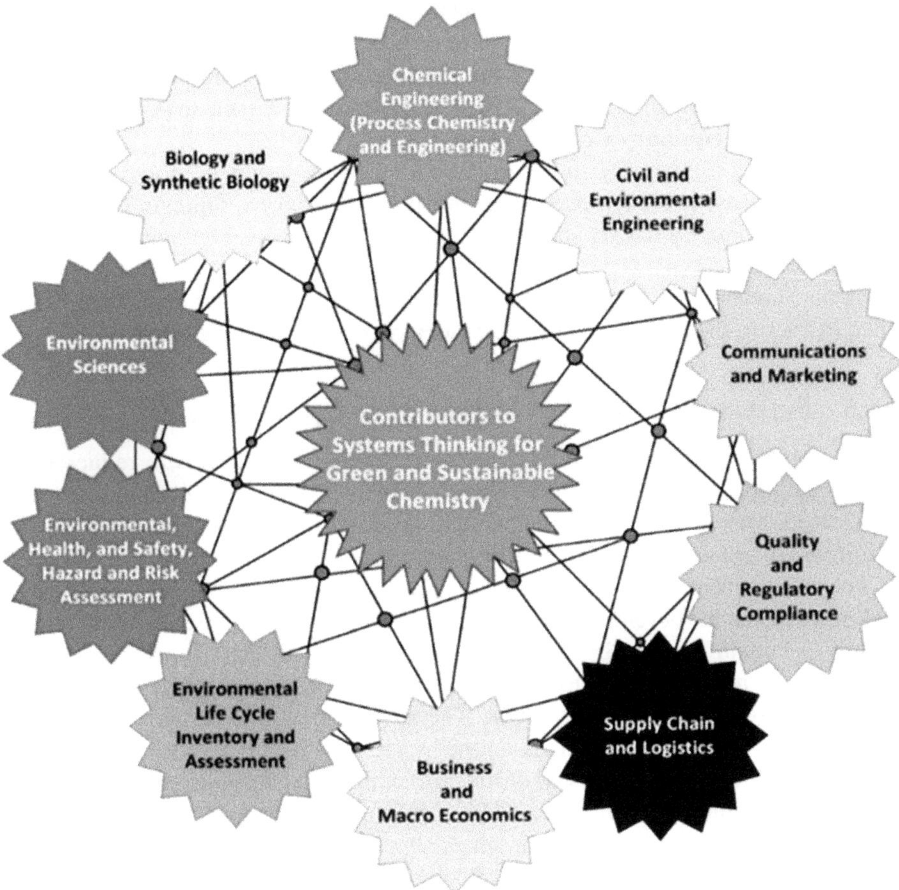

FIGURE 1.6 Contributors to systems thinking (Constable, 2021)(CCC LN 6081681400082).

2. **Designing for Disassembly and Recyclability**
   Materials and molecules are nowadays being designed with their end-of-life in mind. For instance, some polymers with cleavable bonds permit for **controlled depolymerization**, enabling exact chemical recycling rather than downcycling (Yang et al., 2024).
3. **Shifting to Renewable and Regenerative Inputs**
   Circular design inspires the habit of **biobased feedstocks** which regenerate naturally like lignocellulosic biomass or algae while reducing dependence on fossil resources as well as aligning with carbon neutrality goals (Mehmood et al., 2024).
4. **Integrating LifeCycle Thinking**
   Now chemists must consider the **entire life cycle** of a product from used raw material extraction and disposal using tools such as LCA to reduce environmental and social effects (Parvathy et al., 2025).

5. **New Metrics and Evaluation Tools**
   Traditional metrics such as atom or yield economy are being prolonged to include **resource efficiency, energy use, circularity indices**, **and toxicity** pushing chemists to standardize across multiple dimensions.
6. **Cross-Disciplinary Collaboration**
   Circular chemistry insists collaboration with designers, policymakers, economists, and engineers to establish systems where **materials flow in closed loops** which are supported by infrastructure and regulation.

At the fundamental level, circular economy design transforms chemistry from a lined extractive discipline into a **systems-oriented, regenerative science**.

1.1.5 Emerging Technologies– A Promise of Green Chemistry

Green chemistry and sustainability reimagine chemical processes to reduce environmental impact and align with the United Nations Sustainable Development Goals (UNSDGs). They aim to redesign chemical products and processes to minimize or eliminate hazardous substances while enhancing resource efficiency.

The environmental stewardship is overpassed by it with economical viability after making sure that the development of industries does not compromise the heath of ecosystem or human beings. Green chemistry comes up with that 12 principles which range to use a renewable feedstock for the prevention of waste to guide scientists and engineers to direct sustainable solution. The advancement of emerging technologies held with these principles which lessen the by products which are toxic as well as in material effectiveness at every step of lifecycle of product (Mahjoub et al., 2025). The IUPAC, that is, International Union of Pure and Applied Chemistry which organizes the top ten emerging technologies in chemistry annually for showcasing innovation where the transformative progress in circularity even decarbonization and resource efficacy.

Such technologies rely in various stages ranging from lab-scale breakthroughs to initial commercial progress by reducing the waste and safer materials to open every new pathway to green synthesis. The chemistry can decouple economic progress from eco degradation while forging a sustainable future through the integration of tools across industries. These are some emerging green chemistry-dependent technologies which are growing faster and adopted by industries.

1. Biobased Polymers: These polymers are derived from renewable resources tackle the environmental toll based on petroleum plastics (Babu et al., 2013).
   a. Polylactic Acid (PLA): They are formed by fermenting carbohydrates (sugars) from corn or sugarcane to find PLA features biodegradability and a lesser carbon footprint than that of conventional plastics. Revolutions in copolymerization and blending are advancing its thermal and mechanical characteristics for packaging and textile applications.
   b. Polyhydroxyalkanoates (PHAs): Microbial fermentation of lipids and even carbohydrates yield PHAs with a versatile family of biopolymers. Current metabolic engineering efforts have been optimized yield and

decreased production costs while enabling use in medical implants and agricultural films.

The current research focuses on integrating agricultural leftover as feedstock and developing fully compostable composites to wrap up the material loop.

2. Green Solvents: Ionic Liquids and Deep Eutectic Solvents

   Traditional organic solvents give remarkably to VOC emissions and toxicity. Green solvents suggest a safest alternative.

   a. Ionic Liquids: The salt-like liquids which stay liquid at ambient temperatures and show negligible vapor pressure. Their tunable solvation characteristics enable catalysis, selective reactions, and departures in pharmaceuticals and materials synthesis.
   b. Deep Eutectic Solvents (DES): Prepared by mixing hydrogen bond donors and acceptors. DES are biodegradable, having lesser cost, and nontoxic. They facilitate extraction of bioactive compounds using biomass and increase rates of reaction in biomass valorization.

   Task-specific solvent systems were concentrated by researchers which combine high performance with straightforward recovery and recycling (Vanda et al., 2018).

3. Photocatalysis for Solar-Driven Synthesis (Šuligoj et al., 2022).
   a. Capturing solar energy to drive chemical transformations illustrates green chemistry's energy-efficiency ethos.
   b. Researchers work with engineering semiconductor photocatalysts like titanium dioxide modified with dopants to absorb visible light. These kinds of catalysts facilitate water splitting to generate hydrogen and CO_2 reduction into fuels and selective organic oxidations by eliminating harsh reagents.
   c. Integrating solar concentrators with photocatalytic reactors and flow systems promises on-demand production and scalable production of fine chemicals and energy carriers.
4. Supercritical Fluid (SCF) Technologies (Chakravarty et al., 2019).
   a. The gap between liquids and gases is bridged by SCFs that are offering unique solvating abilities.
   b. CO_2 at supercritical conditions for instance 31°C and 74 bar is tunable, non-flammable and leaves zero solvent residues. Supercritical CO_2 extraction extracts provide fragrances, flavors, and lipids with lesser thermal degradation.
   c. Emerging designs utilize SCFs in continuous flow reactors to accomplish green hydrogenation and polymerizations for improving mass transfer and dropping energy consumption.
5. Enzymatic Catalysis and Biocatalysis (Khafaga et al., 2023).
   a. Enzymes transport, unparalleled selectivity and operate beneath mild conditions are slashing byproducts and energy needs.
   b. Current improvements in protein engineering and directed evolution have been formed enzyme variants with increased stability and substrate scope. Selective oxidations, esterifications, and pharmaceutical intermediate syntheses are their industrial applications.

c. Continuous reactor systems streamline downstream processing and enable catalyst reuse, particularly through the application of immobilized enzymes.

6. Microreactor and Flow Chemistry (Gioria et al., 2019).
   a. Batch processes frequently experience with poor heat and mass transfer leads to inefficiencies and safety hazards. Flow chemistry in microreactors offense these issues by channeling reagents from narrow channels.
   b. This technology provides precise rapid mixing, temperature control, and high surface-to-volume ratios. It strengthens green syntheses of active pharmaceutical ingredients, fine chemicals, and nanomaterials with decreased reagent excess and waste.
   c. Modular flow platforms were developing which integrate inline analytics and automated optimization for real-time process intensification.
7. Carbon Capture, Utilization, and Storage (CCUS) (Yan & Zhang, 2019).
   a. Advancing the materials and processes that not only capture CO_2 but also convert it into value-added products is driven by the green chemistry.
   b. Metal Organic Frameworks (MOFs): MOFs are the crystalline porous solids which selectively adsorb CO_2 under ambient conditions and release it into modest pressure or temperature swing.
   c. Electrochemical Conversion: When paired with renewable electricity the electrochemical reactors reduce captured CO_2 into carbon monoxide, format, or ethylene serving as feedstocks for fuels and polymers.

   Maximizing energy efficiency, cycle stability, and integration with industrial point sources have been focused by scientists and engineers.
8. Artificial Photosynthesis and Solar Fuels (Mori et al., 2025).
   a. Mirroring nature's strategy to change the forms like sunlight, water, and CO_2 into chemical energy the artificial photosynthesis could deliver carbon-neutral fuels.
   b. Light-absorbing catalysts with membranes are combined by integrated photoelectrochemical cells to produce hydrogen or hydrocarbons directly from sunlight.
   c. Innovations such as molecular catalysts inspired by natural enzymes, hybrid inorganic–organic assemblies, and scalable photoelectrodes have significantly advanced the field. Improvement in stability and solar-to-fuel efficacy drives pilot scale demonstrations.
9. Green Hydrogen and Sustainable Energy Carriers (Segovia-Hernández et al., 2025).
   a. Sustainably produced hydrogen holds promise as a clean energy vector.
   b. Water electrolysis unable by renewable electricity by using advanced catalysts based on earth-abundant metals has been decreased premiums compared to noble metal systems.
   c. Alternative phenomena such as photoelectrochemical water splitting and biomass-derived hydrogen fabrications are emerging. Linking green hydrogen with fuel cells with hydrogenation processes assists decarbonizing transport and chemical manufacturing.

10. AI (Artificial Intelligence) and Green Chemistry: An article written by Ghasemlou et al. (Ghasemlou et al., 2025) explores the potential of AI to revolutionize sustainable chemistry and give contribution toward a green future. Though current chemical industries face significant environmental challenges which are relying on time-consuming trial-and-error procedures for discovery. ML and AI can dramatically decrease experimental design time, predict optimal reaction conditions, and help to identify new synthetic routes. Though, there are some significant challenges in adopting AI such as the need for extensive, training data, reliable, substantial computational resources, and the environmental effect of AI itself because of high consumption of energy. Meanwhile AI in chemistry is still under development where increased focus on data training and rigorous validation is essential to harness its ability to create the most environmentally compatible products.

1.2 CONCLUDING REMARKS

Green chemistry has emerged as a distinct area within chemistry, providing a responsible framework for achieving the Sustainable Development Goals (SDGs) and advancing the broader concept of sustainable chemistry. Following the Industrial Revolution, its formal foundation was established with the Pollution Prevention Act of 1990 in the United States, which shifted the focus from end-of-pipe pollution control to prevention at the source. The 12 principles of green chemistry have been published by John Warner and Paul Anastas in 1998, served as a foundational blueprint for designing safe and sustainable chemical processes, emphasizing prevention at source, atom economy, and safer syntheses. These aspects are gaining importance specially looking at pressing global challenges such as climate change, pollution, and resource depletion. Apart from the original 12 principles of green chemistry, the modern green chemistry integrates life cycle and systems thinking approach. In this the approach is moving toward "cradle-to-grave" analysis and circular economy designs that underscores importance of waste as feedstock. Modern green chemistry toolkits and metrics such as Atom Economy, E-Factor, and LCAs are crucial for quantifying environmental performance and driving improvement in chemical practices making them more sustainable. Despite significant progress and successful industrial implementations, there are some prevailing challenges such as economic pressures, technical complexity, data gaps, and regulatory fragmentation. But in coming years, the green chemistry will witness accelerated growth with converging trends such as bio-enabled manufacturing, carbon circularity, and digitalization using AI. This is clearly reflated in emerging technologies identified by IUPAC such as biobased polymers, green solvents, photocatalysis, enzymatic catalysis, and microreactor technology. These technologies are enabling promising greener syntheses and waste minimization. This evolving science is supported by interdisciplinary collaborations among chemists, policymakers, engineers, and civil society. This is crucial for reinvention of the chemical enterprise for a sustainable and environmentally conscious civilization. All in all, green chemistry has been evolved from a foundational set of principles into a transformative and dynamic framework that enable scientific innovation to go hand in hand with environmental responsibility. Appropriate integration of green chemistry into industrial practice and academic research underscores its relevance in addressing climate change and attaining sustainable development. It provides a framework of compelling pathway for harmonizing economic progress with ecological stewardship as it matures into the broader paradigm of sustainable chemistry. The widespread adoption of green chemistry tools, metrics, and emerging technologies reflects a shift toward systems thinking and lifecycle awareness ensuring a path toward sustainable future for betterment of humankind.

REFERENCES

Allen, R. C. (2009). The industrial revolution in miniature: The spinning Jenny in Britain, France, and India. *The Journal of Economic History*, *69*(4), 901–927. https://doi.org/10.1017/S0022050709001326

Anastas, P., & Eghbali, N. (2010). Green chemistry: Principles and practice. *Chemical Society Reviews*, *39*(1), 301–312. https://doi.org/10.1039/B918763B

Anastas, P. T., & Kirchhoff, M. M. (2002). Origins, current status, and future challenges of green chemistry. *Accounts of Chemical Research*, *35*(9), 686–694. https://doi.org/10.1021/ar010065m

Anastas, P. T., & Warner, J. C. (2000). *Green chemistry*. Oxford University Press. https://doi.org/10.1093/oso/9780198506980.001.0001

Babu, R. P., O'Connor, K., & Seeram, R. (2013). Current progress on bio-based polymers and their future trends. *Progress in Biomaterials*, *2*(1), 8. https://doi.org/10.1186/2194-0517-2-8

Bernard, R. M. (1983). The healthiest city: Milwaukee and the politics of health reform. By Judith Walzer Leavitt. *Environmental Review*, *7*(4), 382–383. https://doi.org/10.2307/3984183

Boyce, J., Sacchi, R., Goetheer, E., & Steubing, B. (2024). A prospective life cycle assessment of global ammonia decarbonisation scenarios. *Heliyon*, *10*(6), e27547. https://doi.org/10.1016/j.heliyon.2024.e27547

Caicedo, M., Barros, J., & Ordás, B. (2016). Redefining agricultural residues as bioenergy feedstocks. *Materials*, *9*(8), 635. https://doi.org/10.3390/ma9080635

Chakravarty, P., Famili, A., Nagapudi, K., & Al-Sayah, M. A. (2019). Using supercritical fluid technology as a green alternative during the preparation of drug delivery systems. *Pharmaceutics*, *11*(12), 629. https://doi.org/10.3390/pharmaceutics11120629

Chiarappa, M. J. (2011). Managed annihilation: An unnatural history of the Newfoundland Cod Collapse. *Environmental History*, *16*(4), 715–716. https://doi.org/10.1093/envhis/emr100. https://doi.org/10.1093/envhis/emr100

Clark, J. H. (1999). Green chemistry: Challenges and opportunities. *Green Chemistry*, *1*(1), 1–8. https://doi.org/10.1039/a807961g

Clark, J., & Macquarrie, D. (n.d.). *Handbook of green chemistry and technology*. Blackwell Publishing.

Constable, D. J. C. (2021). Green and sustainable chemistry – The case for a systems-based, interdisciplinary approach. *iScience*, *24*(12), 103489. https://doi.org/10.1016/j.isci.2021.103489

Din, M. I., Ghaffar, T., Najeeb, J., Hussain, Z., Khalid, R., & Zahid, H. (2020). Potential perspectives of biodegradable plastics for food packaging application-review of properties and recent developments. *Food Additives & Contaminants: Part A*, *37*(4), 665–680. https://doi.org/10.1080/19440049.2020.1718219

Ganesh, K. N., Zhang, D., Miller, S. J., Rossen, K., Chirik, P. J., Kozlowski, M. C., Zimmerman, J. B., Brooks, B. W., Savage, P. E., Allen, D. T., & Voutchkova-Kostal, A. M. (2021). Green chemistry: A framework for a sustainable future. *Organometallics*, *40*(12), 1801–1805. https://doi.org/10.1021/acs.organomet.1c00343

Ghasemlou, M., Nguyen, H. C., Talekar, S., Pfeffer, F. M., & Barrow, C. J. (2025). Artificial intelligence (AI) for more sustainable chemistry and a greener future. *ACS Sustainable Chemistry & Engineering*, *13*(10), 3830–3833. https://doi.org/10.1021/acssuschemeng.5c00853

Gioria, E., Wisniewski, F., & Gutierrez, L. (2019). Microreactors for the continuous and green synthesis of palladium nanoparticles: Enhancement of the catalytic properties. *Journal of Environmental Chemical Engineering*, *7*(3), 103136. https://doi.org/10.1016/j.jece.2019.103136

Ivanova, M. (2007). Designing the United Nations environment programme: A story of compromise and confrontation. *International Environmental Agreements: Politics, Law and Economics*, *7*(4), 337–361. https://doi.org/10.1007/s10784-007-9052-4

Johnson, N. W., Valenzuela-Ortega, M., Thorpe, T. W., Era, Y., Kjeldsen, A., Mulholland, K., & Wallace, S. (2025). A biocompatible Lossen rearrangement in *Escherichia coli*. *Nature Chemistry*, *17*(7), 1020–1026. https://doi.org/10.1038/s41557-025-01845-5

Khafaga, D. S. R., Radwan, M. G., Muteeb, G., Aatif, M., & Farhan, M. (2023). Green synthesis of biocatalysts based on nanocarriers promises an effective role in pharmaceutical and biomedical fields. *Catalysts*, *13*(11), 1448. https://doi.org/10.3390/catal13111448

Kümmerer, K., Clark, J. H., & Zuin, V. G. (2020). Rethinking chemistry for a circular economy. *Science*, *367*(6476), 369–370. https://doi.org/10.1126/science.aba4979

Li, J., Goerlandt, F., & Reniers, G. (2020). Trevor Kletz's scholarly legacy: A co-citation analysis. *Journal of Loss Prevention in the Process Industries*, *66*, 104166. https://doi.org/10.1016/j.jlp.2020.104166

Liu, H. (2014). Ammonia synthesis catalyst 100 years: Practice, enlightenment and challenge. *Chinese Journal of Catalysis*, *35*(10), 1619–1640. https://doi.org/10.1016/S1872-2067(14)60118-2

Mahjoub, B., Fersi, C., Bouteffeha, M., & Kümmerer, K. (2025). Green chemistry and sustainable chemistry related to water challenges: Solutions and prospects in a changing climate. *Current Opinion in Green and Sustainable Chemistry*, *52*, 101000. https://doi.org/10.1016/j.cogsc.2025.101000

Matlin, S. A., Mehta, G., Hopf, H., Krief, A., Keßler, L., & Kümmerer, K. (2020). Material circularity and the role of the chemical sciences as a key enabler of a sustainable post-trash age. *Sustainable Chemistry and Pharmacy*, *17*, 100312. https://doi.org/10.1016/j.scp.2020.100312

Mehmood, M. A., Amin, M., Haider, M. N., Malik, S., Malik, H. A., Alam, M. A., Xu, J., Alessa, A. H., Khan, A. Z., & Boopathy, R. (2024). Wastewater-grown algal biomass as carbon-neutral, renewable, and low water footprint feedstock for clean energy and bioplastics. *Current Pollution Reports*, *10*(2), 172–188. https://doi.org/10.1007/s40726-024-00294-x

Molina, M. J., & Rowland, F. S. (1974). Stratospheric sink for chlorofluoromethanes: Chlorine atom-catalysed destruction of ozone. *Nature*, *249*(5460), 810–812. https://doi.org/10.1038/249810a0

Mori, S., Hashimoto, R., Hisatomi, T., Domen, K., & Saito, S. (2025). Artificial photosynthesis directed toward organic synthesis. *Nature Communications*, *16*(1), 1797. https://doi.org/10.1038/s41467-025-56374-z

Mossman, B. T., & Gee, J. B. L. (1989). Asbestos-related diseases. *New England Journal of Medicine*, *320*(26), 1721–1730. https://doi.org/10.1056/NEJM198906293202604

Needleman, H. L. (2000). The removal of lead from gasoline: Historical and personal reflections. *Environmental Research*, *84*(1), 20–35.https://doi.org/10.1006/enrs.2000.4069

Parvathy, S. U., Kolil, V. K., Raman, R., Vinuesa, R., & Achuthan, K. (2025). Integrating sustainable development goals into life cycle thinking: A multidimensional approach for advancing sustainability. *Environment, Development and Sustainability*. https://doi.org/10.1007/s10668-024-05810-z

Poliakoff, M., Fitzpatrick, J. M., Farren, T. R., & Anastas, P. T. (2002). Green chemistry: Science and politics of change. *Science*, *297*(5582), 807–810. https://doi.org/10.1126/science.297.5582.807

Reyes, K. M. D., Bruce, K., & Shetranjiwalla, S. (2023). Green chemistry, life cycle assessment, and systems thinking: An integrated comparative-complementary chemical decision-making approach. *Journal of Chemical Education*, *100*(1), 209–220. https://doi.org/10.1021/acs.jchemed.2c00647

Rochester, J. R. (2013). Bisphenol A and human health: A review of the literature. *Reproductive Toxicology*, *42*, 132–155. https://doi.org/10.1016/j.reprotox.2013.08.008

Russell, C. A. (1987). The changing role of synthesis in organic chemistry. *Ambix*, *34*(3), 169–180. https://doi.org/10.1179/amb.1987.34.3.169

Segovia-Hernández, J. G., Hernández, S., Cossío-Vargas, E., Juarez-García, M., & Sánchez-Ramírez, E. (2025). Green hydrogen production for sustainable development: A critical examination of barriers and strategic opportunities. *RSC Sustainability*, *3*(1), 134–157. https://doi.org/10.1039/D4SU00630E

Sheldon, R. A. (2016). Green chemistry and resource efficiency: Towards a green economy. *Green Chemistry*, *18*(11), 3180–3183. https://doi.org/10.1039/C6GC90040B

Šuligoj, A., CercKorošec, R., Žerjav, G., Novak Tušar, N., & LavrenčičŠtangar, U. (2022). Solar-driven photocatalytic films: Synthesis approaches, factors affecting environmental activity, and characterization features. *Topics in Current Chemistry*, *380*(6), 51. https://doi.org/10.1007/s41061-022-00409-2

Tamaru, K. (1991). *The history of the development of ammonia synthesis* (pp. 1–18). https://doi.org/10.1007/978-1-4757-9592-9_1

Tundo, P., Anastas, P., Black, D. StC., Breen, J., Collins, T. J., Memoli, S., Miyamoto, J., Polyakoff, M., & Tumas, W. (2000). Synthetic pathways and processes in green chemistry. Introductory overview. *Pure and Applied Chemistry*, *72*(7), 1207–1228. https://doi.org/10.1351/pac200072071207

Turusov, V., Rakitsky, V., & Tomatis, L. (2002). Dichlorodiphenyltrichloroethane (DDT): Ubiquity, persistence, and risks. *Environmental Health Perspectives*, *110*(2), 125–128. https://doi.org/10.1289/ehp.02110125

Vanda, H., Dai, Y., Wilson, E. G., Verpoorte, R., & Choi, Y. H. (2018). Green solvents from ionic liquids and deep eutectic solvents to natural deep eutectic solvents. *ComptesRendus Chimie*, *21*(6), 628–638. https://doi.org/10.1016/j.crci.2018.04.002

Vargesson, N. (2015). Thalidomide-induced teratogenesis: History and mechanisms. *Birth Defects Research Part C: Embryo Today: Reviews*, *105*(2), 140–156. https://doi.org/10.1002/bdrc.21096

Ventola, C. L. (2015). The antibiotic resistance crisis: Part 1: Causes and threats. *P & T : A Peer-Reviewed Journal for Formulary Management*, *40*(4), 277–283.

Warner, J. C., Cannon, A. S., & Dye, K. M. (2004). Green chemistry. *Environmental Impact Assessment Review*, *24*(7–8), 775–799. https://doi.org/10.1016/j.eiar.2004.06.006

Yan, J., & Zhang, Z. (2019). Carbon capture, utilization and storage (CCUS). *Applied Energy*, *235*, 1289–1299. https://doi.org/10.1016/j.apenergy.2018.11.019

Yang, S., Du, S., Zhu, J., & Ma, S. (2024). Closed-loop recyclable polymers: From monomer and polymer design to the polymerization–depolymerization cycle. *Chemical Society Reviews*, *53*(19), 9609–9651. https://doi.org/10.1039/D4CS00663A

Zuin, V. G., Eilks, I., Elschami, M., & Kümmerer, K. (2021). Education in green chemistry and in sustainable chemistry: Perspectives towards sustainability. *Green Chemistry*, *23*(4), 1594–1608. https://doi.org/10.1039/D0GC03313H

2 Principles of Green Chemistry

Camila de Moura Strieder, Domingos Lusitâneo Pier Macuvele, Natan Padoin, Humberto Gracher Riella, and Cíntia Soares

2.1 INTRODUCTION

Contemporary societies are facing various challenges. Most of them are related to anthropogenic actions. For these reasons, it is imperative to ask important questions: (1) How can we increase the productivity of foods to feed the growing population with less or minimum environmental impact? (2) How can we deal with energetic challenges aligned with sustainability? (3) How to produce chemicals for diverse applications that are environmentally friendly without compromising their performance? These questions and others can be challenged by a multidisciplinary approach where green chemistry is a central science (Anastas & Eghbali, 2010; Chen & Yang, 2024). Green chemistry guides the production of chemicals and processes by applying strategies that reduce the generation or reduction of hazardous substances. This discipline is supported by 12 principles: (1) prevention, (2) atom economy, (3) less dangerous chemical synthesis, (4) designing safer chemicals and products, (5) safer solvents and auxiliaries, (6) design for energy efficiency, (7) use of renewable feedstocks, (8) reduce of derivatives, (9) catalysis, (10) design for degradation, (11) real-time analysis for pollution prevention, and (12) inherent chemistry for accident prevention (Anastas & Eghbali, 2010).

Since the first worries about environmental impacts, which later led to the development of green chemistry, most advances have been made. These include consolidating products and processes to obtain biodegradable polymers, developing novel materials (nano and micro) using green routes, and innovating biodegradable pharmaceuticals, etc. (Bryan *et al.*, 2018; Dhurjad *et al.*, 2022; Grzebieniarz *et al.*, 2023). The road to innovate with more sustainable products and processes is still open, because although countless efforts have been made, there is still a real dominance of petroleum-based chemicals. Therefore, this chapter aims to provide a general outlook of green chemistry, focusing on its principles. The chapter starts by offering concepts and a historical perspective of green chemistry. The second section analyzes the elaboration and examples of each green chemistry application. Next, the advances and challenges in the implementation and innovation in green synthesis will be presented. Finally, the concluding remarks will be provided.

DOI: 10.1201/9781003642589-2

2.2 GREEN CHEMISTRY: GENERAL ASPECTS

2.2.1 Historical Development of Green Chemistry

Green chemistry originated in 1990 as a role model for guiding environmentally friendly products and processes. However, Rachel Carson inspired the origin of green chemistry in 1960. Rachel Carson dedicated most of her life to researching and advocating regarding the role of some chemicals on local ecosystems (Sheldon, 2008). These chemicals included DDT (dichlorodiphenyltrichloroethane), a chemical that was synthesized for the first time in 1874. DDT is a potent insecticide with severe side effects on the environment and human health. The concerns raised by Rachel Carson and endorsed in their book, "Silent Spring," led to the ban of DDT in the late 60 (Carson, 1962). In 70 decades, the scientists and activists pressed the government, resulting in the addition of environmental issues related to chemical activities as a part of state policy. President Richard Nixon's establishment of the US Environmental Protection Agency (EPA) accelerated efforts to impose a definitive ban on DDT (Jarman & Ballschmiter, 2012).

Since 1980, several efforts have been made to address the chemical industry's environmental issues globally, including international meetings. Importantly, more than 30 industrialized countries met more than once to discuss and elaborate on several documents regarding environmental protection. However, the 1990s were the best decade regarding accepting environmental protection and intensifying prevention actions. Furthermore, it was in this decade that several regulatory issues were established, and the first graduate programs in green chemistry were established at the University of Massachusetts. Furthermore, Paul Anastas and John C. Warner published a book about green chemistry in 1998 (Green Chemistry: Theory and Practice), which became the philosophical guide for the green chemistry movement (Anastas & Eghbali, 2010). The initial years of the 21st century consolidated chemistry globally. This consolidation was evidenced by creating a scientific committee dealing with chemistry outside the United States. Furthermore, in 2001, the Nobel Prize in chemistry was awarded for research in green chemistry (Figure 2.1) (Lipshutz *et al.*, 2012).

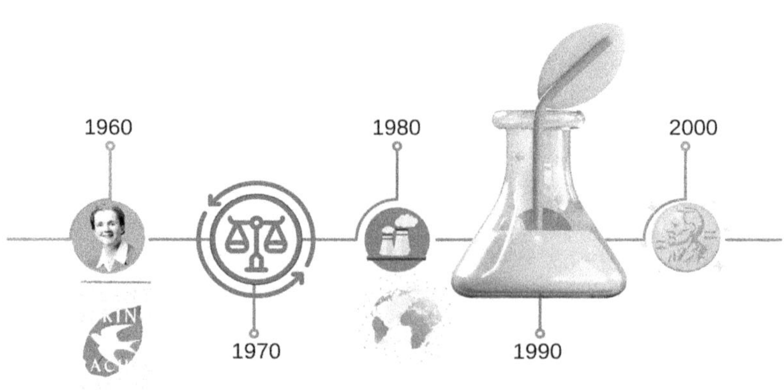

FIGURE 2.1 General illustration of the history of green chemistry.

Several efforts have been made to develop chemicals and processes that align with green chemistry principles. However, several issues remain open, and an in-depth discussion will be provided in the next section.

2.2.2 The Intersection of Sustainability and Economic Efficiency

Economic efficiency and sustainable development are two key issues surrounding the modern productive process (Figure 2.2). However, traditionally, economic efficiency was seen as the opposite of sustainability. Conceptually, economic efficiency is related to balancing resource allocation, increasing productivity, and profitability. Sustainability involves balancing productive processes and environmental benefits (Chen *et al.*, 2019; Dräger & Letmathe, 2023). Another concept related to sustainability and economic efficiency is the circular economy, defined as a productive model that ensures waste reduction, resource efficiency, and resource regeneration (Corvellec *et al.*, 2022; Morseletto, 2020).

Nowadays, the productive process connects sustainability and economic efficiency, with numerous success cases. The synergy of sustainability and economic efficiency can provide long-term benefits to companies. Implementing green chemistry principles can boost these benefits (Chen *et al.*, 2019; Shuai & Fan, 2020). For example, biomass residues can reduce dependence on petroleum-based feedstock, increasing the industry's profitability. However, before profitability, long-term innovation and research are necessary to customize the process initially designed to use petroleum-based feedstock. In the energy field, the use of biomass from agriculture is promising for switching from crude oil to green energy (Ajala *et al.*, 2021; Saleem, 2022; Wu *et al.*, 2025). Data from China, Canada, Poland, and the United States were used to examine the feasibility of utilizing agricultural biomass in the energy sector. A techno-economic analysis revealed economic benefits to using biomass from agriculture as a competitive energy source. Importantly, the biomass was used as a direct energy source by burning

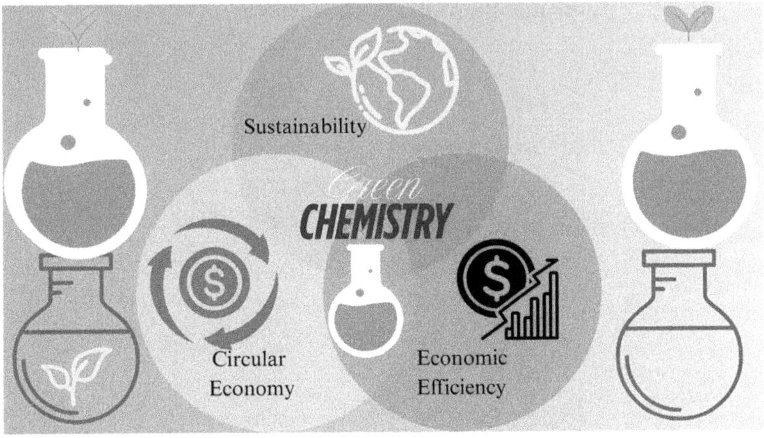

FIGURE 2.2 Intersection of circular economy, economic efficiency, green chemistry, and sustainability.

it to generate electricity and heat, as well as by converting it into biofuels. Challenges such as regulatory requirements and metrics, technological improvements, and public concerns were cited (Saleem, 2022). A more recent work analyzes the availability of corn straw, controlled pyrolysis, environmental impact, net value, and essential costs associated with inputs regarding hydrogen energy and utilization. The work not only shows the feasibility of using corn straw to produce hydrogen that aligns with economic efficiency and sustainability, but also provides critical aspects. These drawbacks include issues related to regulators, biomass, and catalyst costs (Wu *et al.*, 2025).

2.3 THE 12 PRINCIPLES OF GREEN CHEMISTRY

The principles of green chemistry serve as guidelines for developing and improving chemical processes. These principles serve as an essential source of inspiration and reference for researchers to expand their understanding of how their processes and products are designed. More efficient strategies can be developed to minimize environmental impact in this context. The definitions presented for each of the principles in the work are based on those provided by the authors, who selected and published these definitions (Anastas & Warner, 2000). Beyond the environmental issue, the principles of green chemistry are also linked to economic and social aspects (Lane *et al.*, 2023). Additionally, using products of natural origin, rather than relying solely on chemicals, stimulates new production chains. Therefore, promoting green chemistry principles represents a milestone for the planet's future sustainability (Figure 2.3). This is achieved through reducing dependence on fossil fuels, minimizing the environmental impact of chemical synthesis and manufacturing, designing inherently safer chemicals, ensuring the sustainable use of water resources, adopting circular economy practices such as closing material loops, and integrating lifecycle and systems thinking. Furthermore, continued research efforts are crucial to advancing these goals and supporting the development of

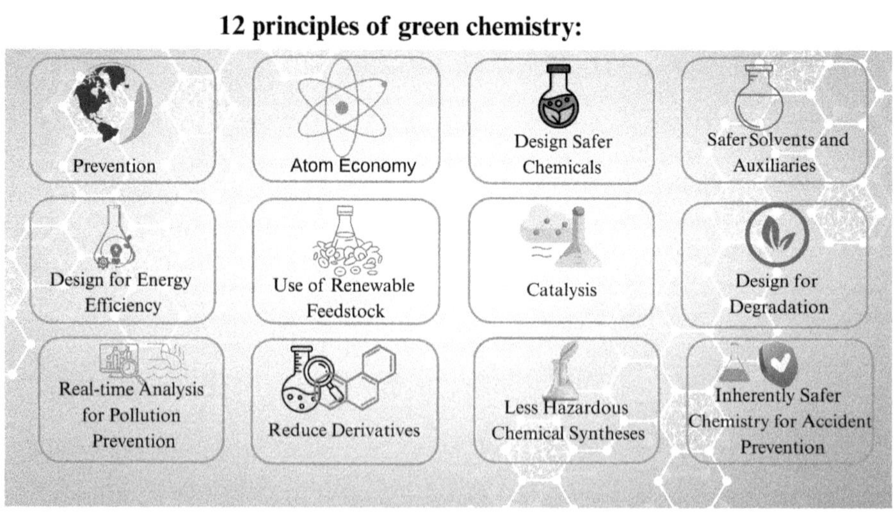

FIGURE 2.3 Twelve principles of green chemistry.

a truly circular economy (Ganesh *et al.*, 2021). Given the broad applicability of green chemistry principles, they will be addressed below to illustrate how each can be incorporated into the day-to-day activities of researchers and society.

2.3.1 Waste Prevention

Waste prevention, the first principle of green chemistry, emphasizes the importance of designing chemical processes and products to minimize or eliminate waste generation at the source, rather than managing or disposing of it after it has formed. This factor is especially relevant because waste is a common denominator of several environmental problems (Sheldon, 2023). Waste prevention involves designing chemical processes and products that minimize the production of by-products, impurities, and waste throughout the substance's life cycle. This is environmentally and economically important, as chemical waste can pose risks to human health and the environment, requiring specialized treatment, safe storage, and proper disposal. Additionally, waste represents material and energy losses, thereby reducing the overall efficiency of processes.

However, despite being ideally desirable, completely avoiding waste is not always practical. Some chemical reactions inevitably generate by-products due to thermodynamic or kinetic limitations or complex reaction mechanisms. Also, alternative technologies or raw materials may not be available or accessible. Restructuring existing industrial processes to eliminate waste can require substantial initial investments and a significant amount of time for development and validation. Some conventional industrial processes are examples of generating large quantities of waste. These include the pyrometallurgical extraction of metals, as seen in the production of aluminum, where bauxite is electrolyzed using the Hall-Héroult process, resulting in the generation of large quantities of greenhouse gases (CO_2, CF_4), cryolite slag, and red mud (Ratvik *et al.*, 2022). Furthermore, the Portland cement production process is another example of high CO_2 emissions, accounting for up to 7% of all global CO_2 generation in the limestone calcination process, as the decomposition of $CaCO_3$ is essential for forming clinker (Wu *et al.*, 2024). In the context of Portland cement production, some alternatives have been studied, as shown in Table 2.1.

Complementary to the points mentioned above, using alternative fuels – such as waste-derived biomass, used tires, and sewage sludge – can significantly reduce industrial waste and fossil fuel consumption. Internal process residues like dust, slag, and by-products can be reused, reducing waste and lowering raw material costs. Moreover, adopting low-emission clinkers (e.g., belite, calcium sulfoaluminate) involves modifying the mineralogical composition of the material to require less heat during production, further enhancing energy efficiency and reducing emissions. Furthermore, metal ions are precipitated with alkalis in synthesizing nanoparticles (NPs) by co-precipitation and generate insoluble salts and excess base in the process residue. In this context, the challenge is to separate the by-products, especially since many products use excess reagents to ensure better yields. One of the alternatives in this case is the green synthesis of NPs, which has been widely studied for different materials (Ag, Au, ZnO, Fe_3O_4/Fe_2O_3, TiO_2, CeO_2, CuO, carbon dots, etc.) and applications (Kazemi et al., 2023).

In the pharmaceutical industry, to prevent waste, high-selectivity reactions are potential, as well as one-pot reactions and process intensification (Kar *et al.*, 2022).

TABLE 2.1
Strategies to Minimize Waste in the Production of Portland Cement

Strategy	How It Works	Environmental Benefit	Reference
Partial replacement of clinker	Use of calcined clay, slag, and fly ash instead of clinker	Reduction of CO_2 and limestone consumption	Shamseldeen Fakhri and Thanon Dawood (2023)
CO_2 Capture and Utilization (CCU)	Capture of CO_2 generated and use as a raw material	Direct reduction of carbon emissions	Meijssen *et al.* (2024)
Accelerated carbonation	Curing concrete with CO_2 to form carbonates	Permanent fixation of CO_2 in the material	Hanifa *et al.* (2023)
3D printing with cement	Additive manufacturing of optimized structures	Less waste of material and forms	Moeini *et al.* (2022)
Nanostructured additives	Incorporation of NPs to improve performance	Reduction in the amount of cement required	Farzadniaand Khayat (2023)
Alternative cements (geopolymers)	Use of ash and alkalized residues as binders	Eliminates clinker and reuses industrial waste	Panda *et al.* (2024)

This can be done with enantioselective catalysts in pharmaceutical synthesis, allowing products with high optical purity to be obtained, minimizing by-products, and reducing purification steps (Yang *et al.*, 2023). Metal-organic frameworks (MOFs) are also a strategy for separating chiral drugs, and MOFs can be obtained via green synthesis routes (Dhurjad *et al.*, 2022; Głowniak *et al.*, 2021). Additive manufacturing techniques also have the potential to meet the first cornerstone of green chemistry, as they use only the material needed to produce parts, unlike conventional machining, which generates leftovers (Javaid *et al.*, 2021). By applying biodegradable or recyclable materials, solid waste generation is practically eliminated. Through 3D printing, some materials are already conventionally used, and their biodegradable characteristics are emphasized, as with polylactic acid (PLA), a polymer made from renewable sources such as corn and cassava (Dananjaya *et al.*, 2024). In agriculture, formulating controlled-release fertilizers through encapsulation or slow release reduces over-application and leaching losses, reducing the load of chemical residues in agricultural ecosystems. This also reduces the need for frequent reapplications, optimizing the use of nutrients and resources. Controlled-release fertilizers, for example, can be made with biopolymers such as chitosan, starch, and cellulose, which contribute to the circular economy model (Eddarai *et al.*, 2024).

2.3.2 Atom Economy

The atom economy (AE) measures the efficiency of a chemical reaction based on the ratio between the mass of the atoms of the desired product ($M_{product}$) and the total mass of the reagents used ($\sum M_{reactants}$), as shown in Equation (2.1) (Trost, 1991).

$$AE = \left(\frac{M_{product}}{\sum M_{reactants}}\right) \cdot 100 \qquad (2.1)$$

This principle guides the development of chemical reactions in which ideally all or almost all-of the reactants' atoms are incorporated into the product. The atomic economy provides a direct quantitative measure of a process's synthetic and material efficiency. Unlike percentage yield, which only evaluates how much product has been isolated, atomic economy analyzes the rational use of the atoms involved. Processes with high atomic economy minimize the generation of by-products and waste, save raw materials, and reduce separation steps, contributing to cleaner, more sustainable, and economically viable processes.

Despite the advantages of this fundamental, some efficient reactions from an energy or economic point of view can have low atom economy, especially when they involve leaving groups or protective agents. These aspects reinforce how the principles interrelate and how finding a balance point is necessary. In many cases, achieving high atomic economy can also require profound re-engineering of the synthetic route, like preventing waste, which is not always feasible from a technical or economic point of view.

About chemical reactions, the direct activation and functionalization of C-H bonds, without intermediate functional groups, is a theoretical possibility for saving atoms. However, numerous factors still hinder its applicability on a large scale (Dalton *et al.*, 2021). In the polymer industry, addition polymerization reactions, such as the formation of polyethylene from ethylene, show atomic savings of close to 100%, as they do not generate by-products. This contrasts with condensation polymerizations, which often release small molecules (e.g., water, HCl), reducing the atomic economy (Coates & Getzler, 2020). In organic synthesis, reactions such as the Claisen rearrangement or the Beckmann rearrangement are examples of transformations that keep all the atoms of the reactants in the products (García-Álvarez *et al.*, 2013). In catalysis, processes like Heck, Suzuki, or Sonogashira couplings, reactions catalyzed by transition metals, when well optimized, allow carbon-carbon bonds to form with minimal atom loss and high selectivity (Ahmed, 2024; Veerakumar *et al.*, 2017).

2.3.3 Less Hazardous Chemical Syntheses

Less hazardous syntheses employ reagents and conditions that produce substances with lower toxicity, flammability, reactivity, or environmental impact during the reaction and subsequent treatment. This principle emphasizes replacing hazardous substances and conditions with safer alternatives and using methodologies that minimize risks to human health and the environment. Reducing hazards in chemical syntheses protects workers, communities, and the environment. Hazardous processes require strict safety protocols, special equipment, and containment measures, which increase operational costs and the risk of accidents. By designing safer synthetic routes, it is possible to reduce the need for intensive risk control, facilitate industrial scale-up, and improve regulatory acceptance. In addition, reducing the toxic potential of products and intermediates contributes to more sustainable life cycles and less problematic disposal and bioaccumulation.

Despite the advantages of this aspect, replacing hazardous reagents with safer alternatives can lead to lower reactivity, lower yields, or increased reaction complexity. Some compounds (e.g., alkylating agents, strong acids, powerful oxidants) are inherently hazardous, and finding less toxic substitutes that maintain efficacy can be challenging. Furthermore, less dangerous methods are not always available for all types of chemical transformation, especially in the pharmaceutical and advanced materials industries, which require a high degree of structural control (Bryan *et al.*, 2018). In the same way that green synthesis of NPs is mentioned for preventing waste, it is an alternative to reduce the hazards of the material synthesis process (Kumar *et al.*, 2021).

In biocatalysis, using enzymes is a safe alternative because they promote reactions under mild conditions (neutral pH, room temperature, absence of toxic solvents) and with high selectivity. In organic and industrial syntheses, enzymes significantly reduce the use of toxic metal catalysts or aggressive reagents (Kumar & Singh, 2024). In synthesis, the development of solid acid catalysts replaces the use of H_2SO_4 or concentrated HCl in esterification reactions, reducing the risks of corrosion, toxicity from vapors, and generation of liquid acid waste. These catalysts, such as zeolites or sulfonated resins, are reusable and safer (Kaur & Banik, 2024). Enzymes are also relevant in the pharmaceutical industry, and mutagenesis is an opportunity for developing enzyme-catalyzed syntheses to produce chiral therapeutic drugs (Reetz *et al.*, 2024). In the pharmaceutical industry, replacing hazardous oxidizing reagents, such as potassium permanganate or sodium dichromate, with benign alternatives such as hydrogen peroxide, considered a green oxidant, or atmospheric oxygen has been widely applied in selective oxidations (Martin *et al.*, 2017; Poursaitidis *et al.*, 2024). This reduces both toxicity and the amount of metal waste generated.

In reactions that traditionally use chlorine gas or phosgene, alternatives such as carbonates or carbonic acid derivatives have been used to obtain carbonates and urethanes, drastically reducing the toxicological risk. For example, the synthesis of polycarbonates via transesterification avoids the direct use of phosgene (Tamura *et al.*, 2022). In analytical chemistry, the development of methods based on stoichiometrically balanced reactions, with non-toxic products and without the generation of liquid or gaseous waste, contributes to the practice of more sustainable analytical chemistry and with greater atomic economy. Product analysis techniques support the sustainable development of materials because they are fundamental to the characterization of new materials and can be intensively used in the research process. In this context, the academic community has widely accepted some methodologies to classify the new analytical methods developed according to environmental aspects, as the AGREE – Analytical GREEnness Metric Approach and Software (Pena-Pereira *et al.*, 2020). Also, there are indicators that evaluate the analytical technique and the sample preparation system, as with AGREEprep (Wojnowski *et al.*, 2022).

2.3.4 Design of Safer Chemicals

The fourth principle is the intentional development of compounds that perform the desired function with the lowest possible risk to human health and the environment. This involves considering toxicity, environmental persistence, bioaccumulation, nonspecific reactivity, and safe degradation from the molecular design stage. This way, it is possible to incorporate intrinsic safety into the compound structure. This aspect is important because the toxicity of chemical substances is at the heart of several environmental, occupational, and health problems. Bioaccumulative, carcinogenic, mutagenic, or persistent compounds represent chronic risks even at low concentrations (Anastas, 2016).

Developing molecules with better safety profiles reduces exposure to hazardous agents, facilitates regulation and disposal, and improves the overall sustainability of the chemical life cycle. In addition, market demand and regulatory frameworks, such as REACH (EU) or TSCA (USA), push for replacing hazardous substances with safe alternatives. However, designing less toxic molecules without compromising their technical efficacy can be complex. Many desired properties, such as high biological activity, are often associated with lipophilicity and stability, favoring bioaccumulation. Also, despite the evolution of computational power, toxicity is multifactorial and not always predictable by computational modeling or initial testing alone (Jia *et al.*, 2023). Developing pesticides with lower collateral toxicity is an applied approach to this principle. This can be achieved by using molecules with greater selectivity for target pests and lower environmental persistence, which have been prioritized in agriculture. For example, compounds degrade rapidly in sunlight (controlled photolability), reducing prolonged ecological exposure and protecting non-target organisms like bees and fish (Huang *et al.*, 2018).

In cosmetics, replacing ethoxylated or petroleum-derived surfactants with sulfated fatty alcohols of plant origin, such as biodegradable sodium lauryl sulfate, reduces skin and aquatic toxicity without compromising cleaning and foaming efficacy (Asio *et al.*, 2023; Bom *et al.*, 2020). Meanwhile, medicinal chemistry prioritizes the design of drugs that avoid toxic bioactivation in the body, such as those that do not generate reactive intermediates in the liver (e.g., epoxides or free radicals). Structures that avoid unstable functional groups or polycyclic aromatics reduce the risk of hepatotoxicity and carcinogenicity (Driscoll *et al.*, 2020). Within engineering materials, the development of non-endocrine-disrupting plasticizers is a potential. The development of alternative plasticizers to di-2-ethylhexyl phthalate (DEHP), such as triethyl citrate or adipates, allows the flexibility of PVC without adverse effects on the human endocrine system. These compounds have significantly more benign toxicological profiles (Park *et al.*, 2019).

2.3.5 Use of Safer Solvents and Auxiliaries

The fifth principle states that auxiliary substances – such as solvents, extraction agents, stabilizers, drying agents, and carriers – should be eliminated whenever feasible. However, when their use is unavoidable, they must be selected based on their minimal impact on human health and the environment. Preferable solvents are those that exhibit low toxicity and flammability, are non-volatile, biodegradable, and ideally sourced from renewable feedstocks. This is important because solvents and auxiliaries represent a significant fraction of the total mass used in chemical processes, especially in organic synthesis, pharmaceutical formulations, cosmetics, and industrial processes. Many conventional solvents (dichloromethane, benzene, toluene, and acetonitrile) are volatile, flammable, carcinogenic, or environmentally dangerous. Replacing them with safer alternatives reduces occupational risks, facilitates disposal, reduces environmental control costs, and improves the sustainability of the process. Despite advances, "green" solvents do not always offer the same performance regarding solubility, selectivity, or kinetics/reaction control. In addition, some alternative solvents can be more expensive or difficult to recycle. Substitution requires process re-optimization, which can affect yields and reproducibility. Hazardous solvents are sometimes still indispensable due to specific polarity, temperature, or reagent compatibility requirements.

This criterion can be prioritized in the pharmaceutical industry using ethanol or water instead of halogenated solvents. Ethanol, especially of renewable origin, is widely used as an alternative to chlorinated solvents such as chloroform or dichloromethane in extraction, crystallization, and synthesis reactions (Usman *et al.*, 2023). Water, when possible, is the ideal solvent because it is non-toxic, cheap, and abundant, especially in enzymatic or biotechnological processes. The use of supercritical processes in chemical engineering also holds significant potential for enhancing efficiency and sustainability. Supercritical carbon dioxide ($scCO_2$) is used as a reaction and extraction medium in industrial and laboratory methods. It is non-flammable, non-toxic, and recyclable. It allows reactions under moderate pressure and facilitates the separation of the product after the reaction by simple decompression (Han *et al.*, 2023). Toxic surfactants and preservatives can be eliminated in cosmetics and cleaning products. This application also closely relates to the fourth fundamental of safer design products and processes. Using aqueous bases with natural preservatives (such as sorbates, organic acids, or natural phenolic compounds) replaces organic solvents and preservatives such as parabens or formaldehyde, which pose potential health risks. This is essential in formulations with a "natural" or "green" appeal (Nowak *et al.*, 2021).

2.3.6 Design for Energy Efficiency

Design for energy efficiency involves developing chemical processes with the lowest possible energy consumption, preferably under ambient conditions (room temperature and pressure). This principle encourages the reduction of energy demand during all stages of production – from synthesis and purification to transport and disposal – to minimize the environmental impacts associated with the generation and use of energy. This criterion is relevant since energy consumption is directly related to greenhouse gas emissions, dependence on fossil fuels, and the economic cost of industrial processes.

More energy-efficient chemical projects significantly reduce the carbon footprint, increase economic viability, and contribute to climate change mitigation. In addition, processes that operate at mild temperatures or without heating are safer and often more compatible with sensitive biomolecules and catalysts (Gabrielli *et al.*, 2023). Despite this, the search for energy efficiency can imply compromises with reaction time, selectivity, yield, or even reaction viability. Not all reactions occur efficiently under mild conditions; some require high temperatures or specific pressures. In this case, using tools to measure and compare different production routes in terms of energy is important. Technologies that allow for greater efficiency (such as microwaves, ultrasound, or continuous systems) can require specialized equipment and considerable investment. In addition, the energy assessment needs to consider the complete life cycle, including the production, transport, and disposal of the materials involved. Replacing conventional heating with microwaves allows for rapid, homogeneous, and direct heating of reagents, with lower-energy consumption and significantly reduced reaction times. It is widely used in organic synthesis, the preparation of nanomaterials, and heterogeneous reactions (Martina *et al.*, 2021).

Catalysis, in turn, can reduce the activation energy of some compounds. This technology also communicates with other fundamentals of green chemistry. Homogeneous and heterogeneous catalysts allow reactions at lower temperatures with less energy input, significantly increasing energy efficiency (Tway & Filip, 2025). For example, catalysis by transition metals in C-C couplings or by enzymes in biocatalysis avoids extreme conditions and reduces reaction times (Zhong *et al.*, 2024). Enzymes are commonly used in aqueous media as single-use catalysts, which are subsequently discarded. This practice is neither economically viable nor aligned with the principles of a circular economy. This shortcoming can be overcome by immobilizing the enzyme as an insoluble, recyclable solid, that is, as a heterogeneous catalyst (Sheldon *et al.*, 2021). Continuous flow reactors, including those with enzymes, offer greater thermal control, allowing efficient use of the heat generated, reducing conduction losses, and facilitating automation (Sheldon, 2024). In this scenario, some conventional technologies are being revisited. This is the case with mechano-chemistry, which promotes reactions without solvents and heating. Reactions promoted by ball milling use mechanical energy to initiate and sustain chemical reactions, eliminating the need for solvents and heat. This method has been successfully exploited to prepare inorganic materials, synthesize organic compounds, and process pharmaceutical compounds (Głowniak *et al.*, 2021; Krusenbaum *et al.*, 2022).

2.3.7 Use of Renewable Feedstock

Modern societies are deeply dependent on feedstocks that are not renewable. In this group, feedstocks from crude oil are mainly used in all productive sectors. These include fuels, fine chemicals for biomedical, aerospace, automotive, food, and many other industries. Therefore, it is urgent to start using and researching renewable feedstock to reduce the environmental implications of using non-renewable feedstock. Briefly, the green chemistry principle 7, states, "Whenever it is economically and technically feasible, renewable raw materials should be used instead of non-renewable" (Anastas & Eghbali, 2010). Adopting this seventh principle of green chemistry has numerous advantages, including sustainability, reducing environmental impact, a circular economy, and innovation (Anastas & Warner, 2000). Renewable feedstocks include biomass, and more recently, carbon dioxide has been considered a renewable feedstock because it is possible to capture it and produce value-added products (Figure 2.4).

Several attempts have been made to obtain chemical products from renewable feedstocks. These contributions have revolutionized the obtaining of several essential products for diverse applications, such as medicine, agriculture, aerospace, and automobility. Among several products, NP synthesis has been highly beneficial, and because of the advances, green nanotechnology is emerging as a new branch of green chemistry (Ali, 2025; De Gregori da Rocha *et al.*, 2025; Strieder *et al.*, 2024). In green nanotechnology, renewable feedstock is used in different ways (Cechinel *et al.*,

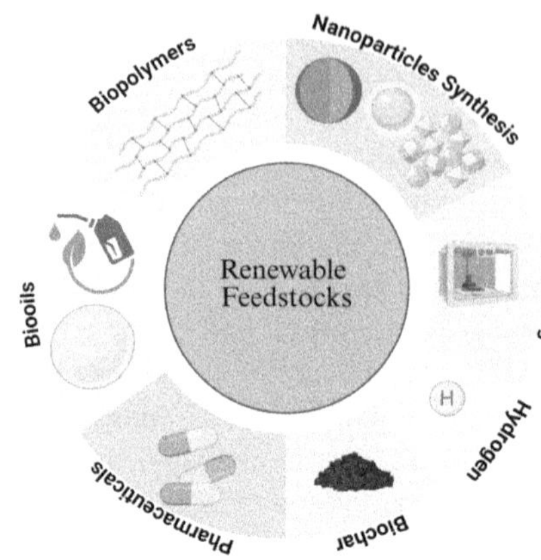

FIGURE 2.4 General outlook of the potential of renewable feedstocks in producing green chemical products.

2025; De Gregori da Rocha et al., 2025; Strieder et al., 2024). The first one involves using biomass as a direct precursor to obtain NPs. In this case, the biomass is mixed with water – hydrothermal process, for example (De Gregori da Rocha *et al.*, 2025) or can be directly converted in a solid-state approach (pyrolysis) (Cechinel *et al.*, 2025). A more recent study produced carbon dots from rice husks by applying a hydrothermal approach. The results showed that the obtained carbon dots presented a spherical-like shape and were sensitive to the pesticide atrazine. Notably, the work provides new insights regarding the carbon dots formation mechanism from rice husk (De Gregori da Rocha *et al.*, 2025). In another work, carbon dots were obtained using a solid-state approach. Briefly, malt bagasse was used to produce carbon dots using pyrolysis. The obtained carbon dots have an average size between 1 nm and 3nm and a spherical shape. Interestingly, presented a sensitivity to the antibiotic tetracycline, a contaminant with emerging concerns (Cechinel *et al.*, 2025).

Renewable feedstocks in green nanotechnology are also used as templates and/or reducing/oxidizing agents (Strieder *et al.*, 2024). Renewable molecules are mixed with precursors as templates to direct the shape and control the size of NPs. This approach is mainly applied in the hydroxyapatite NP synthesis (Ali *et al.*, 2021). Hydroxyapatite NPs were obtained using licorice (*Glycyrrhiza glabra*) herb extracts as a template. The hydroxyapatite NPs obtained in the presence of plant extracts presented a more uniform rod-like shape, compared to those obtained in the absence of plant extracts. These results show the role of plant extracts in controlling the morphology of NPs (Ali *et al.*, 2021). Hydroxyapatite NPs were obtained using *Euclea natalensis* root extracts, popularly known as mulala (in Mozambique). Without *E. natalensis* root extracts, the hydroxyapatite NPs obtained presented a rod-like shape with nano grooves on their surface. However, in the presence of plant extract, the hydroxyapatite NPs presented a spherical-like shape with smooth surfaces (Aguilar *et al.*, 2021).

A recent study reported using plant extracts as an oxidizing agent and stabilizer to obtain ZnO NPs (Mousa *et al.*, 2024). The work used beetroots, pomegranates, and seder for green synthesis of ZnO, applying a simple approach without pH adjustment for $Zn(OH)_2$ precipitation. The obtained NPs presented morphological features dependent on the plant extract used. ZnO NPs prepared without plant extracts presented mixed shapes, including spherical and rod-shaped. However, plant extracts formed mostly spherical for seder extract, hexagonal for beetroots, and hexagonal for pomegranates (Mousa *et al.*, 2024).

In another work, *E. natalensis* root extracts were used as stabilizers and reducing/oxidizing agents to synthesize zirconium oxide NPs (da Silva *et al.*, 2019). Zirconium oxide NPs presented spherical shapes and sizes between 5.90 nm and 8.54 nm. Furthermore, the crystallite size agreed well with the particle size measured by TEM, indicating the formation of single-crystal zirconium NPs (da Silva *et al.*, 2019).

Environmentally friendly feedstocks are being used to obtain other chemicals, such as polymers and low-molecular-weight organic molecules, for a wide range of applications (Alhodaib *et al.*, 2024; Kings *et al.*, 2025). Cellulose, a well-known biopolymer, was obtained using the aquatic plant *Pistia stratiotes.* The product presents a crystallite size of 8.2 nm and a particle size of 23.253 μm ± 6.55 μm. Furthermore, microcrystalline cellulose presented a thermal stability comparable to that of conventional raw materials (Kings *et al.*, 2025).

A recent study produced phenolic monomers from lignin without a catalyst. The process consisted of optimization of the solvent/solid ratio and lignin depolymerization. The depolymerization approach was performed using a stirred and high-pressure reactor. The work showed that at 300°C, 120 minutes, and in water/acetone/formic acid medium, the maximum yield of phenolic monomers and biooil was achieved (Zhao *et al.*, 2024). Renewable feedstocks have provided innovation, sustainability, and a circular economy. However, replicating process performance is challenging due to their variability in composition. Therefore, a customizable innovation to tackle this variability is still needed. Despite this limitation, novel products and processes are being developed from renewable resources, providing innovation, less environmental impact, and a circular economy.

2.3.8 Reduce Derivatives

The derivatives application in chemistry, mainly in organic and analytical chemistry, has been an exciting strategy to protect some groups from undesirable reactions. However, when these compounds are used, additional reactants are necessary to remove them from the reaction mixture. Although of great importance, this approach results in waste generation. For these reasons, the green principle "reduce derivatives" has been formulated to reduce waste generation by avoiding or lowering derivatives in chemistry, mainly in organic chemistry and synthetic approaches (Anastas & Warner, 2000). Briefly, the eighth principle of green chemistry is *that Unnecessary derivatization (e.g., installation/removal of use-protecting groups) should be minimized or avoided, if possible, because such steps require additional reagents and can generate waste* (Anastas & Eghbali, 2010).

In organic synthesis, using derivative molecules to protect some groups is a consolidated practice (Schelhaas & Waldmann, 1996). Traditionally, the derivatization approach consists of forming a covalent bond between the protective molecule and a certain group that is a target for protection. For example, the synthesis of some peptides with biomedical interest involves the protection of non-participating amino groups. In this case, protecting groups are used, resulting in waste (Figure 2.5). Therefore, the search for alternatives is an object of study in organic chemistry (Liu *et al.*, 2024).

A more recent study performed an inverse chemical peptide synthesis using transiently protected amino acids (Liu *et al.*, 2024). The strategy briefly consists of activation, transient protection, aminolysis, and *in situ* deprotection (Figure 2.6). All these steps were performed in a one-pot system.

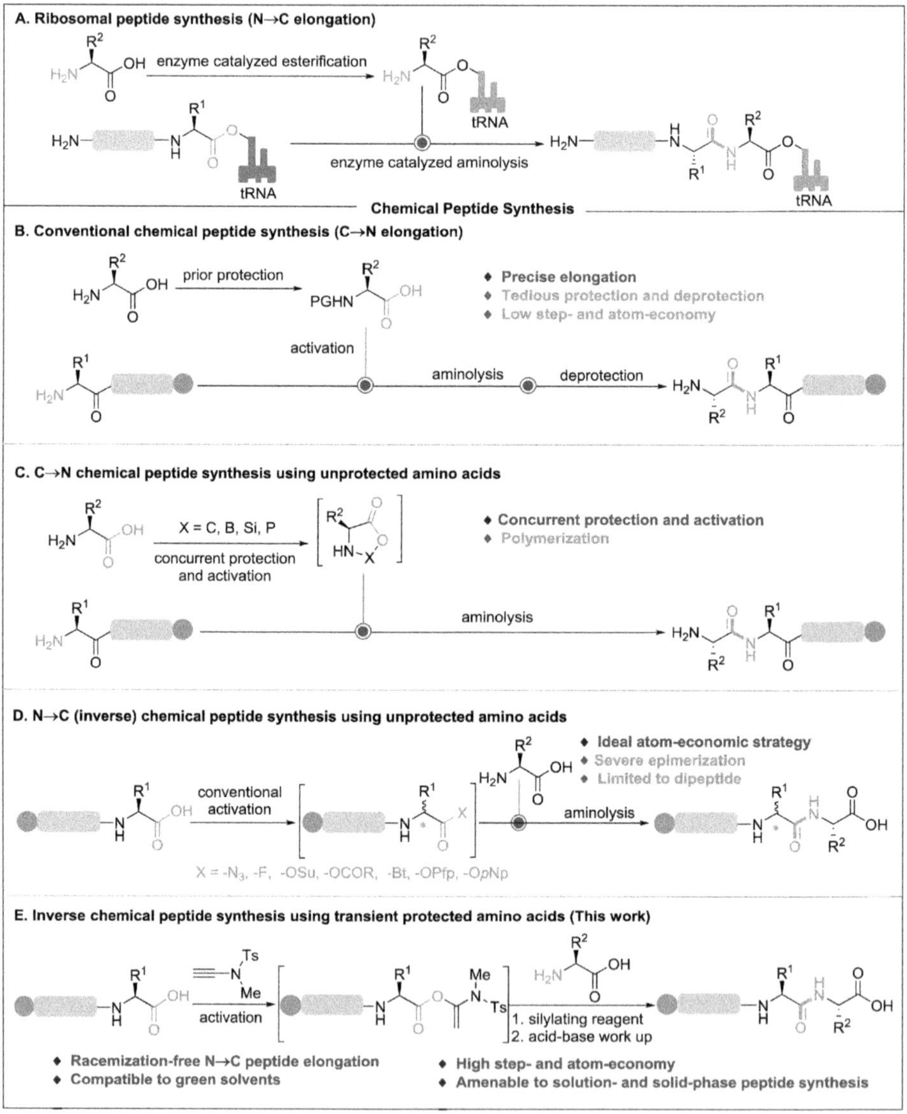

FIGURE 2.5 Ribosomal and chemical peptide synthesis (Liu et al., 2024).

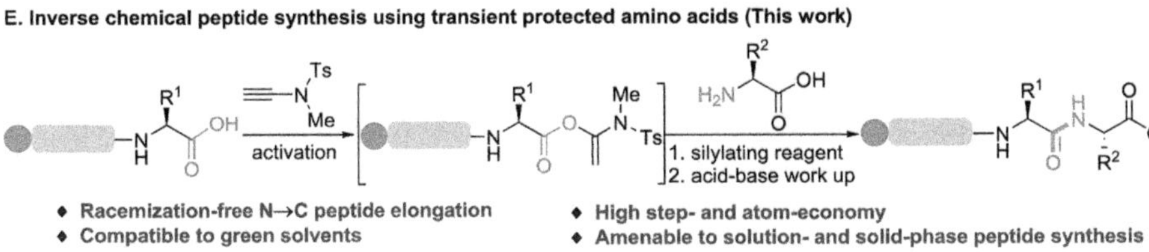

FIGURE 2.6 Chemical peptide synthesis using transient protected amino acids (Liu et al., 2024).

Beyond mitigating the environmental impact of waste generation, the proposed strategy addresses a century-old synthetic challenge in organic chemistry: the efficient production of peptides without the use of protecting groups (Liu *et al.*, 2024).

Another strategy involves non-covalent derivatization. Briefly, this strategy consists of self-assembly and molecular recognition. In this approach, the target molecule interacts using intermolecular forces such as hydrogen bonds. Derivation is guided by a fundamental understanding of molecular behavior and intrinsic chemical tendencies. Understanding this phenomenon allows scientists to modify the properties of target molecules by reducing or increasing their polarity, for example. The same aim is conventionally performed by forming a covalent bond with a derivatizing agent (Cannon & Warner, 2002). A non-covalent derivatization was applied to produce dicarboxylic acids with urea. The main goal of this work was to provide a controlled release of urea, a major nitrogen source for plant nutrition. The co-crystals were obtained by combining dicarboxylic acid with urea in various stoichiometric ratios (2:1, 1:1, and 1:2). In both co-crystals, the peak of urea disappeared, and a new peak appeared at $2\theta=6$ and between $2\theta= 10$ and $2\theta = 20$, indicating the formation of a new phase. According to the authors, the co-crystal obtained in their work was novel. More importantly, the co-crystals released the urea slowly, providing a promising approach for sustainable agriculture (Faulks *et al.*, 2025).

In analytical chemistry, derivatization is commonly employed, particularly in techniques such as gas chromatography coupled with mass spectrometry (GC-MS). It is used to enhance the volatility and thermal stability of analytes, which are essential properties for effective analysis in GC-MS.A greener derivatization approach for patulin analysis was developed. The approach involved reacting the patulin with sodium benzene-1,2-bis(thiolate) (a derivatizing agent) to generate the adduct. This strategy provided a clean and efficient approach to analyze the patulin with higher sensitivity using liquid chromatography coupled with tandem time-of-flight mass spectrometry (LC-QTOF-MS) (Duncan *et al.*, 2023). This approach is greener because the conventional methods involve laborious processes and the generation of large amounts of waste to derivatize the patulin before the analysis.

Several approaches have been implemented to make the derivatization process greener. These approaches include microwaves, radiation, enzymatic approaches, electrochemical derivatization, etc. All these approaches do not remove the derivative molecules entirely, but provide paths that make the processes more analytically greener.

2.3.9 Catalysis

Catalysis is a central principle in green chemistry. Traditionally and didactically, catalysts are used to accelerate the rate of chemical reactions. However, catalysts present benefits such as reducing the use of hazardous substances, improving efficiency and selectivity, and saving energy (Anastas *et al.*, 2001). The green chemistry principle of catalysis emphasizes that catalytic reagents, particularly those exhibiting high selectivity, are superior to stoichiometric counterparts due to their ability to increase reaction efficiency and minimize by-product formation (Anastas & Warner, 2000). Catalysts are revolutionizing the synthesis of several compounds, including pharmaceuticals, polymers, and novel catalysts. Traditional catalysts include solid catalysts, which are mainly composed of noble metals as the active part of the catalyst. On the other hand, a new and abundant substance is being used as a catalyst, carbon dioxide (CO_2) (Font *et al.*, 2018).

In the pharmaceutical industry, the use of catalysts represents a routine process. However, some molecules are to obtain under moderate conditions, even with catalysts. Daclatasvir is an antiviral drug conventionally obtained with higher quantities of catalyst, homocoupling ligands, and organic solvents. A research obtained this antiviral using water and ligand-free homocoupling in the presence of Pd/C catalysts (Mhate *et al.*, 2025).

In recent years, research has emerged in the literature that uses carbon dioxide as a promoter/mediator/catalyst in synthesizing pharmaceutical-relevant substances. CO_2 is an abundant, well-known molecule. For the first time, a selective *meta*-functionalization of fluoroarenes using carbon dioxide as a transient directing group was reported (Figure 2.7) (Font *et al.*, 2018). The approach involves lithiation, carboxylation, and tandem arylation and decarboxylation of fluoroarene and aryl iodide coupling partners. More importantly, the synthesis occurs in one pot. This method not only provides a sustainable strategy for meta-arylation of fluoroarenes, but is also applicable to other groups.

The Das group published an in-depth review article on the use of CO_2 in the synthesis of fine chemicals and pharmaceuticals in 2021 (Sahoo et al., 2021). In this review, the authors summarized all the possibilities of using carbon dioxide to obtain naturally occurring products and relevant molecules in the pharmaceutical industry (Sahoo *et al.*, 2021).

Catalysts are core in green chemistry, and several advances have been achieved. These advances include intensifying the use of biocatalysts, designing more selective catalysts with a high reusability index, and applying catalysts in environmental remediation. However, the innovation window is still open, and more research is still needed.

2.3.10 Design for Degradation

Many products from the chemical industry are discarded in nature after their useful life. Most of these products are difficult to degrade by environmental conditions (biotic and abiotic conditions). Among these hardly degraded products, plastics represent the top of the list. The lack of degradation of these products or the degradation

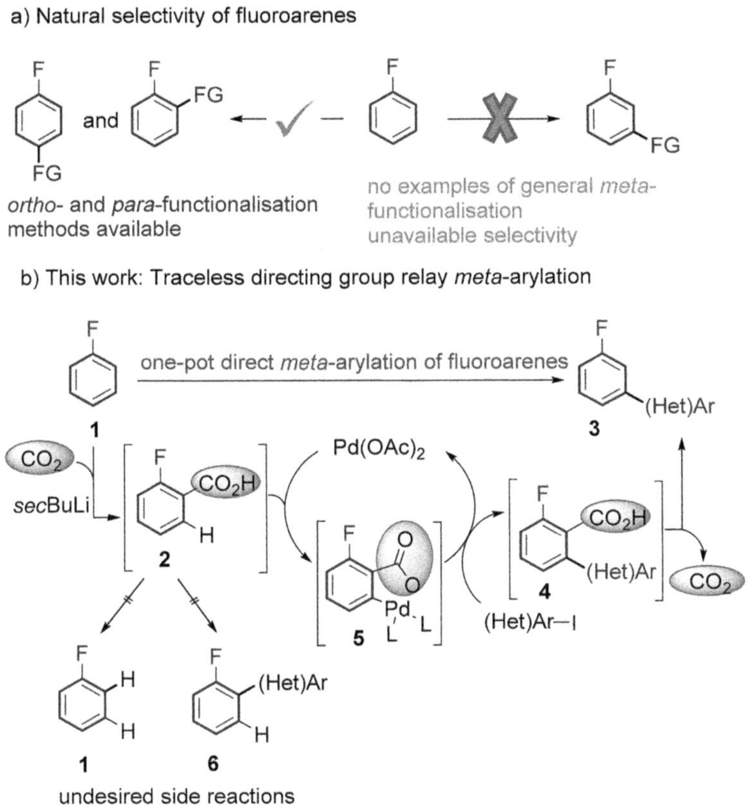

FIGURE 2.7 Proposed approach for regioselective meta-arylation of fluoroarenes (Font et al., 2018).

resulting in more toxic chemical species has been a challenge. For these reasons, the green chemistry principle, which is designed for degradation, is fundamental. The green chemistry principle, design for degradation, states that *chemical products should be designed so that at the end of their function, they break down into innocuous degradation products and do not persist in the environment* (Anastas & Warner, 2000). Several products that follow the design for degradation have been developed. These products include polymers and low-molecular-weight molecules for diverse applications.

Polymers, mainly from crude oil, represent a plethora of environmental issues. For these reasons, research on polymers that follow the design for degradation has been reported. These biodegradable polymers are generally obtained from biomass. Aiming to tackle the proliferation of non-biodegradable food packaging materials, a biodegradable polymer reinforced with cellulose nanocrystals was developed (Andrade *et al.*, 2022). The composite obtained presented a mechanical performance comparable to commercial films. These results show that biodegradable materials, although providing environmentally friendly outputs, still present a good performance (Andrade *et al.*, 2022). Balancing environmental benefits and performance of biobased plastics has been addressed by recent work (Kwak *et al.*, 2025). Therefore, cellulose nanofiber films modified by esterification with short-chain fatty acids have been obtained. The product presents enhanced mechanical properties and retains biodegradability. These results indicated that the esterification of cellulose nanofibers can balance the mechanical performance and biodegradability (Kwak *et al.*, 2025).

Pharmaceutical products are another group that benefits from implementing design for degradation. Pharmaceutical products, such as antibiotics, are becoming major threats to the environment because they are not biodegradable and cause various concerns. For these reasons, toxicologists and environmentalists coined contaminants of emerging concern (CECs) for contaminants like pharmaceutical products.

Designing drugs by applying design for degradation is not easy. However, results from the past century have shown that modifying the drug molecule increases its biodegradability. This successful result was achieved by changing the well-known antitumor molecules cyclophosphamide (CP) and ifosfamide (IF) by obtaining analogs such as glucosylisophosphoramidmustard (β-d-Glc-IPM; INN=glufosfamide) and β-l-glucosylisophosphoramidmustard (β-l-Glc-IPM). Importantly, the analogs improved their biodegradability without losing the antitumoral effect (Kümmerer *et al.*, 2000). Another approach to implement the benign by design includes the application of computational advances coupled with theoretical/quantum calculations. The *in silico* tools were proposed in 2021. However, there is still more research for the elaboration of real products elaboration (Lorenz *et al.*, 2021).

Design by benignity or design for degradation is an exciting principle in green chemistry. Advances in this direction have been achieved in the formulation of novel catalysts, polymers, pharmaceuticals, etc. However, more research is still needed to provide more biodegradable products.

2.3.11 Real-Time Analysis for Pollution Prevention

Analytical chemistry plays a paramount role in green chemistry. Timely identifying molecules before reaching a dangerous level can avoid environmental concerns. In this way, the chemical process must be equipped with analytical tools to monitor the hazardous compounds in their early formation stages. For these reasons, the green chemistry principle, which addresses these issues, was developed and states that *analytical methodologies must be further developed to allow for real-time, in-process monitoring and control before forming hazardous substances* (Anastas & Warner, 2000). With the advent of computational science, mainly AI, real-time prevention is expected to be boosted. The AI approach can use previous data to train larger models in languages, reducing costs, and consequently increasing sustainability outcomes (Figure 2.8).

Several advances in developing real-time analysis for pollution prevention have been made (Das *et al.*, 2025; Tang *et al.*, 2022; Zeng *et al.*, 2024). A recent study developed a system for monitoring cadmium in real-time using chemiluminescence (Zeng *et al.*, 2024). Another study showed that real-time monitoring data has the potential to reduce air pollution and carbon dioxide in the cement sector (Tang *et al.*, 2022).

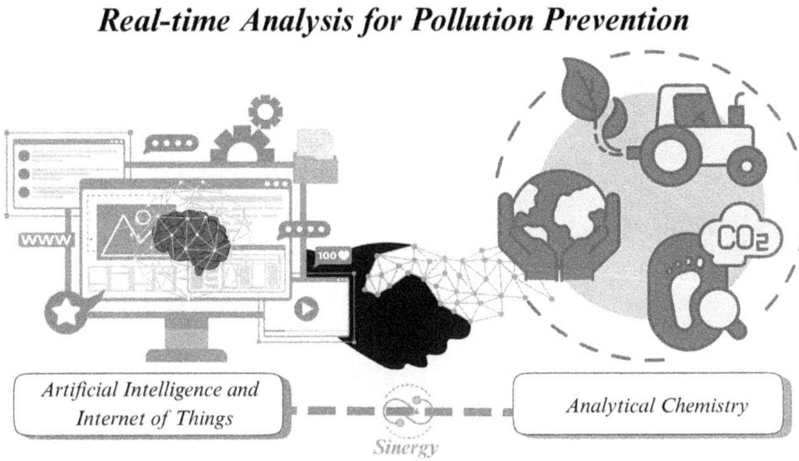

FIGURE 2.8 Synergy of AI and IoT, and analytical chemistry on boosting the real-time analysis for pollution prevention.

More recently, a real-time water pollution monitoring system was developed by integrating computer vision-based AI and large language models. To train the vision model, the main contaminant indicators were used. Furthermore, this information was connected to the ChatGPT API to generate alerts regarding the early-stage contamination (Samuel *et al.*, 2024).

2.3.12 Inherently Safer Chemistry for Accident Prevention

Traditionally, accident prevention was tackled mainly by elaborating security norms to protect workers from exposure to hazardous substances, such as API. However, in light of green chemistry principles, accident prevention can be approached by designing an eco-friendly process that avoids producing hazardous chemicals (Andrade-Guel *et al.*, 2020). Process customization to attend to the green chemistry principle includes substituting raw materials and innovation (Figure 2.9). Inherently Safer Chemistry for Accident Prevention green chemistry principle states that *Substances and the form of a substance used in a chemical process should be chosen to minimize the potential for chemical accidents, including releases, explosions, and fires* (Anastas & Warner, 2000).

Inherent safety in the chemical industry is a complex issue. It can be approached from the perspectives of safety and design/process engineering. Independent of the perspective, several tools and metrics are used. These include inherent safety assessment tools (ISATs), inherent safer design (ISD), and risk management metrics. All these issues must be addressed, considering the environmental aspects, economic, sustainability, and safety.

In the last 15 years, the integration of AI in inherently safety processes has been reported (Karun *et al.*, 2022). By applying machine learning and fuzzy logic, an accuracy of 99.9% in terms of prediction was achieved (Karun *et al.*, 2022).

FIGURE 2.9 Elements of the 12th green chemistry principle.

2.4 ADVANCES IN GREEN CHEMISTRY

Driven by regulatory pressures, societal demands for sustainability, and scientific and technological advances, green chemistry has evolved significantly in the last few decades. This section explores cutting-edge innovations in materials science, sustainable production processes, and cross-cutting applications that are reshaping industries through the perspective of environmentally responsible chemistry.

2.4.1 Innovations in Materials Science

Advances in materials science are fundamental to consolidating the principles of green chemistry, promoting more sustainable technological alternatives in several areas. The development of functionalized nanomaterials through green routes represents the main advances and is a significant achievement in green materials. One of the most promising fields of study is the synthesis of NPs through plant extracts, sugars, amino acids, or microorganisms, which eliminates the need for toxic reagents, acting simultaneously as synthesis agents and stabilizers(Kazemi et al., 2023).

Simultaneously, biopolymer engineering has experienced significant advances, especially in the search for high-performance compostable materials. Polymers such as PLA, polyhydroxyalkanoates (PHAs), and bacterial cellulose are improved through surface modification techniques, copolymerization, and additives with compounds obtained through green routes (Samir *et al.*, 2022). Functionalizing these materials with inorganic or organic fillers of sustainable origin has enabled the development of biopolymers with intelligent properties, such as self-repair, antimicrobial action, and hydrophobicity. These characteristics broaden the range of applications for these materials, which are now used in active packaging, biomedical devices, and disposable electronics with low environmental impact (Nadim *et al.*, 2025).

Another important area of innovation is using green solvents, like deep eutectic solvents (DESs). By replacing conventional organic solvents, which are often toxic and volatile, DESs provide a safer and more environmentally friendly alternative (Perna *et al.*, 2020). Composed of simple combinations of organic salts with proton donors, such as choline and lactic acid, these solvents have been successfully used in processes for extracting bioactive compounds and in producing polymeric materials more cleanly and efficiently (van Erven *et al.*, 2024).

Photocatalytic materials have been playing a crucial role in environmental remediation, especially in the treatment of contaminated water. Semiconductor nanocomposites based on TiO_2, ZnO, and CeO_2, often doped with transition metals or rare earths, have been synthesized by hydrothermal routes or by green methods (Nawaz *et al.*, 2023; Sikiru *et al.*, 2022). These materials are applied in solar photocatalysis processes to degrade persistent organic pollutants, such as textile dyes, antibiotics, and pesticides. Such technologies allow simulating and accelerating natural photodegradation processes, promoting the mineralization of toxic compounds with high efficiency and low formation of undesirable by-products.

2.4.2 Cross-Industry Applications and Success Stories

The principles of green chemistry have been incorporated into various industrial sectors, promoting success stories that combine technological innovation, functional performance, and commitment to sustainability. Pharmaceutical companies like Merck and Pfizer have been reformulating synthetic routes to increase atomic efficiency and reduce process toxicity (Kar *et al.*, 2022). An emblematic case is synthesizing the active ingredient in Sitagliptin, an antidiabetic drug produced by Merck, which was optimized using an enzymatic transaminase in an aqueous medium. This substitution eliminated the need for a toxic metal reagent. It allowed for a wide reduction in waste, representing an advance in clean chemistry and process efficiency (Khobragade *et al.*, 2023).

In the packaging and functional plastics sector, the development of biodegradable materials has gained prominence. Companies such as NatureWorks and BASF have been investing in producing polymers such as PLA, Ecoflex®, and Ecovio®, which have controlled degradation and can be composted under industrial conditions (Siegenthaler *et al.*, 2011). At the same time, there is a growth in active films containing NPs with photoprotective properties. These materials are being used in food packaging with functional properties, and their production prioritizes techniques such as thermal extrusion combined with natural additives and compatibilizers with low environmental impact (Grzebieniarz *et al.*, 2023).

In the agribusiness, the principles of green chemistry have been guiding the development of more sustainable agricultural pesticides. Innovative formulations include pesticides encapsulated in biodegradable matrices with controlled release, reducing leaching and adverse environmental impacts. One example is Eden Research, which develops natural terpene-based microcapsules with fungicidal action that apply to organic crops (Soto *et al.*, 2022). In addition, the extraction of bioactive compounds from biological sources has been enhanced through green technologies – such as ultrasound-assisted extraction and supercritical fluid extraction – which eliminate the use of toxic solvents and significantly reduce energy consumption (Molino *et al.*, 2020).

Furthermore, the paints and coatings sector has also adopted more environmentally friendly strategies. Free of volatile organic compounds (VOCs), water-based formulations have been replacing solvent-based systems. Large companies such as AkzoNobel and Sherwin-Williams are investing in emulsion polymerization technologies to produce acrylic resins and develop environmentally friendly inorganic pigments and additives with low toxicity. Photocatalytic paints containing doped TiO_2 are also applied to external surfaces, such as building facades, to promote air purification by photodegrading atmospheric pollutants such as nitrogen oxides (NO_x) and sulfur dioxide (SO_2) (Vallieres *et al.*, 2023; Yu *et al.*, 2020).

Finally, the electronics and advanced materials industry has been driven by innovations to reduce the environmental impact of electronic devices. The growing demand for sustainable electronics has stimulated the development of devices based on biodegradable materials and organic semiconductors, such as OLEDs (organic light-emitting diodes) and OFETs (organic field-effect transistors), obtained from compounds derived from biomass. These green materials replace traditional components containing bromine or heavy metals, and in some cases are integrated with compostable conductors made with functionalized cellulose, contributing to the creation of more responsible electronic technologies aligned with the circular economy (Islam *et al.*, 2023; Zheng *et al.*, 2021).

2.5 CONCLUSIONS

The principles of green chemistry offer a comprehensive framework for transforming chemical research, production, and application in line with sustainability, safety, and efficiency. While initially associated primarily with environmental and health benefits, growing evidence and innovation have demonstrated that green chemistry enables economic viability and competitiveness. Nevertheless, global geopolitical instability and economic uncertainties, such as potential recessions, may deprioritize sustainability efforts. It is therefore vital to emphasize a key message of this chapter: green chemistry is not a short-term trend but a long-term strategy for achieving true sustainability. Though often focused on environmental outcomes, its principles also address critical economic and social dimensions, making it a holistic approach to sustainable development. In this context, nations and organizations proactively adopting green chemistry principles will be better positioned to lead innovation and competitiveness. Countries previously classified as less developed – yet rich in biodiversity and increasingly investing in science and technology – have a unique opportunity to harness their native resources for the development of high-impact, exponential technologies. Coordinated international action is essential because climate change and resource scarcity pose urgent global threats. Adopting and integrating green chemistry into education, policy, and industrial practice is no longer an option; it is imperative. By practicing the green chemistry principles in all productive process, we establish a robust foundation for a more sustainable, equitable, and resilient global future.

ACKNOWLEDGMENTS

This work was supported by CAPES (finance code 001), FAPESC (processes 2021TR00029, 2022TR001399, and 2023TR000451), and CNPq (processes 313202/2021-4and 312247/2022-2).

REFERENCES

Aguilar, A. E. M., Fagundes, A. P., Macuvele, D. L. P., Cesca, K., Porto, L., Padoin, N., Soares, C., & Gracher Riella, H. (2021). Green synthesis of nano hydroxyapatite: Morphology variation and its effect on cytotoxicity against fibroblast. *Materials Letters, 284*, 129013. https://doi.org/10.1016/j.matlet.2020.129013

Ahmed, N. (2024). Advances and emerging trends in mechanistic insights of Sonogashira-type coupling reactions towards green protocols for sustainable organic transformations. *Journal of Organometallic Chemistry, 1016*, 123243. https://doi.org/10.1016/j.jorganchem.2024.123243

Ajala, E. O., Ighalo, J. O., Ajala, M. A., Adeniyi, A. G., & Ayanshola, A. M. (2021). Sugarcane bagasse: A biomass sufficiently applied for improving global energy, environment and economic sustainability. *Bioresources and Bioprocessing,8*(1). https://doi.org/10.1186/s40643-021-00440-zhttps://doi.org/10.1186/s40643-021-00440-z

Alhodaib, A., Yahya, Z., Khan, O., Equbal, A., Equbal, M. S., Parvez, M., Kumar Yadav, A., & Idrisi, M. J. (2024). Sustainable coatings for green solar photovoltaic cells: Performance and environmental impact of recyclable biomass digestate polymers. *Scientific Reports, 14*(1), 11221. https://doi.org/10.1038/s41598-024-62048-5

Ali, S. M. (2025). Biomass derived carbon/platinum nanoparticles as electrocatalyst for the hydrogen evolution. *Scientific Reports, 15*(1), 4602. https://doi.org/10.1038/s41598-024-84727-z

Ali, A. F., Alrowaili, Z. A., El-Giar, E. M., Ahmed, M. M., & El-Kady, A. M. (2021). Novel green synthesis of hydroxyapatite uniform nanorods via microwave-hydrothermal route using licorice root extract as template. *Ceramics International, 47*(3), 3928–3937. https://doi.org/10.1016/j.ceramint.2020.09.256

Anastas, N. D. (2016). Connecting toxicology and chemistry to ensure safer chemical design. *Green Chemistry, 18*(16), 4325–4331. https://doi.org/10.1039/C6GC00758A

Anastas, P., & Eghbali, N. (2010). Green chemistry: Principles and practice. *Chem. Soc. Rev, 39*(1), 301–312. https://doi.org/10.1039/B918763B

Anastas, P. T., Kirchhoff, M. M., & Williamson, T. C. (2001). Catalysis as a foundational pillar of green chemistry. *Applied Catalysis A: General, 221*(1–2), 3–13. https://doi.org/10.1016/S0926-860X(01)00793-1

Anastas, P. T., & Warner, J. C. (2000). *Green chemistry.* Oxford University Press. https://doi.org/10.1093/oso/9780198506980.001.0001

Andrade, M. S., Ishikawa, O. H., Costa, R. S., Seixas, M. V. S., Rodrigues, R. C. L. B., & Moura, E. A. B. (2022). Development of sustainable food packaging material based on biodegradable polymer reinforced with cellulose nanocrystals. *Food Packaging and Shelf Life, 31*, 100807. https://doi.org/10.1016/j.fpsl.2021.100807

Andrade-Guel, M., Cabello-Alvarado, C., Caicedo, C., Melo, L., & Ávila-Orta, C. (2020). Inherent safer chemistry for accident prevention. In *Green chemistry and applications* (pp. 244–275). CRC Press. Edited By Aide Sáenz-Galindo, Adali Facio, Raul Rodriguez-Herrera. https://doi.org/10.1201/9780429291166-14

Asio, J. R. G., Garcia, J. S., Antonatos, C., Sevilla-Nastor, J. B., & Trinidad, L. C. (2023). Sodium lauryl sulfate and its potential impacts on organisms and the environment: A thematic analysis. *EmergingContaminants, 9*(1), 100205. https://doi.org/10.1016/j.emcon.2023.100205

Bom, S., Fitas, M., Martins, A. M., Pinto, P., Ribeiro, H. M., & Marto, J. (2020). Replacing synthetic ingredients by sustainable natural alternatives: A case study using topical O/W emulsions. *Molecules*, *25*(21), 4887. https://doi.org/10.3390/molecules25214887

Bryan, M. C., Dunn, P. J., Entwistle, D., Gallou, F., Koenig, S. G., Hayler, J. D., Hickey, M. R., Hughes, S., Kopach, M. E., Moine, G., Richardson, P., Roschangar, F., Steven, A., & Weiberth, F. J. (2018). Key green chemistry research areas from a pharmaceutical manufacturers' perspective revisited. *Green Chemistry*, *20*(22), 5082–5103. https://doi.org/10.1039/C8GC01276H

Cannon, A. S., & Warner, J. C. (2002). Noncovalent derivatization: Green chemistry applications of crystal engineering. *Crystal Growth & Design*, *2*(4), 255–257. https://doi.org/10.1021/cg0255218

Carson, R. (1962). *Silent spring*. Readers Union, Hamish Hamilton.

Cechinel, M. A. P., Rocha, L. F., Padoin, N., & Soares, C. (2025). Dual functionality of malt bagasse-based carbonaceous material for obtaining carbon dots and adsorbent: Removing and detecting tetracycline. *Environmental Science and Pollution Research*. https://doi.org/10.1007/s11356-025-36409-0

Chen, W., Huang, X., Liu, Y., Luan, X., & Song, Y. (2019). The impact of high-tech industry agglomeration on green economy efficiency-evidence from the Yangtze River economic belt. *Sustainability (Switzerland)*, *11*(19). https://doi.org/10.3390/su11195189

Chen, S., & Yang, Q. (2024). Renewable energy technology innovation and urban green economy efficiency. *Journal of Environmental Management*, *353*. https://doi.org/10.1016/j.jenvman.2024.120130

Coates, G. W., & Getzler, Y. D. Y. L. (2020). Chemical recycling to monomer for an ideal, circular polymer economy. *Nature Reviews Materials*, *5*(7), 501–516. https://doi.org/10.1038/s41578-020-0190-4

Corvellec, H., Stowell, A. F., & Johansson, N. (2022). Critiques of the circular economy. *Journalof Industrial Ecology*, *26*(2). https://doi.org/10.1111/jiec.13187

Dalton, T., Faber, T., & Glorius, F. (2021). C–H activation: Toward sustainability and applications. *ACS Central Science*, *7*(2), 245–261. https://doi.org/10.1021/acscentsci.0c01413

Dananjaya, S. A. V., Chevali, V. S., Dear, J. P., Potluri, P., & Abeykoon, C. (2024). 3D printing of biodegradable polymers and their composites – Current state-of-the-art, properties, applications, and machine learning for potential future applications. *Progress in Materials Science*, *146*, 101336. https://doi.org/10.1016/j.pmatsci.2024.101336

Das, S., Khondakar, K. R., Mazumdar, H., Kaushik, A., & Mishra, Y. K. (2025). AI and IoT: Supported sixth generation sensing for water quality assessment to empower sustainable ecosystems. *ACS ES&T Water*, *5*(2), 490–510. https://doi.org/10.1021/acsestwater.4c00360

da Silva, A. F. V., Fagundes, A. P., Macuvele, D. L. P., de Carvalho, E. F. U., Durazzo, M., Padoin, N., Soares, C., & Gracher Riella, H. (2019). Green synthesis of zirconia nanoparticles based on eucleanatalensis plant extract: Optimization of reaction conditions and evaluation of adsorptive properties. *Colloids and Surfaces A: Physicochemical and Engineering Aspects*, *583*, 123915. https://doi.org/10.1016/j.colsurfa.2019.123915

De Gregori da Rocha, J., Santana Junior, M. B., Pier Macuvele, D. L., Riella, H. G., Ienczak, J. L., Padoin, N., & Soares, C. (2025). Uncovering engineering and mechanistic insights in green synthesis of carbon dots from rice husks. *Chemical Engineering Journal*, *505*, 159364. https://doi.org/10.1016/j.cej.2025.159364

Dhurjad, P., Dhalaram, C. S., Ali, N., Kumari, N., & Sonti, R. (2022). Metal-organic frameworks in chiral separation of pharmaceuticals. *Chirality*, *34*(11), 1419–1436. https://doi.org/10.1002/chir.23499

Dräger, P., & Letmathe, P. (2023). Who drives circularity?—The role of construction company employees in achieving high circular economy efficiency. *Sustainability (Switzerland)*, *15*(9). https://doi.org/10.3390/su15097110

Driscoll, J. P., Sadlowski, C. M., Shah, N. R., & Feula, A. (2020). Metabolism and bioactivation: It's time to expect the unexpected. *Journal of Medicinal Chemistry*, *63*(12), 6303–6314. https://doi.org/10.1021/acs.jmedchem.0c00026

Duncan, H., Juan, C., Mañes, J., Mercader, J. V., Abad-Somovilla, A., & Abad-Fuentes, A. (2023). Green derivatization strategy coupled to high-resolution mass spectrometry (QTOF-MS) for patulin monitoring in fruit products. *Talanta*, *253*, 124061. https://doi.org/10.1016/j.talanta.2022.124061

Eddarai, E. M., El Mouzahim, M., Ragaoui, B., Eladaoui, S., Bourd, Y., Bellaouchou, A., & Boussen, R. (2024). Review of current trends in chitosan based controlled and slow-release fertilizer: From green chemistry to circular economy. *International Journal of Biological Macromolecules*, *278*, 134982. https://doi.org/10.1016/j.ijbiomac.2024.134982

Farzadnia, N., & Khayat, K. H. (2023). Modification of nanomaterials for nanostructured cement-based materials. In *Nanotechnology for civil infrastructure* (pp. 5–37). Elsevier. Edited by: Kamal H. Khayat and Weina Meng. https://doi.org/10.1016/B978-0-12-817832-4.00008-0

Faulks, T., Huang, R., Forsyth, C. M., Blair, V. L., Turney, T., Warner, J. C., & Patti, A. (2025). Control of urea release through noncovalent derivatization with aliphatic dicarboxylic acids. *Green Chemistry Letters and Reviews*, *18*(1). https://doi.org/10.1080/17518253.2025.2478881

Font, M., Spencer, A. R. A., & Larrosa, I. (2018). *Meta* -C–H arylation of fluoroarenes*via* traceless directing group relay strategy. *Chemical Science*, *9*(35), 7133–7137. https://doi.org/10.1039/C8SC02417K

Gabrielli, P., Rosa, L., Gazzani, M., Meys, R., Bardow, A., Mazzotti, M., & Sansavini, G. (2023). Net-zero emissions chemical industry in a world of limited resources. *One Earth*, *6*(6), 682–704. https://doi.org/10.1016/j.oneear.2023.05.006

Ganesh, K. N., Zhang, D., Miller, S. J., Rossen, K., Chirik, P. J., Kozlowski, M. C., Zimmerman, J. B., Brooks, B. W., Savage, P. E., Allen, D. T., & Voutchkova-Kostal, A. M. (2021). Green chemistry: A framework for a sustainable future. *Environmental Science & Technology*, *55*(13), 8459–8463. https://doi.org/10.1021/acs.est.1c03762

García-Álvarez, J., García-Garrido, S. E., & Cadierno, V. (2013). Nanocatalysts for rearrangement reactions. In *Nanocatalysis synthesis and applications* (pp. 251–285). Wiley. Edited by: Vivek Polshettiwar and Tewodros Asefa. https://doi.org/10.1002/9781118609811.ch8

Głowniak, S., Szczęśniak, B., Choma, J., & Jaroniec, M. (2021). Mechanochemistry: Toward green synthesis of metal–organic frameworks. *Materials Today*, *46*, 109–124. https://doi.org/10.1016/j.mattod.2021.01.008

Grzebieniarz, W., Biswas, D., Roy, S., & Jamróz, E. (2023). Advances in biopolymer-based multi-layer film preparations and food packaging applications. *Food Packaging and Shelf Life*, *35*, 101033. https://doi.org/10.1016/j.fpsl.2023.101033

Han, Y., Zhou, X., Fang, R., Lu, C., Wang, K., Gan, Y., He, X., Zhang, J., Huang, H., Zhang, W., Xia, X., & Xia, Y. (2023). Supercritical carbon dioxide technology in synthesis, modification, and recycling of battery materials. *Carbon Neutralization*, *2*(2), 169–185. https://doi.org/10.1002/cnl2.49

Hanifa, M., Agarwal, R., Sharma, U., Thapliyal, P. C., & Singh, L. P. (2023). A review on CO_2 capture and sequestration in the construction industry: Emerging approaches and commercialised technologies. *Journal of CO_2 Utilization*, *67*, 102292. https://doi.org/10.1016/j.jcou.2022.102292

Huang, B., Chen, F., Shen, Y., Qian, K., Wang, Y., Sun, C., Zhao, X., Cui, B., Gao, F., Zeng, Z., & Cui, H. (2018). Advances in targeted pesticides with environmentally responsive controlled release by nanotechnology. *Nanomaterials*, *8*(2), 102. https://doi.org/10.3390/nano8020102

Islam, A., Usman, K., Haider, Z., Alam, M. F., Nawaz, A., & Sonar, P. (2023). Biomass-derived materials for interface engineering in organic/perovskite photovoltaic and light-emitting devices. *Advanced Materials Technologies*, *8*(7). https://doi.org/10.1002/admt.202201390

Jarman, W. M., & Ballschmiter, K. (2012). From coal to DDT: The history of the development of the pesticide DDT from synthetic dyes till silent spring. *Endeavour*, *36*(4), 131–142. https://doi.org/10.1016/j.endeavour.2012.10.003

Javaid, M., Haleem, A., Singh, R. P., Suman, R., & Rab, S. (2021). Role of additive manufacturing applications towards environmental sustainability. *Advanced Industrial and Engineering Polymer Research*, *4*(4), 312–322. https://doi.org/10.1016/j.aiepr.2021.07.005

Jia, X., Wang, T., & Zhu, H. (2023). Advancing computational toxicology by interpretable machine learning. *Environmental Science & Technology*, *57*(46), 17690–17706. https://doi.org/10.1021/acs.est.3c00653

Kar, S., Sanderson, H., Roy, K., Benfenati, E., & Leszczynski, J. (2022). Green chemistry in the synthesis of pharmaceuticals. *Chemical Reviews*, *122*(3), 3637–3710. https://doi.org/10.1021/acs.chemrev.1c00631

Karun, B., Renjith, V. R., & Elayidom, S. (2022). Application of fuzzy logic and machine learning techniques to improve inherently safer design in process safety management: A brief study. *Process Safety Progress*, *41*(S1). https://doi.org/10.1002/prs.12331

Kaur, N., & Banik, N. (2024). Industrial applications of solid base catalysis. In *Solid base catalysts* (pp. 169–231). Wiley. Edited by: K. K. Pant, Ravi Tomar and Ramesh Chandra. https://doi.org/10.1002/9783527846719.ch7

Kazemi, S., Hosseingholian, A., Gohari, S. D., Feirahi, F., Moammeri, F., Mesbahian, G., Moghaddam, Z. S., & Ren, Q. (2023). Recent advances in green synthesized nanoparticles: From production to application. *Materials Today Sustainability*, *24*, 100500. https://doi.org/10.1016/j.mtsust.2023.100500

Khobragade, T. P., Pagar, A. D., Giri, P., Sarak, S., Jeon, H., Joo, S., Goh, Y., Park, B.-S., & Yun, H. (2023). Biocatalytic cascade for synthesis of sitagliptin intermediate employing coupled transaminase. *Biotechnology and Bioprocess Engineering*, *28*(2), 300–309. https://doi.org/10.1007/s12257-023-0085-5

Kings, A. J., Suyambulingam, I., Narayanaperumal, S., Miriam, L. R. M., Divakaran, D., Murali, A., & Han, S. S. (2025). Exploring biomass derived microcrystalline cellulose from the waste aquatic plant *Pistia stratiotes*: A comprehensive characterization for polymer composite reinforcement. *International Journal of Biological Macromolecules*, *300*, 140217. https://doi.org/10.1016/j.ijbiomac.2025.140217

Krusenbaum, A., Grätz, S., Tigineh, G. T., Borchardt, L., & Kim, J. G. (2022). The mechanochemical synthesis of polymers. *Chemical Society Reviews*, *51*(7), 2873–2905. https://doi.org/10.1039/D1CS01093J

Kumar, A., Duhan, S., Kumari, S., Devi, S., & Dhayal, S. S. (2021). Green nanomaterials: Synthesis, characterization, and their industrial applications. In *Handbook of smart materials, technologies, and devices* (pp. 1–20). Springer International Publishing. Edited by: Chaudhery Mustansar Hussain and Paolo Di Sia. https://doi.org/10.1007/978-3-030-58675-1_121-1

Kumar, P., & Singh, J. (2024). Exploring enzyme biotechnology's role in green chemistry and advancements in pharmaceutical technologies. In *Enzyme biotechnology for environmental sustainability*(pp. 465–495). Elsevier. Edited by: Praveen Dahiya, Joginder Singh and Ajay Kumar. https://doi.org/10.1016/B978-0-443-22072-2.00015-2

Kümmerer, K., Al-Ahmad, A., Bertram, B., & Wießler, M. (2000). Biodegradability of antineoplastic compounds in screening tests: Influence of glucosidation and of stereochemistry. *Chemosphere*, *40*(7), 767–773. https://doi.org/10.1016/S0045-6535(99)00451-8

Kwak, H., Jung, H., Park, S.-A., Park, S. B., Shin, G., Choi, S., Kim, H. J., & Jeon, H. (2025). Balancing performance and eco-friendliness: Esterified cellulose nanofiber films for next-generation food packaging. *Carbohydrate Polymers, 360*, 123612. https://doi.org/10.1016/j.carbpol.2025.123612

Lane, M. K. M., Rudel, H. E., Wilson, J. A., Erythropel, H. C., Backhaus, A., Gilcher, E. B., Ishii, M., Jean, C. F., Lin, F., Muellers, T. D., Wang, T., Torres, G., Taylor, D. E., Anastas, P. T., & Zimmerman, J. B. (2023). Green chemistry as just chemistry. *Nature Sustainability, 6*(5), 502–512. https://doi.org/10.1038/s41893-022-01050-z

Lipshutz, B. H., Taft, B. R., Abela, A. R., Ghorai, S., Krasovskiy, A., & Duplais, C. (2012). Catalysis in the service of green chemistry: Nobel prize-winning palladium-catalysed cross-couplings, run in water at room temperature. *Platinum Metals Review, 56*(2), 62–74. https://doi.org/10.1595/147106712X629761

Liu, T., Peng, Z., Lai, M., Hu, L., & Zhao, J. (2024). Inverse peptide synthesis using transient protected amino acids. *Journal of the American Chemical Society, 146*(6), 4270–4280. https://doi.org/10.1021/jacs.4c00314

Lorenz, S., Amsel, A.-K., Puhlmann, N., Reich, M., Olsson, O., & Kümmerer, K. (2021). Toward application and implementation of *in silico* tools and workflows within benign by design approaches. *ACS Sustainable Chemistry & Engineering, 9*(37), 12461–12475. https://doi.org/10.1021/acssuschemeng.1c03070

Martin, B., Sedelmeier, J., Bouisseau, A., Fernandez-Rodriguez, P., Haber, J., Kleinbeck, F., Kamptmann, S., Susanne, F., Hoehn, P., Lanz, M., Pellegatti, L., Venturoni, F., Robertson, J., Willis, M. C., & Schenkel, B. (2017). Toolbox study for application of hydrogen peroxide as a versatile, safe and industrially-relevant green oxidant in continuous flow mode. *Green Chemistry, 19*(6), 1439–1448. https://doi.org/10.1039/C6GC02899C

Martina, K., Cravotto, G., & Varma, R. S. (2021). Impact of microwaves on organic synthesis and strategies toward flow processes and scaling up. *The Journal of Organic Chemistry, 86*(20), 13857–13872. https://doi.org/10.1021/acs.joc.1c00865

Meijssen, M., Becattini, V., & Mazzotti, M. (2024). Integrated carbon capture and utilization in the cement industry: A comparative study. *ACS Sustainable Chemistry & Engineering, 12*(7), 2709–2718. https://doi.org/10.1021/acssuschemeng.3c07081

Mhate, M., Khanra, M., Saharia, N., Radhakrishnanand, P., Ravichandiran, V., & Swain, S. P. (2025). Pd/C as a heterogeneous catalyst for ligand-free homocoupling reaction of halo compounds in water and total synthesis of daclatasvir. *Applied Organometallic Chemistry, 39*(2). https://doi.org/10.1002/aoc.8006

Moeini, M. A., Hosseinpoor, M., &Yahia, A. (2022). 3D printing of cement-based materials with adapted buildability. *Construction and Building Materials, 337*, 127614. https://doi.org/10.1016/j.conbuildmat.2022.127614

Molino, A., Mehariya, S., Di Sanzo, G., Larocca, V., Martino, M., Leone, G. P., Marino, T., Chianese, S., Balducchi, R., & Musmarra, D. (2020). Recent developments in supercritical fluid extraction of bioactive compounds from microalgae: Role of key parameters, technological achievements and challenges. *Journal of CO_2 Utilization, 36*, 196–209. https://doi.org/10.1016/j.jcou.2019.11.014

Morseletto, P. (2020). Targets for a circular economy. *Resources, Conservation and Recycling, 153*. https://doi.org/10.1016/j.resconrec.2019.104553

Mousa, S. A., Wissa, D. A., Hassan, H. H., Ebnalwaled, A. A., & Khairy, S. A. (2024). Enhanced photocatalytic activity of green synthesized zinc oxide nanoparticles using low-cost plant extracts. *Scientific Reports, 14*(1), 16713. https://doi.org/10.1038/s41598-024-66975-1

Nadim, E., Major, I., Devine, D., & Paraskar, P. (2025). Biobased self-healing functional composites and their applications. *Journal of Materials Science: Composites, 6*(1), 3. https://doi.org/10.1186/s42252-025-00065-x

Nawaz, R., Ullah, H., Ghanim, A. A. J., Irfan, M., Anjum, M., Rahman, S., Ullah, S., Abdel Baki, Z., & Kumar Oad, V. (2023). Green synthesis of ZnO and black TiO_2 materials and their application in photodegradation of organic pollutants. *ACS Omega*, *8*(39), 36076–36087. https://doi.org/10.1021/acsomega.3c04229

Nowak, K., Jabłońska, E., & Ratajczak-Wrona, W. (2021). Controversy around parabens: Alternative strategies for preservative use in cosmetics and personal care products. *Environmental Research*, *198*, 110488. https://doi.org/10.1016/j.envres.2020.110488

Panda, S., Nanda, A., & Panigrahi, S. K. (2024). Potential utilization of waste plastic in sustainable geopolymer concrete production: A review. *Journal of Environmental Management*, *366*, 121705. https://doi.org/10.1016/j.jenvman.2024.121705

Park, J., Park, C., Gye, M. C., & Lee, Y. (2019). Assessment of endocrine-disrupting activities of alternative chemicals for bis(2-ethylhexyl)phthalate. *Environmental Research*, *172*, 10–17. https://doi.org/10.1016/j.envres.2019.02.001

Pena-Pereira, F., Wojnowski, W., & Tobiszewski, M. (2020). AGREE—Analytical GREEnness metric approach and software. *Analytical Chemistry*, *92*(14), 10076–10082. https://doi.org/10.1021/acs.analchem.0c01887

Perna, F. M., Vitale, P., & Capriati, V. (2020). Deep eutectic solvents and their applications as green solvents. *Current Opinion in Green and Sustainable Chemistry*, *21*, 27–33. https://doi.org/10.1016/j.cogsc.2019.09.004

Poursaitidis, E. T., Gkizis, P. L., Triandafillidi, I., & Kokotos, C. G. (2024). Organocatalytic activation of hydrogen peroxide: Towards green and sustainable oxidations. *Chemical Science*, *15*(4), 1177–1203. https://doi.org/10.1039/D3SC05618J

Ratvik, A. P., Mollaabbasi, R., & Alamdari, H. (2022). Aluminium production process: From Hall–Héroult to modern smelters. *ChemTexts*, *8*(2), 10. https://doi.org/10.1007/s40828-022-00162-5

Reetz, M. T., Qu, G., & Sun, Z. (2024). Engineered enzymes for the synthesis of pharmaceuticals and other high-value products. *Nature Synthesis*, *3*(1), 19–32. https://doi.org/10.1038/s44160-023-00417-0

Sahoo, P. K., Zhang, Y., & Das, S. (2021). CO_2-promoted reactions: An emerging concept for the synthesis of fine chemicals and pharmaceuticals. *ACS Catalysis*, *11*(6), 3414–3442. https://doi.org/10.1021/acscatal.0c05681

Saleem, M. (2022). Possibility of utilizing agriculture biomass as a renewable and sustainable future energy source. *Heliyon*,*8*(2). https://doi.org/10.1016/j.heliyon.2022.e08905

Samir, A., Ashour, F. H., Hakim, A. A. A., & Bassyouni, M. (2022). Recent advances in biodegradable polymers for sustainable applications. *Npj Materials Degradation*, *6*(1), 68. https://doi.org/10.1038/s41529-022-00277-7

Samuel, D. J., Sermet, Y., Cwiertny, D., & Demir, I. (2024). Integrating vision-based AI and large language models for real-time water pollution surveillance. *Water Environment Research*, *96*(8). https://doi.org/10.1002/wer.11092

Schelhaas, M., & Waldmann, H. (1996). Protecting group strategies in organic synthesis. *Angewandte Chemie International Edition in English*, *35*(18), 2056–2083. https://doi.org/10.1002/anie.199620561

Shamseldeen Fakhri, R., & Thanon Dawood, E. (2023). Limestone powder, calcined clay and slag as quaternary blended cement used for green concrete production. *Journal of Building Engineering*, *79*, 107644. https://doi.org/10.1016/j.jobe.2023.107644

Sheldon, R. A. (2008). Why green chemistry and sustainability of resources are essential to our future. *Journal of Environmental Monitoring*, *10*(4), 406. https://doi.org/10.1039/b801651h

Sheldon, R. A. (2023). The e factor at 30: A passion for pollution prevention. *Green Chemistry*, *25*(5), 1704–1728. https://doi.org/10.1039/D2GC04747K

Sheldon, R. A. (2024). Green chemistry and biocatalysis: Engineering a sustainable future. *Catalysis Today*, *431*, 114571. https://doi.org/10.1016/j.cattod.2024.114571

Sheldon, R. A., Basso, A., & Brady, D. (2021). New frontiers in enzyme immobilisation: Robust biocatalysts for a circular bio-based economy. *Chemical Society Reviews*, *50*(10), 5850–5862. https://doi.org/10.1039/D1CS00015B

Shuai, S., & Fan, Z. (2020). Modeling the role of environmental regulations in regional green economy efficiency of China: Empirical evidence from super efficiency DEA-Tobit model. *Journal of Environmental Management*, *261*. https://doi.org/10.1016/j.jenvman.2020.110227

Siegenthaler, K. O., Künkel, A., Skupin, G., & Yamamoto, M. (2011). *Ecoflex® and Ecovio®: Biodegradable, Performance-Enabling Plastics*(pp. 91–136). https://doi.org/10.1007/12_2010_106

Sikiru, S., Abiodun, O. J. A., Sanusi, Y. K., Sikiru, Y. A., Soleimani, H., Yekeen, N., & Haslija, A. B. A. (2022). A comprehensive review on nanotechnology application in wastewater treatment a case study of metal-based using green synthesis. *Journal of Environmental Chemical Engineering*, *10*(4), 108065. https://doi.org/10.1016/j.jece.2022.108065

Soto, E. R., Rus, F., & Ostroff, G. R. (2022). Yeast particles hyper-loaded with terpenes for biocide applications. *Molecules*, *27*(11), 3580. https://doi.org/10.3390/molecules27113580

Strieder, C., de, M., Macuvele, D. L. P., Soares, C., Padoin, N., & Riella, H. G. (2024). Plant-mediated green synthesis of cerium oxide nanoparticles: A critical perspective of some unclear issues. *Journal of Materials Research and Technology*, *30*, 6376–6388. https://doi.org/10.1016/j.jmrt.2024.05.022

Tamura, M., Nakagawa, Y., & Tomishige, K. (2022). Direct CO_2 transformation to aliphatic polycarbonates. *Asian Journal of Organic Chemistry*, *11*(11). https://doi.org/10.1002/ajoc.202200445

Tang, L., Ruan, J., Bo, X., Mi, Z., Wang, S., Dong, G., & Davis, S. J. (2022). Plant-level real-time monitoring data reveal substantial abatement potential of air pollution and CO_2 in China's cement sector. *One Earth*, *5*(8), 892–906. https://doi.org/10.1016/j.oneear.2022.07.003

Trost, B. (1991). The atom economy - a search for synthetic efficiency. *Science*, *254*(5037), 1471–1477. https://doi.org/10.1126/science.1962206

Tway, C. L., & Filip, S. V. (2025). Catalysis at the crossroads. *Science*, *388*(6742), 29–30. https://doi.org/10.1126/science.adw5529

Usman, M., Cheng, S., Boonyubol, S., & Cross, J. S. (2023). Evaluating green solvents for bio-oil extraction: Advancements, challenges, and future perspectives. *Energies*, *16*(15), 5852. https://doi.org/10.3390/en16155852

Vallieres, M., Jones, S. H., Schwartz-Narbonne, H., & Donaldson, D. J. (2023). Photochemical renoxification on commercial indoor photoactive paint. *Scientific Reports*, *13*(1), 17835. https://doi.org/10.1038/s41598-023-44927-5

van Erven, G., Boerkamp, V. J. P., van Groenestijn, J. W., & Gosselink, R. J. A. (2024). Choline and lactic acid covalently incorporate into the lignin structure during deep eutectic solvent pulping. *Green Chemistry*, *26*(12), 7101–7112. https://doi.org/10.1039/D4GC00909F

Veerakumar, P., Thanasekaran, P., Lu, K.L., Lin, K.C., & Rajagopal, S. (2017). Computational studies of versatile heterogeneous palladium-catalyzed suzuki, heck, and Sonogashiracoupling reactions. *ACS Sustainable Chemistry & Engineering*, *5*(10), 8475–8490. https://doi.org/10.1021/acssuschemeng.7b00922

Wojnowski, W., Tobiszewski, M., Pena-Pereira, F., & Psillakis, E. (2022). AGREEprep – Analytical greenness metric for sample preparation. *TrAC Trends in Analytical Chemistry*, *149*, 116553. https://doi.org/10.1016/j.trac.2022.116553

Wu, S., Shao, Z., Andrew, R. M., Bing, L., Wang, J., Niu, L., Liu, Z., & Xi, F. (2024). Global CO_2 uptake by cement materials accounts 1930–2023. *Scientific Data*, *11*(1), 1409. https://doi.org/10.1038/s41597-024-04234-8

Wu, W., Wang, Y., Alofaysan, H., Junjie, X., & Shuo, C. (2025). Hydrogen energy and utilization: Economic and technological perspectives in China's transition to sustainability transition to sustainability. *International Journal of Hydrogen Energy, 127*, 737–749. https://doi.org/10.1016/j.ijhydene.2025.04.168

Yang, H., Yu, H., Stolarzewicz, I. A., & Tang, W. (2023). Enantioselective transformations in the synthesis of therapeutic agents. *Chemical Reviews, 123*(15), 9397–9446. https://doi.org/10.1021/acs.chemrev.3c00010

Yu, H., Dai, W., Qian, G., Gong, X., Zhou, D., Li, X., & Zhou, X. (2020). The NO_x degradation performance of nano-TiO_2 coating for asphalt pavement. *Nanomaterials, 10*(5), 897. https://doi.org/10.3390/nano10050897

Zeng, S., Zhu, H., Sohan, A. S. M. M. F., Liu, J., Wan, X., Lin, X., & Yin, B. (2024). A remote-controlled portable workstation for highly sensitive and real-time chemiluminescent detection of cadmium. *Food Chemistry, 452*, 139549. https://doi.org/10.1016/j.foodchem.2024.139549

Zhao, X., You, C., Li, X., Zhang, Y., & Wang, F. (2024). The depolymerization of lignin in water/acetone/formic acid synergistic solvents to produce high-value added phenolic monomers without external hydrogen and catalyst. *Fuel Processing Technology, 261*, 108102. https://doi.org/10.1016/j.fuproc.2024.108102

Zheng, Q., Li, H., Zheng, Y., Li, Y., Liu, X., Nie, S., Ouyang, X., Chen, L., & Ni, Y. (2021). Cellulose-based flexible organic light-emitting diodes with enhanced stability and external quantum efficiency. *Journal of Materials Chemistry C, 9*(13), 4496–4504. https://doi.org/10.1039/D1TC00019E

Zhong, Y., Sun, Y., Meng, L., & Li, X. (2024). Bimetallic Cu/Pd(ppm)-catalyzed suzuki-miyaura couplings: DFT studies on mechanism, roles of Pd(ppm) and ligands. *Molecular Catalysis, 553*, 113776. https://doi.org/10.1016/j.mcat.2023.113776

3 Design, Synthesis, and Analysis of Green Chemical Processes

Omer Faruk Sarioglu

3.1 INTRODUCTION

Nowadays, there is a strong emphasis on sustainable chemical manufacturing, and green chemistry is wanted to be integrated into the process design. Anastas and Warner (1998) defined green chemistry apparently, it is not merely a new take on the older, slightly more eco-friendly approaches from the past, instead, it is a complete departure from the bad habit of using dangerous chemicals simply because they work. Pollution prevention is at the core of this new perspective.

In the past, chemists aimed for "less toxic," "more biodegradable," or "friendlier" chemicals. That is a great advance, yet green chemistry goes further. The current goal is to generate "nonhazardous" substances from the beginning. That means designing processes that use and produce substances not dangerous, and making chemicals safe from the beginning. In fact, it is more than just a minor adjustment, it is a fundamental change in our perspective.

When we look at the potentials of green chemistry, the benefits are obvious. Maximizing atom economy (AE) and using renewable feedstocks are not just buzzwords, they are practical tools that yield real consequences. These concepts may have complexities, but the overall picture is obvious, green chemistry shifts our focus from concentrating on single molecules to embracing the entire process. This represents a transformation, but it also provides a toolkit. Green chemistry offers both principles and practical methods to help establish a sustainable chemical industry.

This chapter takes a close look at how to architect, whittle down, and make sense of very green chemical processes in a way that also assures a return on investment, a must for anything one might call sustainable.

3.2 FUNDAMENTAL PRINCIPLES OF GREEN CHEMISTRY

The 12 Principles of Green Chemistry (Anastas & Warner, 1998) serve as the foundation for sustainable process development. The items of 12 Principles of Green Chemistry are given in Table 3.1. Key focus areas selected to describe in further parts include:

- Waste minimization/Prevention
- Energy efficiency

DOI: 10.1201/9781003642589-3

TABLE 3.1
The 12 Principles of Green Chemistry According to Anastas and Warner(1998)

1. Prevent Waste
2. Atom Economy
3. Less Hazardous Synthesis
4. Design Benign Chemicals
5. Benign Solvents and Auxiliaries
6. Design for Energy Efficiency
7. Use of Renewable Feedstocks
8. Reduce Derivatives
9. Catalysis
10. Design for Degradation
11. Real-time Analysis for Pollution Prevention
12. Inherently Benign Chemistry for Accident Prevention

- Use of renewable feedstocks
- Design for degradation
- Inherently safer chemistry

The first key focus area is prevention/waste minimization. A common metric for waste is Roger Sheldon's E-factor (Sheldon, 2007), which compares the weight of coproduced waste to the desired product's weight. More recently, the ACS Green Chemistry Institute Pharmaceutical Roundtable (ACS GCIPR) has offered process mass intensity (PMI) (Cespi et al., 2015). PMI expresses the ratio of all materials used (water, organic solvents, raw materials, reagents, process aids) to the weight of the active pharmaceutical ingredient (API) produced. This focus is particularly important for the pharmaceutical industry, given the historically enormous amounts of wastes produced during pharmaceutical manufacturing (often exceeding 100 kilograms per kilogram of API manufacturing). However, applying green chemistry principles to API process design frequently leads to dramatic waste reduction (in some conditions approaching to ten-fold decrease). It is of utmost importance to extend the outstanding outcomes from the ACS GCIPR to the entire pharmaceutical sector, which encompasses biopharma and generics, and including other chemical industries that rely on synthetic chemistry

The second key focus area can be given as energy efficiency. Most of the energy isn't consumed during the reactions themselves, instead, it's consumed in the processes like solvent removal to prepare for subsequent steps, switch of solvents, isolation of the desired product, or elimination of the impurities. While most reactions, aside from hydrogenations or those sensitive to oxygen or moisture, occur at atmospheric pressure, this does not eliminate the significance of energy. It simply highlights that energy consumption is often significant in areas where chemists typically do not focus normally. Therefore, a holistic approach is crucial for design of new molecules, considering the need of more than just a single reaction step or process are required for those. This broader perspective is not just focus on energy, such that energy should be viewed as another crucial design parameter for synthetic and process chemists. Hence, historically overlooked, chemists can no longer pursue designing new molecules without a detailed and comprehensive consideration of their energy footprint (ACS, 12 Principles of Green Chemistry, accessed 06/28/2025).

The third key focus area is use of renewable feedstocks. The idea of sourcing all our future fuels, chemicals, and materials from non-depleting feedstocks might initially seem unrealistic, though it will become a must in the future eventually. Humanity currently extracts fossil fuels (e.g., oil, natural gas) and minerals till exhaustion, primarily for profit. Our reliance on fossil fuels for carbon-based chemicals and materials results in their rapid and foreseeable depletion, which is a natural outcome of the rising global populations and expanding energy-intensive economies worldwide. This has significant implications for human health and the environment, posing prominent challenges for next decades. However, by using renewable feedstocks, humanity may have a chance to reduce global warming gases and mitigate climate change. Even though renewable substances have a high potential for chemical processes, their safety should be ensured that these new chemicals and materials derived from renewable resources are non-toxic or harmless to human health and the planet. In recent years, remarkable progresses have been made in development of fuels, chemicals, and materials from renewable feedstocks, including biodiesel production from plant oils and green algae, bioethanol and butanol production from sugars and lignocellulose, plastics, foams, and thermosets production from lignin and plant oils, and even production of electronic materials from chicken feathers. In line with 12 Principles of Green Chemistry's Principle 7, use of renewable feedstocks, the future seems promising, due to the ongoing, worthy collaborations across diverse disciplines like biotechnology, agronomy, physics, and engineering, and these collaborations are successfully yielding new fuels, chemicals, and materials from renewable feedstocks "out of thin air" without causing noticeable harm on people's health and the environment (ACS, 12 Principles of Green Chemistry, accessed 06/28/2025).

The fourth key focus area is design for degradation. If a chemical possess risk due to its inherent toxicity, then there is a risk due to potential contact of this chemical with a living organism. Even though the chemicals possess this risk, degradation of these chemicals can dramatically minimize potential exposure, thereby eliminating risk irrespective of the chemical's intrinsic toxicity. A core principle for green chemistry is considering the molecular features responsible for hazardous characteristics and risk during design, and if molecular features of the desired chemicals may have an increased risk of intrinsic hazard, the best way is termination of their commercial production and seeking for alternative approaches and designs. Tools for degradation including biodegradation, hydrolysis, and photolysis can be intentionally designed into chemical products. In addition, a deep understanding of degradation and persistence mechanisms is also required to engineer chemical features that promote degradation and eliminate others which involved in persistence. For instance, a number of persistent compounds are heavily chlorinated, though halogens like chlorine can have a negative impact on microbial enzyme systems by inhibiting them, due to the nature of aerobic microbial degradation which favors presence of electron-rich structures in general. For design of degradation, prediction methods might be utilized for guiding the molecular designs with intended degradability, and these methods include models that assess biodegradability, databases of existing knowledge, and experimental testing. Here, one of the issue that should be considered is occurrence of degradation using relevant environmental compartments with a meaningful rate. A common example is remediation of domestic wastewater, which passes through effective bioreactors in wastewater treatment plants during remediation. For effective degradation of chemicals within domestic wastewater, companies have designed new molecules with higher degradability for removal within these bioreactors, and a renowned example is the shift of non-biodegradable branched surfactants to biodegradable linear alkyl benzene sulfonate-based surfactants, which still has positive contributions to today's society (ACS, 12 Principles of Green Chemistry, accessed 06/28/2025).

The fifth and last key focus area to be mentioned here is inherently safer chemistry. Safety, as its core focus, is about managing known hazardous substances to reach an admissible level of risk. Even though it is often overlooked, the "Safety Principle" of 12 Principles of Green Chemistry is a natural consequence of other green chemistry principles. Since green chemistry is strictly associated with reducing or eliminating hazardous substances during use or as a result of the process, it is inherently tied to laboratory safety as well. With only some exceptions, most green chemistry principles will lead to a safer outcome. Intrinsically, green chemistry aims to enhance environmental safety, by assuring substances and processes with higher safety for the environment and general society, and the positive results of higher safety can be first directly distinguished on the workers involved in manufacturing and laboratory settings, who will experience improved safety in their daily tasks (ACS, 12 Principles of Green Chemistry, accessed 06/28/2025).

3.3 GREEN PROCESS DESIGN STRATEGIES

Green process design strategies are crucial to reach sustainability in the chemical industry, and these strategies involve a comprehensive and integrated approach, hence covering environmental considerations from the earliest stages of chemical and process development to final stages. In addition, green process design strategies focus on inherent safety via ensuring utilization of less hazardous chemistries and processes, providing design of products with higher degradability that break down into benign substances after completing their lifecycle. This holistic perspective, often benefitted from strategies as Process Intensification (PI) and tools like Life Cycle Assessment (LCA), directing toward cleaner and more resource-efficient, hence more sustainable chemical manufacturing after all.

3.3.1 Process Intensification

PI is highly important in chemical engineering for a very good reason. It focuses on improving efficiency, shrinking equipment size, reducing energy use, and minimizing waste. PI does not simply adjust old setups, but it also introduces new techniques and smart designs that can achieve strict efficiency goals. For instance, microreactors allow precise control over heat and mass transfer, providing a better control than traditional setups. Everything occurs at the microscale, which brings new opportunities. Another development in this field is continuous manufacturing. Instead of using stop-and-go batches, this approach keeps production flowing, making the overall process much smoother. Hybrid separation processes are another smart approach, which combine several steps into one unit for streamlining operations later on. These changes do not just speed up production or reduce operation expenses, also maintain or even improve product quality and safety. The next few paragraphs explore these techniques in more detail and demonstrate how they work in practical use.

3.3.1.1 Microreactor Technology

Microreactor technology involves shrinking traditional chemical reactors down to sizes generally in the millimeter or even below millimeter range. The benefits are remarkable. The high surface area-to-volume ratio in microreactors improves heat and mass transfer rates, allowing reactions to happen faster and with more control. As a result, it reduces unwanted side reactions and hot spots, which are common in larger traditional reactors. The small volumes also enhance safety, especially when dealing with hazardous or exothermic reactions, since the amount of material at any given time is much lower in these compact systems. By keeping tight control over the conditions of a reaction generally leads to a higher yield and a more selective formation of the desired product is also achievable.

3.3.1.2 Continuous Manufacturing Processes

Most chemical processes traditionally work in batch mode. This means that certain amounts of raw materials are processed in stages. Each batch is followed by a clean-up period, during which materials from the last batch are adequately removed before the next round of processing begins. Conversely, continuous manufacturing flows materials steadily and uninterruptedly through a series of operations connected in line. This method achieves the same end as batch mode but more quickly. Continuous processes also have better control, which is important for maintaining product quality. A well-controlled continuous process will have less variability than a well-controlled batch process and will almost certainly have less variability than a badly controlled batch process. Finally, a continuous process is more efficient because it minimizes resource use and production of waste. Continuous processes don't need a clean-up cycle between batches because they don't have any batches to begin with.

3.3.1.3 Hybrid Separation Processes

In traditional chemical plants, various types of separation equipment, such as distillation columns, extractors, and crystallizers, are arrayed in sequence. Chemical engineers, however, are increasingly turning to hybrid separation processes that offer a more integrated approach to separation. These techniques are a relatively new development, though they are based on well-understood principles of separation. Hybrid separation processes generally use two types of separation units that are closely linked together. A good example is reactive distillation. It is not as simple as it sounds since a chemical reaction and distillation occur simultaneously in the same column in this process. This integration can overcome the thermodynamic limits, improve reaction efficiency, and reduce energy use and costs by eliminating the need for a separate equipment. In addition, other examples of hybrid separation processes may include the closely linked combination of membrane separation with distillation and the combination of adsorption with crystallization, both of which are in accordance with the goal to obtain a cleaner product from a chemical reaction.

3.3.2 Life Cycle Assessment (LCA)

The method serving as the basis for determination of the effects on the environment for a product or process is called as Life Cycle Assessment (LCA). It is used to consider the lifespan of a product, from raw material collection to the moment when the product is no longer needed. For Green Chemistry, LCA is a vital way of getting a handle on what the real impact is of a chemical development that could potentially go to scale. By aligning perfectly with Green Chemistry's fundamental principles, conducting a complete LCA can be quite simple and it always provides absolute and definitive answers about a product's "greenness." LCA helps to identify where additional improvements can be made, for instance, by calculating the carbon footprint of products. LCA is also helpful for identification of potential causes of some burdens of chemical industry, including high-energy consumption, production of significant amount of waste, or the use of hazardous materials, at any stage of a chemical's existence. By systematically assessing multiple environmental impact categories (e.g., global warming, ozone depletion, eutrophication), LCA ensures that improvements in one area do not cause unintentional cause of environmental problems to another. Although, conducting a full and comprehensive LCA can be very complex and data-intensive due to requirement of robust methodologies and assumptions, it can provide invaluable and quantifiable insights that guide chemists and engineers to design processes and products more efficiently with higher sustainability, following the principles of green chemistry.

3.4 SOLVENT SELECTION AND REPLACEMENT

Solvent selection and replacement are critical components of green chemistry strategies, as solvents often contribute the majority of process waste and they can lead to significant environmental and health hazards. Green solvent selection approach focuses on minimizing these impacts by use of alternative solvent systems to traditional volatile organic compounds (VOCs), involving the use of bio-based solvent systems which are derived from renewable biomass, known to be mostly biodegradable and having lower toxicity. Another exciting development is the use of supercritical fluids, especially supercritical CO_2, which has tunable solvent properties, is recyclable, and is of lower toxicity. Nonetheless, the most ideal approach is pursuing solvent-free reactions which do not need presence of a solvent, thereby achieving maximum economy and dramatically reducing production of wastes and their related purification costs.

Solvents contribute significantly to process waste. Green solvent selection criteria involve use of bio-based solvents, use of supercritical fluids (e.g., CO_2), and pursuit of solvent-free reactions. The following paragraphs describe these green solvent selection criteria in more detail.

3.4.1 Use of Bio-Based Solvents

Using bio-based solvents strategically is a big step forward for green chemistry. Bio-based solvents have a clear advantage over petroleum-based ones. They are more sustainable because we can produce them from renewable materials like crops or even waste. We do not need to depend on large fields growing a single crop just for biofuels that give poor energy returns. Instead, we can use various renewable sources to create these solvents. In many cases, the production processes for these solvents consume less energy and produce far fewer harmful by-products than the processes used to make solvents from petrochemicals, which contributes to their safety and supports the Green Chemistry principles.

3.4.2 Use of Supercritical Fluids

Using supercritical fluids in green chemistry is very promising and offers a substantial approach to the problem of solving bio-based solvents. Compared to traditional organic solvents, using a supercritical fluid offers greater benefits and flexibility in solution properties. The term supercritical fluid represents a stage of a substance, where it becomes supercritical fluid upon heating and pressurizing it beyond its critical point, providing gain of unique properties that are intermediate between those of a gas and a liquid, hence it can diffuse through solids like a gas while dissolving materials like a liquid. Supercritical carbon dioxide ($scCO_2$) is the most widely employed type of SCF due to its mild critical conditions, non-toxicity, non-flammability, and high availability (Armenta et al., 2022), which often sourced as a by-product from other industrial processes, making it a highly sustainable SCF. Applications of SCFs in green chemistry are diverse and effective, these are extensively used for extraction processes (e.g., decaffeinating coffee), extraction of essential oils from plants, and purification of APIs. Utilization of SCFs in chemical processes can result in production of cleaner, greener, and solvent-free products. In addition, SCFs have a potential to be utilized as an excellent reaction media, to improve reaction rates and selectivity in various chemical syntheses (e.g., hydrogenation and polymerization reactions), by offering tunability in solvent power. SCFs can be utilized in particle formation (e.g., micronization), environmental remediation for extraction of contaminants from soil, cleaning and degreasing applications (i.e., replacing harsh chemicals) as well. Nevertheless, there are also limitations for SCFs, and the major ones are the requirement of complex equipment, and high-energy consumption during the process, which is also directly proportional to the high-energy costs (Armenta et al., 2022). You can adjust the pressure and temperature of SCFs to change how the solvent works. This leads to better selectivity and efficiency. By producing less waste and eliminating harmful solvent leftovers, it fits well with several green chemistry principles.

3.4.3 Solvent-Free Reactions

Preference of solvent-free reactions is really concurrent with the Principle 5 of Green Chemistry. This principle focuses on avoiding unnecessary extras, like auxiliary substances. Even when a reaction does not produce other by-products, use of unwanted solvents can still harm the environment. The solvent-free approaches that are discussed here take sustainability to another level since they eliminate what is often the biggest hazard in the process. Indeed, if you are not adding extra materials, you are likely not wasting extra energy either. Most reactions require a boost of energy to get started, however, without a suitable solvent, it is impossible to increase the heat and conduct the reactions. Therefore, preference of solvent-free reactions provides the reaction to find a more direct and efficient path, in terms of energy use, and supports the goal of green chemistry principles.

3.5 REACTION ENGINEERING FOR SUSTAINABILITY

Ensuring sustainability in processes which involve chemical reactions on an industrial scale is impossible to achieve without improvements in the energy and material efficiency of these processes. Synthesis of chemical reactions must be integrated with separation steps to sufficiently simplify the overall processes and to notably cut down on the number of required unit operations. High selectivity and yield of synthesis reactions must be ensured if the desired products are to be obtained in good amounts and with good quality. Favorable conditions (i.e., pressure, temperature, etc.) must be maintained during the reaction, and any feedback that can be coupled in a control scheme to stabilize the synthetic process will also be helpful in maintaining overall selectivity.

Any process that is to be sustainable must also be economically viable. This implies that the costs associated with any catalyst design or reactor configuration must be sufficiently low to ensure overall process profitability. Finally, despite all of these conditions that have been specified, the overall process must also be safe, not only for the operators involved in the process but also for anyone who might come into contact with the products obtained from the process.

3.5.1 Catalyst Design

Catalyst design strategies are essential for meeting the sustainability goals of green chemistry. Catalysts play a key role in creating more efficient, selective, and sustainable chemical processes. The main goal of catalyst design is to accelerate reactions while creating better conditions that lower energy use and produce fewer hazardous by-products and waste. By using highly selective catalysts, such as biocatalysts, we can steer reactions mainly toward the desired product. Green catalyst design strategies can lower costs and reduce the need for energy-intensive separations. Additionally, these strategies emphasize using abundant, readily available, non-toxic, and renewable materials for making catalysts. This approach reduces the dependence on hazardous noble metals. Designing green catalysts is not just about improving their performance. It also involves making them easier to separate and reuse. For instance, catalysts with magnetic properties can be removed with a magnet and reused, which helps reduction of waste production along with environmental impact (Wang et al., 2025). Moreover, we can apply computational methods and AI to speed up the search for new catalysts and refine them for even better results. This pushes the chemical industry toward more sustainable and cost-effective practices (Su et al., 2024).

3.5.2 Microwave and Ultrasound-Assisted Synthesis

Microwave and ultrasound-assisted synthesis are two effective methods that are important in green chemistry. Microwave irradiation heats the reactants directly using dielectric heating. This raises the temperature quickly and evenly, reaching levels that traditional heating cannot achieve. As a result, microwave assistance can greatly reduce reaction times, often from hours to minutes or even seconds. It can also lead to higher yields and better selectivity, while reducing or eliminating the need for solvents. This helps minimizing the waste production and energy use. Likewise, ultrasound-assisted synthesis (sonochemistry) is based on use of high-frequency sound waves to induce acoustic cavitation, leading to the formation, growth, and explosive collapse of microscopic bubbles in a liquid, and this cavitation can generate localized hot spots with extreme temperatures and pressures, facilitating enhanced mass transfer, enhanced reactivity, and enhanced yields, even in heterogeneous systems (Suslick et al., 1999). Both microwave and ultrasound-assisted synthesis techniques accelerate reactions, reduce or eliminate the need for harsh reaction conditions, enable the use of greener solvents or promote solvent-free reactions, and contribute to inherently safer chemical processes by improvement of controlling and minimizing the scale of chemical reactions. Thus, both synthesis methods have valuable potential for sustainable chemical synthesis.

3.5.3 Photocatalysis

The method of carrying out photocatalysis makes it possible to perform very clean transformations in organic syntheses. The method is based on the use of a photocatalyst, which is a substance that, when illuminated with light of certain wavelengths, is able to drive a chemical reaction. In this case, we used a titanium dioxide (TiO_2) semiconductor, which, when dry, is a white powder. Under ultraviolet light, the TiO_2 drives the transformation of an organic molecule to its corresponding product. During this process, the photocatalyst does not become consumed and is able to drive the reaction over and over again, with the light energy enabling the necessary changes in the structure and bonding of the reacting molecules. In this textbook, we discover many such structures and transformations that can occur under the right conditions and by using the right (non)photonic (i.e., TiO_2 in sunlight) catalyst system.

3.6 GREEN METRICS FOR PROCESS EVALUATION

For those aiming to make processes and products in the chemical realm more sustainable, green metrics are truly essential. What are green metrics? Altogether, they offer a broader view, a bird's-eye view, if you will, of just how sustainable or unsustainable a given chemical process is. This is something that can definitely be applied to the kinds of difficult reactions that are conducted in the lab. Bolstering the appearance of sustainability in product plans is important because a broad view of sustainability allows chemists, chemical engineers, and other scientists to see not just the yield of a product, but also to see how much waste is being produced, the resource efficiency of the process, and the level of hazard found in the chemicals and solvents being used. The following paragraphs describe those different green metrics for process evaluation.

3.6.1 Atom Economy

Barry M. Trost introduced the idea of AE back in 1991, and honestly, it changed the way chemists think about reaction efficiency. Before AE came along, people mostly cared about yield, basically, how much of the target product you got compared to the amount of starting material you used. But that approach overlooked all the extra material you end up discarding. Trost's idea changed the focus. Instead of just measuring how much product you get, AE asks how many atoms from your starting materials actually make it into your final product instead of turning into waste. The fewer by-products you produce, the better your AE, which means you are making your process more efficient. When chemists discuss efficiency today, AE is the main topic. It is a simple concept: take all the atoms you start with and see how many end up in the end product you actually want. If your reaction has a high AE, this means that you are using your materials wisely, with very little going to waste. The goal is really simple; maximizing what you keep and minimizing what you lose.

3.6.2 The E-Factor

The E-factor serves as a vital environmental measurement tool in green chemistry. Roger A. Sheldon presented the E-factor concept for the first time in 1992 in his "Organic Synthesis: Past, Present and Future" paper, before expanding this concept in his subsequent publications. The E-factor serves as a basic yet powerful tool which evaluates the waste output from chemical manufacturing operations. The E-factor represents the waste-to-product mass ratio which equals kilograms of waste per kilogram of product. The E-factor helps chemists evaluate process efficiency through its measurement of all substances which fail to become part of the final product including reagents and solvents and process aids. The E-factor reveals unacceptable waste levels that exceed acceptable limits by ten times according to Sheldon's method when fine chemical and pharmaceutical industries use their typical multistep synthesis with stoichiometric reagents.

3.6.3 Process Mass Intensity (PMI)

PMI is a highly useful metric in green chemistry, especially in the pharmaceutical industry. It assesses how efficiently not just space but also materials are utilized in a chemical process. Most existing environmental metrics focus on what's coming out of the process, like waste and emissions (e.g., the E-factor, which has been a really useful and important tool for many people). PMI, however, focuses on what went into the whole process, even the materials that went into any pre- or post-reaction work. A lower PMI number is good and indicates an efficient, "green" process. PMI has become a common metric and many synthetic chemists, especially in the field of eco-pharmaceutics, gauge their work.

3.6.4 Reaction Mass Efficiency (RME)

Reaction mass efficiency (RME) is an important measure of green chemistry that counts not only the atoms that end up in the product (the yield), but also the atoms that don't end up in the product (the wasted mass). Happily, the calculation is simple; you just weigh the product and then weigh the starting materials. Actually, you should use the starting amounts of the starting materials, as you may need excess reagents to get a decent yield, or to ensure that just one product is made and not several look-alikes. Therefore, inclusion in RME is the concept that there may be "stewardship" as regards the appearance of the different species in the reaction mixture. To make the result of the reaction more environmentally benign, you want a high RME. A higher RME means fewer throwaway atoms and a much greener process.

3.7 ENABLING TECHNOLOGIES FOR GREEN CHEMISTRY

The enabling technologies of green chemistry are vital tools for putting green chemistry's principles into sensible industrial application. Of these tools, biocatalysis is perhaps the best understood and most frequently used. Biocatalysis employs enzymes or whole cells to produce reactions with high specificity and efficiency—often in dilute aqueous solutions and under mild reaction conditions. This allows biocatalytic reactions to avoid many of the pressing issues associated with chemical synthesis—such as the use of hazardous reagents and solvents, and the generation of toxic by-products. Moreover, when coupled with other chemical operations in a flow system (see below), biocatalysis can be part of a safe, efficient, and sustainable manufacturing system. This is because flow systems offer much better control over heat and mass transfer, and are easier to design for safe scale-up than batch systems. Even if one were to use microwave-assisted biocatalysis, which is currently not done, there would still be far less risk of employing highly dangerous and toxic chemicals in a biocatalytic reactor than in a traditional reactor for the same purpose. Indeed, biocatalysis is a safe way to produce chemicals in a very high yield, with no energy problems associated with working up the reaction (i.e., purification of the product), or with the way one puts the product for use.

3.7.1 Biocatalysis

Biocatalysis is a key part of developing green chemistry. It uses enzymes and living bacterial and fungal cells to carry out chemical changes (De Regil & Sandoval, 2013). Biocatalysis in chemical synthesis offers a safe and eco-friendly method that works under mild conditions. The natural catalysts in biocatalysis work well at normal temperatures and pressures, using water as a solvent instead of harmful, flammable ones. The process of choosing enzymes is very precise. These enzymes target specific parts of large molecules to do their work. They function at room temperature and below, using water as their solvent, and they do not produce harmful by-products (Li et al., 2024). Biocatalysis shows great flexibility, effectively performing many chemical transformations that were once thought impossible. It allows for the creation of green chemicals through straightforward and efficient methods for synthesizing drugs, fuels, and more.

3.7.2 Flow Chemistry

The chemical synthesis industry has moved away from conventional batch manufacturing methods to adopt continuous production techniques. The process of continuous manufacturing uses non-stop fluid pumping to supply reactants directly to synthesis equipment which operates as autonomous chemical agents in an upcoming advanced production system. The narrow tubing and channels function as controlled reaction vessels which enable us to achieve reaction parameter optimization that was impossible in batch processing. The implementation of flow reactors enables us to achieve better safety outcomes and enhanced yield and selectivity with reduced by-products and faster production rates and smaller reaction volumes and environmentally friendly operations without solvents. The chemoselective revolution of flow chemistry presents these and additional episodes which form its effusive nature. The technique shows great potential for sustainable chemical production according to Capaldo et al. (2023) and Heretsch (2023).

3.7.3 Microwave-Assisted Synthesis

Microwave-assisted synthesis of chemical compounds is becoming more popular as a greener way to make chemicals. It is considered as a better heating protocol with more eco-friendliness. When you heat a chemical reaction, you don't just raise the temperature; you're doing it in a specific way. With traditional heating, energy moves from an external source to the walls of the reaction container, which then heats the mixture inside. This process can be uneven. In contrast, when you use a microwave, the energy interacts directly with the molecules in the reaction. This means the atoms and electrons, which need energy the most, get energized immediately. This focused heating makes the reaction happen in a shorter time, often leading to better results. It speeds up the process and usually reduces waste, making it a cleaner option. In many situations, this method is safer for those conducting the experiments and more beneficial for the environment, aligning well with the 12 Principles of Green Chemistry. Overall, microwave-assisted synthesis is an exciting advancement in sustainable chemical synthesis.

3.8 INDUSTRIAL AND ACADEMIC CASE STUDIES

Green chemistry carries a significance that extends well beyond theory. It serves as a tool to guide industrial applications and academic research toward systematic environmental friendliness. In simpler terms, this means promoting design, synthesis, and analysis that seeks to minimize harm to people and the planet. This goal is not just an ideal, it closely connects with sustainable economic development for clear reasons. Application of green chemistry in production situations lowers energy consumption and the amount of harmful by-products associated with synthesis. The pure substances promised by next-step research in academia will make those result reductions even more substantial. These are ensuring that the move toward greener production methods will also include new reaction pathways and sustainable materials, resulting in transition to a greener, more responsible series of chemical products and pathways across a range of industrial applications. The following paragraphs describe some of these exemplary case studies from industry and academia.

3.8.1 Case Studies from Pharmaceutical Industry

Although there is an increased public interest for manufacturing eco-friendly products, still the chemical industry has not changed significantly, and the rising global population boosts the demand for drug production, which further impelling limited resources (Ahmad et al., 2024). Here, a general overview for the attempt of transition of traditional chemical processes to greener, sustainable ones in pharmaceutical manufacturing in recent years is presented. Some existing examples, like ethanol fermentation and Ibuprofen synthesis, already follow green chemistry principles (Zimmerman et al., 2020). However, there is still a considerable need for applying green chemistry principles to minimize generation of pollution due to utilization of hazardous solvents and toxic reagents. Although the pharmaceutical industry has a number of alternative strategies for green manufacturing, the process developers need to be more educated to truly adopt these safer, more efficient methods (Ahmad et al., 2024). This includes demonstration of successful real-world examples to pharmaceutical industry representatives, like those in Table 3.2, which highlight how techniques like biocatalysts, supercritical fluids, solvent-free reactions, and energy-efficient methods facilitate drug production more sustainable and less harmful to the environment, by following the principles of green chemistry.

Pfizer successfully applied green chemistry principles to significantly improve how they produce the drug sertraline. They carefully looked at every step of the process and brought in a new, greener technology to create a very pure product. This made things much safer for both workers and the environment. The new "combined" process is a great success for pollution prevention, which provides handling materials safer, reducing energy, solvent (*about 80% reduction in solvent use*) and water use, while doubling the amount of product synthesis instead. This success motivated Pfizer's green chemistry approach to find other manufacturing processes to make greener as well.

In 2002, Pfizer earned the prestigious Green Chemistry Award for revolutionizing the manufacturing of sertraline, which is the key ingredient for the antidepressant Zoloft. Actually, this was not their first attempt, instead it was the third major redesign of the process for sertraline production since 1985. Each redesign aimed to shrink the number of solvents used in the chemical synthesis, making the process simpler, much safer for workers, more energy-efficient, and more eco-friendly due to reduction of waste production. The biggest acquisition came from ditching a highly problematic chemical called titanium tetrachloride, which is highly toxic, corrosive, and can react with air to generate harmful hydrochloric acid as a by-product. In the old process, titanium tetrachloride was used for removal of water, which otherwise ruin the reaction. However, this step was leading to production of a mix of hazardous by-products like hydrochloric acid and titanium wastes, which were costly and

TABLE 3.2
Exemplary Case Studies for Application of Different Green Chemistry Approaches in Drug Synthesis, Representing Which Green Chemistry Approach Is Applied in Each Case Along With the Benefit(s) Obtained by Applying This Approach

Case Study	Which Green Chemistry Approach Is Applied?	Benefit	Reference
Green Synthesis of Ibuprofen	Use of renewable resources	Reducement of the environmental impact for ibuprofen synthesis	Karaman and Qtait (2021)
Eco-friendly Synthesis of Paracetamol	Waste minimization	Minimization of waste for paracetamol synthesis	Amina et al. (2018)
Eco-friendly Synthesis of Aspirin	Solvent-free reactions	Minimization of waste and the demand for hazardous solvents for aspirin synthesis	Hussain (2020)
Green Synthesis of Taxol	Catalytic processes	Improvement in reaction efficiency and decreasing reaction by-products for taxol synthesis	Sabzehzari et al. (2020)
Green Synthesis of Metformin	Energy-efficient methods	Decrease of energy consumption and improvement in chemical yield for metformin synthesis	Patel (2017)
Green Synthesis of Bupropion	Biocatalysis	Improvement in sustainability by avoiding toxic reagent use for bupropion synthesis	Andrew et al. (2022)
Green Synthesis of Sildenafil	Use of green solvents	Reducement of the environmental impact for sildenafil synthesis	Ahsan et al. (2020)
Green Synthesis of Donepezil	Microwave-assisted synthesis	Reducement in reaction times and energy consumption for donepezil synthesis	Munir et al. (2021)
Eco-friendly Synthesis of Amlodipine	Green chemistry metrics	Improvement in overall sustainability for amlodipine synthesis	Jurjeva and Koel (2022)
Eco-friendly Synthesis of Erythromycin	Use of supercritical fluids	Reducement for the need of harmful solvents during erythromycin synthesis	Athparia et al. (2023)

complicated to dispose of. Here, Pfizer's alternative greener approach was switching ethanol with titanium tetrachloride, a much safer solvent which provided complete elimination of the hazardous titanium tetrachloride and its associated wastes, making a huge difference in terms of safety for both workers and environment. Using ethanol as the solvent also leads to a dramatic decrease in the demand for one of the starting material, and it could be even recycled back, which boosts the overall efficiency. Pfizer's scientists have also discovered a smarter catalyst for this process, where the old catalyst led to unwanted impurities that were hard to remove, wasting product and requiring lots of solvent and energy input for purification. On the other hand, the new, more selective catalyst produced fewer impurities, meaning less of another expensive ingredient (i.e., mandelic acid) is needed, and the catalyst itself could be recovered and reused further. These green chemistry improvements, called as "combined" process, brought both economic and environmental benefits, by significantly reducing the manufacturing costs (*which made up a significant portion of the wholesale price*), thus the global price of the drug could be reduced and this provided a crucial advantage for Pfizer. Building on this success, Pfizer has later attempted to make this process even more efficient. In traditional process, the starting material, tetralone, was in a half-to-half mix, with only half of it turning into sertraline, while the other half was waste. By using a clever separation method called multiple-column chromatography (MCC), Pfizer scientists could isolate the useful part of tetralone, and the previously unusable part could then be recycled and reprocessed, thereby essentially doubling the sertraline yield from the same amount of starting material and halving the required manufacturing plant capacity. This advanced, FDA-approved process provided a significant contribution to Pfizer and making it a leading sustainable drug manufacturer in pharmaceutical industry (ACS, Pfizer Pharmaceuticals: Green Chemistry Innovation and Business Strategy, accessed 06/28/2025).

3.8.2 Case Studies from Agrochemical Manufacturing

An early success in making agrochemical production greener was by Ciba-Geigy (now Syngenta), for making the herbicide prosulfuron. They developed a clever, "one-pot" method for a key intermediate (intermediate 13) of the chemical process, by combining three steps, diazotization, Heck-Matsuda coupling, and hydrogenation in a single sequence, thereby achieving an impressive 93% overall yield from the starting material of Ciba-Geigy process, 2-aminobenzenesulfonic acid to intermediate 13. This significantly reduced waste, reportedly generating only 2 kilograms of waste per kilogram of intermediate 13, largely due to high reaction yields and the use of catalytic transformations instead of less efficient stoichiometric ones. Here, a major green chemistry application is the use of 1-pentanol as the sole solvent for this three-step sequential reaction. The solvent with which we are working is chosen because it has very low toxicity and it is easily recyclable, and so right there you've got two advantages, hence "solvent selection." Because what followed was kind of a "cookbook" when it comes to synthesizing prosulfuron. Prosulfuron is an herbicide. So we have these two advantages from solvent selection, plus what in synthetic organic chemistry is considered very advantageous, which is the use of sustained yields and catalytic methods. Where we run into trouble, though, is with phosgene. Everything else in this synthesis is fairly mild, fairly easy, does not produce a lot of waste, and yet, at the same time, is producing something that is a highly toxic chemical powder. Which, of course, contradicts everything that was mentioned above about these synthesizing methods being advantages in terms of environmental chemistry.

Corteva Agriscience offers another apt illustration. The company recently introduced Rinskor active, a novel herbicide formulated for rice that takes aim at very tough weeds, including resistant ones (Epp et al., 2015). What sets Rinskor apart, and what won it an EPA Green Chemistry Challenge Award in 2018 (Drahl, 2018), is its extremely low application rate—just 5–50 grams per hectare. To help put that in context, consider that the EPA considers a "low use rate" to be 500 grams per hectare (Drahl, 2018). Using a fraction of that amount, Rinskor has an exceptionally safe toxicological profile, which derives from its structure. Studies show it has no harmful effects on mammals in terms of reproduction, development, or any signs of causing cancer, even at very high doses. Its safety led the EPA to give it a "Reduced Risk Pesticide" designation. A similar herbicide, Arylex active, can also work at very low rates for weeds in cereal crops and share a very high safety profile, demonstrating the potential of these greener agricultural chemicals, compatible with the Principle 4 of 12 Principles of Green Chemistry (Whiteker, 2019).

While transition metal catalysts are very common in agrochemical production, still biocatalysis is emerging as a powerful, greener alternative, with promising commercial applications and success. An important example comes from DuPont's work on the production of azafenidin herbicide. To make a crucial intermediate for azafenidin (which is called as 96), DuPont's scientists need addition of water to only one of two similar chemical groups in a molecule called adiponitrile (Shapiro et al., 2001). Nevertheless, traditional chemical methods were imprecise, and hydrating both groups, resulting in a mixture of products with low conversion. On the other hand, by using an enzyme from immobilized *Pseudomonas chlororaphis* bacteria, DuPont has achieved a remarkable precision in the azafenidin synthesis. This biocatalytic approach selectively added water to only one of the desired groups, resulting in a 93% selective yield at 97% conversion. This highly efficient, clean and green reaction is now used commercially to produce metric tons of this intermediate. Even though harnessing more traditional chemistry, the subsequent steps ultimately result in production of the final herbicide, yet the contribution of biocatalysis-based production of intermediate 96 to "greenness" of overall azafenidin manufacturing is undeniable. This case study highlights how biocatalysis can dramatically improve selectivity and efficiency, and remarkably reduce production of waste when compared with traditional chemical routes, even for large-scale industrial processes in agrochemical manufacturing (Whiteker, 2019).

3.8.3 Case Studies from Fine Chemicals Manufacturing

Vitamin C is currently made commercially by two main routes: the Reichstein process and the two-step fermentation process. The Reichstein process, used by companies like Roche and BASF (Liese et al., 2006), blends traditional chemical steps with a biological one, biocatalysis, which starts by using the bacterium *B. xylinum* to convert sorbitol into sorbose (Reichstein & Grussner, 1934), then this sorbose undergoes several chemical modifications and oxidations before finally rearranging in acid to form crude vitamin C, which is eventually purified. The two-step fermentation process, pioneered in China, is more biological and has become the dominant method due to its lower cost, which involves two separate fermentation steps: first, glucose is transformed into 2,5-diketo-gluconate using strains of *Erwinia* or *Acetobacter* bacteria, then, a *Corynebacterium* strain converts this 2,5-diketo-gluconate into

2-ketogulonic acid. Since this two-step fermentation is very efficient and cost-effective method, China has become the world's leading producer of vitamin C.

Additionally, there are recent efforts from BASF to improve the production of biocatalytic processes. In one case study, BASF, the Austrian Research Centre of Industrial Biotechnology (acib), and the University of Graz worked together to develop a new computer-assisted model for this purpose. They have developed a groundbreaking statistical computer model that significantly simplifies the optimization of enzyme performance in chemical reactions, which covers production of fine chemicals like vitamins, flavorings, and other related chemicals. This new regression model moves beyond traditional methods by allowing scientists to predict the best conditions for an enzyme with far fewer lab tests. Here, instead of extensive trial-and-error, only a small amount of initial data, like how the enzyme unfolds, is needed. The newly created computer model accurately determines the ideal reaction temperature and solvent concentration necessary to optimize enzyme activity. The model takes in the relevant reaction information and produces in silico predictions of how the reactions will behave. This is happening at a time when divine-catalysis research and development must speed up its transition from the lab to large-scale production because catalyzed reactions are the basis for chemical synthesis. BASF stated that the model represents a significant step in biocatalysis development because it is leading to a reduction in resource input and production cost.

3.9 FUTURE PERSPECTIVES AND CHALLENGES

Green chemistry's future looks bright, however still some tough problems to battle. Turning lab discoveries into factory production? That's a tough climb. When they push to scale, therefore the project just stays stuck at the prototype, never moving beyond that early demo. That trick works fine in lab, however on a larger scale it costs a fortune and is hard to copy. The field moves fast, new trends pop up all the time. For instance, AI is changing how we develop and conduct chemical reactions. It might even shift the entire design process. It finds the best conditions and identifies key data quickly. This speed helps us discover new catalysts and keeps waste low. Currently, green chemistry is more linked to the circular economy. We are moving away from the old "use it once, toss it" mindset. We aim to choose products that minimize waste. We keep reusing materials and assist nature in restoring itself. These breakthroughs suggest that green chemistry has a bright future, which is promising. We are dedicated to sustainability, making the future look even brighter. Additionally, innovative ideas help the planet and create new ways to earn money. More engineering and inventive approaches are needed to tackle these challenges, leading to the development of green methods that reduce costs and safety issues. Development of smarter materials and catalysts is another key focus. It is important to search for catalysts that speed up reactions, and if they can be made from common materials, that is much more efficient. Quick and accessible options mean a higher chance of success. It keeps us away from costly rare metals and saves money. We need to design new materials that are safe from the start. Corteva Agriscience's Rinskor Active does a solid job by safely degrading at the end of its life, proving that this approach can work. Why do not we prioritize pollution prevention and making it the first step instead of the last? Indeed, currently, cutting pollution is the top priority. It is not a last-minute fix, it is the main mission. Future scientists and engineers should incorporate green ideas right from the beginning. However, convincing industries and investors to support new green technologies is challenging. It can feel like a hard sell. One reason is that traditional methods appear cheaper. In the long run, however, green methods tend to offer better outcomes. Progress in green technology is fueled by ongoing research, teams linking universities and factories, and a commitment to protect the planet while pursuing profit.

Together, these elements shape the next chapter of green science. Anastas and Warner have offered the 12 rules of green chemistry. These rules serve as a starting point for chemists and engineers. By following them, we can generate a clearer plan to reduce pollution and enhance safety. This is crucial for industries that aim to operate for decades (Whiteker, 2019).

3.10 CONCLUSION

In conclusion, application of green chemistry on a large scale is a major and complicated task requiring a team work for success. We need to focus on new molecular designs and reconsider chemical processes to make sure that they are more efficient and environmentally friendly. It is important to establish reliable metrics for measurement and evaluation of the performance for these new processes and chemicals. Using all these advancements in a clear and contemporary way must also be a part of this plan. By following this roadmap with a clear vision and a strong motivation, it is possible to achieve the changes we are looking for.

REFERENCES

12 Principles of Green Chemistry [Internet]. American Chemical Society [cited 2025 June 28].

Ahmad, S., Jaiswal, R., Yadav, R., & Verma, S. (2024). Recent advances in green chemistry approaches for pharmaceutical synthesis. *Sustainable Chemistry One World*, *4*, 100029.

Ahsan, H., Islam, S. U., Ahmed, M. B., Lee, Y. S., & Sonn, J. K. (2020). Significance of green synthetic chemistry from a pharmaceutical perspective. *Current Pharmaceutical Design*, *26*(45), 5767–5782.

Amina, M., Alarfaj, N. A., El-Tohamy, M. F., Al Musayeib, N. M., & Oraby, H. F. (2018). Sequential injection-chemiluminescence evaluation of stigmasterol glucoside and luteolin via green synthesis of silver nanoparticles using biomass of *Plectranthus asirensis*. *Green Chemistry Letters and Reviews*, *11*(4), 523–533.

Anastas, P. T., & Warner, J. C. (1998). *Green chemistry: Theory and practice*. Oxford University Press.

Andrew, O. B., Sherwood, J., & Hurst, G. A. (2022). A greener synthesis of the antidepressant bupropion hydrochloride. *Journal of Chemical Education*, *99*(9), 3277–3282.

Armenta, S., Esteve-Turrillas, F. A., Garrigues, S., & De La Guardia, M. (2022). Alternative green solvents in sample preparation. *Green Analytical Chemistry*, *1*, 100007.

Athparia, M., Bora, N., Deka, A., Sohtun, P., Padhi, P., Bhuyan, N., Bordoloi, N. J., Gogoi, L., & Kataki, R. (2023). Non-fuel applications of bio-oil for sustainability in management of bioresources. *Environmental Science and Pollution Research*, *32*(12), 7468–7492.

BASF and the University of Graz Develop Fundamentally New Computer-Assisted Model to Optimize Efficiency of Biocatalytic Production Processes [Internet]. BASF [cited 2025 June 29]. https://www.basf.com/global/en/media/news-releases/2024/07/p-24-247.

Capaldo, L., Wen, Z., & Noël, T. (2023). A field guide to flow chemistry for synthetic organic chemists. *Chemical Science*, *14*(16), 4230–4247.

Cespi, D., Beach, E. S., Swarr, T. E., Passarini, F., Vassura, I., Dunn, P. J., & Anastas, P. T. (2015). Life cycle inventory improvement in the pharmaceutical sector: Assessment of the sustainability combining PMI and LCA tools. *Green Chemistry*, *17*(6), 3390–3400.

De Regil, R., & Sandoval, G. (2013). Biocatalysis for biobased chemicals. *Biomolecules*, *3*(4), 812–847.

Drahl, C. (2018). Green chemistry challenge awards announced. *Chemical & Engineering News*, *96*(46), 35.

Epp, J. B., Alexander, A. L., Balko, T. W., Buysse, A. M., Brewster, W. K., Bryan, K., Daeuble, J. F., Fields, S. C., Gast, R. E., Green, R. A., Irvine, N. M., Lo, W. C., Lowe, C. T., Renga, J. M., Richburg, J. S., Ruiz, J. M., Satchivi, N. M., Schmitzer, P. R., Siddall, T. L., Webster, J. D., Weimer, M. R., Whiteker, G. T., & Yerkes, C. N. (2015). The discovery of Arylex™ active and Rinskor™ active: Two novel auxin herbicides. *Bioorganic & Medicinal Chemistry*, *24*(3), 362–371.

Heretsch, P. (2023). Modern flow chemistry – Prospect and advantage. *Beilstein Journal of Organic Chemistry*, *19*, 33–35.

Hussain, E. A. (2020). Chapter 4 - green solvents for drug synthesis. In *Green sustainable process for chemical and environmental engineering and science* (pp. 55–86). Elsevier. Edited by: Inamuddin, Rajender Boddula, ... Abdullah M. Asiri. https://doi.org/10.1016/B978-0-12-821885-3.00011-6

Jurjeva, J., & Koel, M. (2022). Implementing greening into design in analytical chemistry. *Talanta Open*, *6*, 100136.

Karaman, R., & Qtait, A. (2021). Design of bitterless atenolol prodrugs for the use in aqueous formulations for pediatrics and geriatrics. *Technological Innovation in Pharmaceutical Research*, *4*, 10–27.

Li, G., Zhang, N., Dai, X., & Fan, L. (2024). EnzyACT: A novel deep learning method to predict the impacts of single and multiple mutations on enzyme activity. *Journal of Chemical Information and Modeling*, *64*(15), 5912–5921.

Liese, A., Seelbach, K., & Wandrey, C. (2006). *Industrial biotransformations.* Wiley.

Munir, R., Zia-Ur-Rehman, M., Murtaza, S., Zaib, S., Javid, N., Awan, S. J., Iftikhar, K., Athar, M. M., & Khan, I. (2021). Microwave-assisted synthesis of (piperidin-1-Yl) quinolin-3-Yl)methylene)hydrazinecarbothioamides as potent inhibitors of cholinesterases: A biochemical and in silico approach.*Molecules*, *26*(3), 656.

Patel, R. M. (2017). Biocatalysis for synthesis of pharmaceuticals. *Bioorganic & Medicinal Chemistry*, *26*(7), 1252–1274.

Pfizer Pharmaceuticals: Green Chemistry Innovation and Business Strategy [Internet] American Chemical Society [cited 2025 June 28]. https://www.acs.org/content/dam/acsorg/greenchemistry/industriainnovation/Pfizer-business-case-study.pdf

Reichstein, T., & Grussner, A. (1934). Eine ergiebige synthese der 1-Ascorbinsaure (C-vitamin). *Helvetica Chimica Acta*, *17*, 311–315.

Sabzehzari, M., Zeinali, M., & Naghavi, M. R. (2020). Alternative sources and metabolic engineering of taxol: Advances and future perspectives. *Biotechnology Advances*, *43*, 107569.

Shapiro, R., DiCosimo, R., Hennessey, S. M., Stieglitz, B., Campopiano, O., & Chiang, G. C. (2001). Discovery and development of a commercial synthesis of azafenidin. *Organic Process Research & Development*, *5*, 593.

Sheldon, R. A. (1992). Organic synthesis; past, present and future. *Chemistry and Industry*, *23*, 903–906.

Sheldon, R. A. (2007). The E factor: Fifteen years on. *Green Chemistry*, *9*(12), 1273.

Su, Y., Wang, X., Ye, Y., Xie, Y., Xu, Y., Jiang, Y., & Wang, C. (2024). Automation and machine learning augmented by large language models in a catalysis study. *Chemical Science*, *15*(31), 12200–12233.

Suslick, K. S., Didenko, Y., Fang, M. M., Hyeon, T., Kolbeck, K. J., McNamara, W. B., Mdleleni, M. M., & Wong, M. (1999). Acoustic cavitation and its chemical consequences. *Philosophical Transactions of the Royal Society a Mathematical Physical and Engineering Sciences*, *357*(1751), 335–353.

Trost, B. M. (1991). The atom economy—A search for synthetic efficiency. *Science*, *254*(5037), 1471–1477.

Wang, W., Liu, J., Shi, G., Wu, S., Zhang, S., & Yuan, R. (2025). Preparation of recyclable magnetic catalyst (PD/PDA@FE_3O_4) and the catalytic degradation of 4-nitrophenol and rhodamine B. *Catalysts*, *15*(2), 175.

Whiteker, G. T. (2019). Applications of the 12 principles of green chemistry in the crop protection industry. *Organic Process Research & Development*, *23*(10), 2109–2121.

Zimmerman, J. B., Anastas, P. T., Erythropel, H. C., & Leitner, W. (2020). Designing for a green chemistry future. *Science*, *367*(6476), 397–400.

4 Re-refining Technologies for Circularity and Sustainability in the Oil and Gas Supply Chain

Abdolvahhab Fetanat and Mohsen Tayebi

4.1 INTRODUCTION

The industry of oil and natural gas (O&G) energy continues to maintain its position as one of the world's leading energy sectors that contributes considerably to the worldwide economy. Nevertheless, it has confronted market obstacles over the past few years. As most countries transition away from fossil fuels, there is an increasing shift toward greener, cleaner, and more sustainable energy production. (Fetanat et al. 2023; Lau et al., 2022; Zgheib & Takache, 2021). The O&G energy sector operations emit a significant volume of carbon dioxide (CO_2), which conflicts with policymakers' intention to achieve decarbonization goals. In addition, the public opposes this industry because environmental sustainability issues have emerged from the processing operations of this energy sector. In the O&G processing operations, the waste oil dumping can result in polluting natural resources such as water and disrupting the ecosystem related to the marine. Moreover, in this type of waste, there are oxidation products and worn metals that have high toxicity to humans (Lau et al., 2022; Nour et al., 2021; Saleem & Karim, 2020). Given the dwindling volume of crude oil as a fossil fuel resource that serves as the main feedstock in the O&G supply chain, it underscores the necessity of the processing operations of the O&G energy sector to construct significant reforms with the aim of recovering the unfavorable situation (Lau et al., 2022). Therefore, the current situation provides an opportunity to modify the economy with the target of achieving better utilization of resources and reduction of waste in the studied energy sector (Fetanat & Tayebi, 2022, 2023; Richard et al., 2020).

The issue related to the sustainability in O&G processing operations can be ameliorated by implementing a circular economy approach. It is a valuable approach for achieving the target of resource efficiency. This approach concentrates on closing substance and energy loops with the aim of the mitigatation in producing less waste and consuming fewer resources (Agamuthu & Mehran, 2019). This approach has gained renewed attention over the past few years as concerns in relation tothe issues of sustainability and the environment have increased. Besides its environmental advantages, the circular economy approach can result in economic advantages,

DOI: 10.1201/9781003642589-4

including cost savings and profit production. This is done by eliminating the demand for imported fresh resources and treating waste before the operations related to the disposal stage (Fetanat et al., 2021; Sillanpää & Ncibi, 2019). In the context of a circular economy, incinerating waste lubricant oil as a solution intended for re-refining is problematic, as it reduces system circularity by preventing the product from closing its life cycle and re-entering the value stream. (Ellen MacArthur Foundation 2015; Silva et al., 2022). Owing to the absence of technologies for re-refining and/or supervisory causes, most countries that are developing still transmit waste to be incinerated. In the countries of France and the United States, a portion of the waste lubricant oil accumulated that assembles certain requirements for technical and environmental fields is purified and then incinerated in different industrial sectors (such as thermoelectric plants and cement kilns) (Gonzaga et al., 2021). In this regard, a methodology for managing environmental issues is life cycle assessment. It aims at helping the processes of decision-making (Campitelli et al., 2019; Hassanain et al., 2017). Also, this methodology is a valuable and useful tool in order to identify the ideal solutions (e.g., the disposal of waste lubricant oil)(Silva et al., 2022). Nevertheless, recent researches in relation to this issue show that there is an absence of information and data for a whole life cycle assessment of waste lubricant oil in many countries (Botas et al., 2017; Hassanain et al., 2017; Pires & Martinho, 2013; Tsambe et al., 2019). Therefore, there is a significant restriction to developing the studies related to life cycle assessment, particularly because of the absence of elemental data on the sector of used lubricant oil. The most widely employed indicators for circularity in recent years are those that refer to the assessment at two levels of macro and micro in relation to the systems related to generation (de Oliveira et al., 2021; Hassanain et al., 2017; Pires & Martinho, 2013; Rigamonti & Mancini, 2021), and the study (Ellen MacArthur Foundation, 2015) developed an index, namely materials cyclical index. It is the most widely utilized indicator for the study of technical cycles (Silva et al., 2022). These cycles in the context of the circular economy concentrate upon various strategies. These strategies involve repair, reuse, refurbish, remanufacturing, and recycling in relation to products. Their aim is to return components and substances to the market with quality and as much quality as possible. The cyclical index of materials is an uncomplicated way in order to show the amount of material turnover in the life cycle of a product. Also, this index can be well utilized with the aim of integrating findings with other approaches associated with life cycle, including material flow assessment and life cycle assessment (Campitelli et al., 2019; de Oliveira et al., 2021; Hassanain et al., 2017; Rigamonti & Mancini, 2021).

The indicators associated with the circularity context (such as material flow assessment)are planned to consider the five Rs ((1) repair, (2) reuse, (3) refurbish, (4) remanufacturing, and (5) recycling) and they are defined as input elements in the process of the computation. Hence, the target of these five is to evaluate how products and substances are used as a loop in the technological domain (Ellen MacArthur Foundation 2015; Rigamonti & Mancini, 2021). Lonca et al. (2018) evaluated and modified 16 methods related to circular economy indices. Among these, only material flow assessment accounts for product durability, whereas other indices—such as the reuse potential index and the circular economy index (Di Maio and Rem, 2015; Park and Chertow, 2014)—consider a more limited set of requirements.. Generally, the indices related to circularity suppose some level the necessity for using circular inputs, recovered substances, and extended lifespan as the most related elements for determining the performance of circularity. Also, it is noteworthy that the solutions of the circular economy must reduce the potential of the environmental impacts and help more sustainable production. In addition, minimizing the impacts of the life cycle related to the product, even for cyclical products, is necessary (de Oliveira et al., 2021; Hishammuddin et al., 2018; Pinheiro et al., 2021; Rigamonti & Mancini, 2021).

Increasingly, waste lubricant oil has been the topic of research works, mostly on life cycle assessment. The research work by Pires and Martinho (2013) assessed 16strategies of end-of-life in relation to waste lubricant oil. These involve the recycling of producing expanded clay, recycling and production of electricity, energy recovery, and refining. A comparison of refining waste lubricant oil and combustion can be seen in the research published by Hassanain et al. (2017). Based on this work, the most sustainable approach to managing waste lubricant oil includes two plans. These plans are (1) recycling with re-incineration and (2) recovery with co-incineration, and with calculations carried out on a treatment company of this type of waste, it concluded that oil refining is distinct from other methods and is moving toward cleaner production. Also, this work specified that the most appropriate finding (refining) is assured contingent on the quality of the waste lubricant oil, because if it has high viscosity values and more additives, the plan can alter and it is no longer the most suitable option. A research study by Botas et al. (2017) analyzed the environmental and energy performance of end-of-life strategies for waste lubricant oil using life cycle assessment. This research work aims at measuring the productiveness of a specific plant from a company. This work stated that re-refining is much more environmentally friendly and even proposed more evaluations with the target of optimizing the current process of re-refining, but it does not, for example, examine the refining of oil by incineration for energy. A study by Kanokkantapong et al. (2009) handled a comparative life cycle assessment relates to six plans in order to manage waste lubricant oil. These plans are four boilers and two refineries. The evluations of this study showed that each alternative could have importantly different impacts on the environment. The findings obtained illustrated good performance in relation to the strategies of recycling. Nevertheless, the cement kiln had the lowest negative effects regarding the potential of global warming and the generation of heavy metal, as it allowed for complete combustion of the organic compounds constituting waste lubricant oil, while other pollutants (such as heavy metals), were collected in the form of mortar for external recycling.

The principles of circular economy have been employed in diverse industrial sectors (Fetanat et al., 2021; Islam Sazzad et al., 2024; Lau et al., 2022; Pichler et al. 2023; Pinheiro et al., 2021; Sillanpää & Ncibi, 2019; Syahir et al., 2017). Huang and Zhang (2011) suggested strategies in relation to integrating the circular economy with the upstream operations inthe industry of China's O&G energy sector. They are groupedin five classes: (1) developing a circular economy supporting system, (2) progressing entire resource efficiency, (3) optimizing raw substances utilization, (4) developing technology and knowledge, and (5) improving the chain of industry on raw substance extraction. The study by Hishammuddin et al. (2018) presented the

TABLE 4.1
The Benefits of Re-refining Process

Benefits	Description
Environmental protection	Mitigating the amount of utilized oil, which is improperly disposed, reducing the pollutions related to the environment.
Resource conservation	Conserving useful crude oil resources by recycling used oil.
Economic profits	Being more cost-effective than producing virgin oil, particularly crude oil prices rise.
Mitigated reliance on imported oil	Helpingthe mitigation of a country's dependence on imported crude oil.
Lower emissions of greenhouse gas (GHG)	Resulting in lower emissions of GHG compared to virgin oil generation.

techniques, including the recovery of energy from CO_2 and the treatment of wastewater in order to construct the connections of the circular economy at the stage related to refining. The sector of O&G energy is presently governed by using the approach of linear economy, in which the products related to waste petroleum are disposed of directly into the environment with out implementing the strategies of reuse and recycling for them (Islam Sazzad et al., 2024). Waste lubricant oil is an important waste petroleum product produced by consumers in the industry of O&G energy. Integrating a re-refining stage associated with waste lubricant oil into the value chain of O&G energy could be a solution to the challenges related to the environmental and health issues, as informed by Pinheiro et al. (2021). This stage relates to the process that its aim is to treat waste oils for recovering a valuable and profitable source of mineral-based oil (Islam Sazzad et al., 2024; Mikhail & Riad, 2014; Mohammed et al., 2013). Re-refining has benefits that are given in Table 4.1 (Islam Sazzad et al., 2024; Lau et al., 2022; Oladimeji et al., 2018; Pinheiro et al., 2021; Sarkar et al., 2023). Various technologies in relation to the re-refining issue have been developed. Re-refining technologies are crucial for managing sustainable lubricant oil.

4.2 CONTAMINANTS EFFECTS AND THE SIGNIFICANCE OF RECYCLING RELATED TO WASTE LUBRICANT OIL

The pollutants in waste lubricant oil pose numerous risks to two fields of the environment and human health (Islam Sazzad et al., 2024; Nowak et al., 2019). Waste lubricant oil can become harmful when it is contaminated with decomposed pollutants, additives, sediments, and water. Improper methods of disposal for this kind of waste may negatively influence these fields. The presence of water-soluble substances in lubricants makes them susceptible to leaching into waste streams and subsequent groundwater contamination through runoff, leaching, and infiltration into landfills. Also, the discharge of waste lubricant oil into rivers can affect surface waters and coastal areas (Udonne & Bakare, 2013). By reviewing the studies (Islam Sazzad et al., 2024; Lau et al., 2022; Nowak et al., 2019), improper disposal of waste lubricant oil may lead to significant and persistent unfavorable influences on the environment. This type of waste can obstruct vital chemical components, including oxygen from going in water resources, so jeopardizing the health of diverse organisms that live in aquatic environments. Furthermore, waste lubricant oil consists of harmful compounds. These compounds include polycyclic aromatic hydrocarbons (PAHs). Therefore, ensuring appropriate techniques for the disposal of waste lubricant oil is crucial to reduce these health and environmental consequences (Michael, 2015). The waste lubricant oil may evaporate and pollute the surrounding atmosphere when exposed to high temperatures in the presence of air (Rahman et al., 2008). This waste contains a diversified range in relation to pollutants. They can bring risks to aquatic organisms, human health, and the environment (Ani et al., 2015). These pollutants consist of polychlorinated biphenyls, halogen compounds, hydrocarbons, gums, and varnish. As well, the risk of cancer increases significantly when hazardous components (like PAHs and aliphatic hydrocarbons)are present in waste lubricant oil (Wong & Wang, 2001). An overview of the negative effects related to waste lubricant oil pollutants and oil spills on the public health and environment is provided in Table 4.2 (Anisuzzaman et al., 2021; Islam Sazzad et al., 2024; Kupareva et al., 2013; Nowak et al., 2019; Oladimeji et al., 2018; Sarkar et al., 2023; Shakirullah et al., 2006).

The existence of heavy metals in lubricant oil can result indifferent diseases in animals plants, and humans, containing skin ulcers, cancer, cardiovascular diseases, and anemia. Also, waste lubricant oil consists of additives (Sánchez-Alvarracín et al., 2021).Ecosystems and human health are furthermore at risk from them. Lubricant oils, including PAHs, polychlorinated biphenyls (PCBs), benzene-toluene-xylene (BTEX), and chlorinated solvents have been reported to be associated with tumor growth and cancer. Leakage prevention and improper disposal of these oils are critical to mitigate the destructive effects on ecosystems and human health (Jafari & Hassanpour, 2014; Pichler et al. 2023; Syahir et al., 2017). Owing to the class of waste lubricant oil as a hazardous substance and its important impact on the environment, there is a risk of

TABLE 4.2
Pollutants Impacts Related to the Emissions of Waste Lubricant Oil on the Environment and Human Health

Oil Pollutants	Impacts
Oil mist	Adverse effects on the respiratory and nervous systems.
Base oil	Ecotoxicity, poor biodegradability, and high carcinogenic potential.
	Leading to significant health effects, containing changes in lung, liver, kidney, adrenal, and heart function
Water	Formingrust in metal, whichleads to foaming problems owing tothe formation of water sludge
Sludge and lacquers	Close the valve and inhibit the continuous operation of pumping oil.
Solid particles	Superfluous wear, scratch of the bearing surface and possible failure to absorb metal fatigue
	Narrowness of oil passages and clearances because of sludge deposits

water, air, and soil contaminations by harmful compounds. With regard to the significant quantities produced globally, waste lubricant oil management is crucial. Improper disposal of this waste can result in reclamation, direct reuse, reprocessing, and adverse effects for the environment. The waste lubricant oil production is increasing along with the continuous increase in the number of vehicles. The management and disposal of this waste, especially in urban and industrial areas, is becoming increasingly important. Currently, many countries are facing the problem of environmental pollution caused by waste lubricant oil (Shaikh & Mahanwar, 2018). In addition to protecting the environment, recovering waste lubricant oil helps conserve a highly valuable and non-renewable resource (Ali et al., 2010; Elmehbad, 2017). This type of waste is known as a valuable source from energy on a worldwide scale, as well as is recognized a valuable product due to its carbon content. Nevertheless, conserving these valuable, non-renewable resources is about more than just protecting ecosystems from the effects of this waste. In addition, it contains responsible management and effective exploitation for their economic potential maximum and waste minimum (Islam Sazzad et al., 2024; Nowak et al., 2019; Pichler et al. 2023; Sharma et al., 2021; Syahir et al., 2017).Through the implementation of sustainable practices and promotion of recycling and reprocessing strategies of waste related to lubricant oil, it can help preserve these valuable resources for future generations (Islam Sazzad et al., 2024). When it comes to saving energy, the efficient utilization of this waste preserves invaluable renewable resources. Many companies currently invest significant amounts of money in recovering, exploring, and processing waste lubricant oil into high-quality oil. Furthermore, the waste produced from this oil by customers can be recycled repeatedly into profitable products such as pure filler oil, resulting in overall savings of effort and money (Anisuzzaman et al., 2021; Botas et al., 2017; Lau et al., 2022; Udonne & Bakare, 2013). The equipment of refining can save costs associated with pollution, water treatment, and product storage. In addition, waste lubricant oil can be treated, recycled, or utilized with the aim of keeping water and soil resources instead of being improperly disposed of. As a result, if waste lubricant oil is properly managed, there will be less dependence on imported fossil fuels (Anisuzzaman et al., 2021; Botas et al., 2017; Islam Sazzad et al., 2024).The country will be of advantage to the suitable revitalization and reorganization of the type of waste, which will conserve natural resources, mitigate environmental pollution, and save foreign exchange for purchasing pure oil (Anisuzzaman et al., 2021; Islam Sazzad et al., 2024; Lau et al., 2022; Rahman et al., 2008; Sharma et al., 2021).

4.3 WASTE LUBRICANT OIL MANAGEMENT

Currently, the principles of sustainability are widely recognized as having a significant impact on the industrial sector under study.. Hence, reducing the pollutants, preserving the energy resources, and developing a sustainability approach have become the principal goals. Furthermore, societal and national attention in relation to the issue of producing and recovering lubricants is growing because of their potential to advance targets such as sustainable development and resource conservation (Pinheiro et al., 2021).The linear economy functions according to the take-make-dispose principle. Here, substances pursue a linear path. This path is from extraction to production. Eventually, they end up as waste (Buchmann-Duck & Beazley, 2020; Tan &

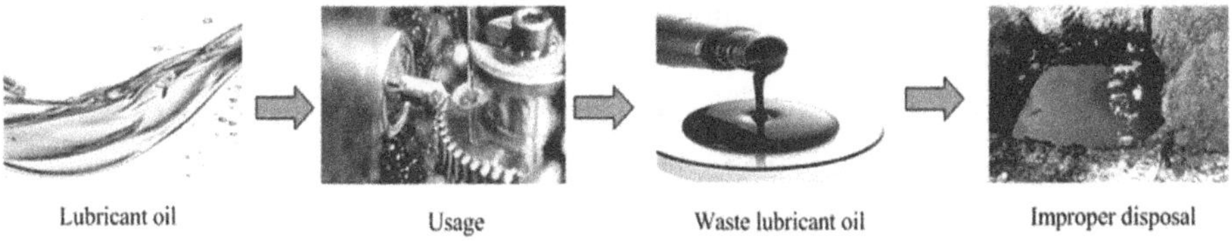

FIGURE 4.1 Linear economy related to lubricant oil.

Lamers, 2021). Conversely, a valuable approach called the circular economy aims at concentrating on strategies such as recycling and reuse related to raw materials. Another aim of this approach is to optimize resource efficiency and mitigate waste production (Fetanat et al., 2021; Grafström & Aasma, 2021; Islam Sazzad et al., 2024; Sharma et al., 2021). This model is shown in Figure 4.1 (Islam Sazzad et al., 2024), allows for transitioning from the dominant procedure that mainly accentuates the concepts related to the linear mode of generation and use (Kirchherr et al., 2017). Figure 4.2 shows the circular economy model for lubricant oil (Lau et al., 2022; Pinheiro et al., 2021). It involves various aspects of economic activity, starting with formulating lubricant oil and ending with refining crude oil to extract base oil, producing the new lubricants, following the use trends, generating waste lubricant oil, and finally collecting and recycling the waste lubricant oil. The aim is to mitigate disposal, which has harmful influences, ideally to achieve the future of zero waste. It is essential to develop appropriate criteria for the circular economy and waste lubricant oil management. In the circular economy of waste lubricant oil, the part

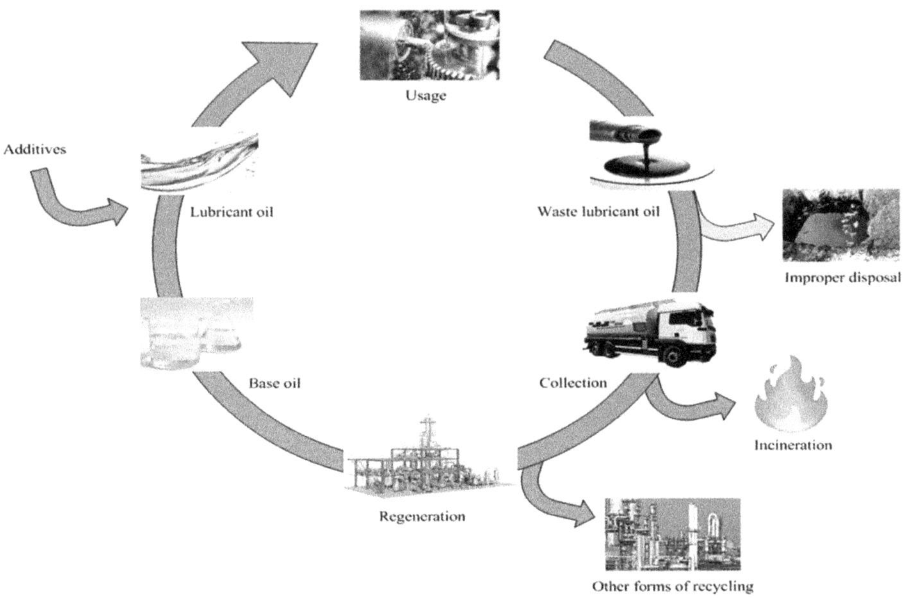

FIGURE 4.2 Circular economy for managing waste lubricant oil.

of regeneration is pivotal, which highlights the importance of recycling and reusing resources (Fetanat et al., 2021; Islam Sazzad et al., 2024; Lau et al., 2022).

4.4 TYPES OF RE-REFINING TECHNOLOGIES

Re-refining is the process of restoring used lubricant oil to a comparable condition to or better than virgin base oils. This process comprises contaminants removal and oil's original properties restoration, making it appropriate for diverse applications. It is an environmentally friendly process that diverts waste and reduces the need for refining crude oil. Re-refining technologies are shown in Figure 4.3 and they are explained as follows (Islam Sazzad et al., 2024; Lau et al., 2022; Pinheiro et al., 2021):

1. Acid-clay treatment: In this technology, an acid solution—such as formic acid, acetic acid, or sulfuric acid—is added to waste lubricant oil to remove impurities, including heavy metals, asphaltenes, and acidic compounds. However, the oil recovery efficiency is relatively low (approximately 50%), and the generated by-products are highly toxic. (Udonne et al., 2016).
2. **Solvent extraction:** The re-refining technology of solvent extraction is superior compared to the acid-clay re-refining technology. Due to the solvent extraction technology does not produce acidic sludge. The efficiency of sludge removal of this re-refining technology is highly dependent on the kind of used solvents, the solvent to oil ratio and the operating temperature. The study by Mohammed et al. (2013) assessed the quality of oil obtained utilizing six different solvents. These six were carbon tetrachloride, acetone, petroleum ether, 1-butanol, 1-hexanol, and n-hexane. The results of this study indicated that 1-butanol has the highest efficiency for sludge removal.
3. **Thin-film evaporation:** The re-refining technology of thin-film evaporation is a process in which a thin layer of liquid is quickly warmed and evaporated. It happens frequently under vacuum states, to form a thin layer of vaporized substances. The thin-film evaporation re-refining technology is

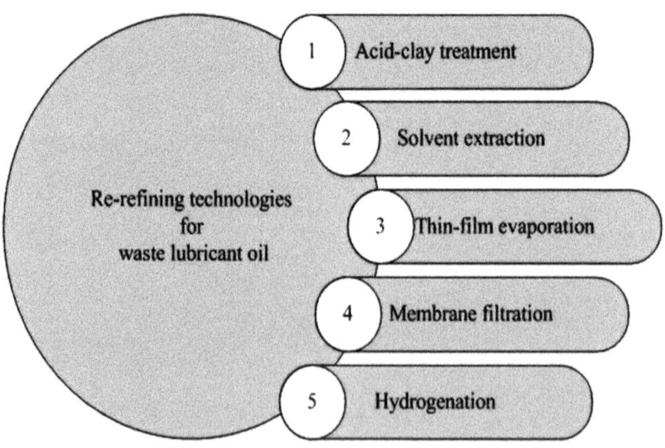

FIGURE 4.3 The used technologies for re-refining waste lubricant oil.

commonly utilized in diverse industries to concentrate, separate, and purify liquids, particularly heat-sensitive or high-boiling point substances (Saleem & Karim, 2020).

4. Membrane filtration: This technology is a process-based physical separation. It is utilized in re-refining lubricant oils with the target of removing pollutants and recovering valuable base oils. Membrane filtration technology employs a semi-permeable membrane. It permits certain components for passing through while keeping others on the basis of size and other features. For membrane filtration, the most widely used is ultrafiltration. Ultrafiltration technology is the most valuable re-refining process for removing suspended solids from waste lubricant oil (Lau et al., 2022).
5. **Hydrogenation:** The hydrogenation re-refining technology uses high-pressure hydrogen gas and catalysts in order to mitigate the metal content, nitrogen, and sulfur of heavy crude oil. The steps of this technology involve: (1) utilizing vacuum distillation for separating gas oil, base oil, and distillation residues; (2) warming the waste oil for removing water and light chemicals; (3) putting the base oil to protective bed filtration for removing any material, which could obstruct the catalytic bed; and (4) putting the resultant base oil for hydrotreating (Chen et al., 2011; Islam Sazzad et al., 2024).

4.5 ADVANTAGES AND DISADVANTAGES OF RE-REFINING TECHNOLOGIES

In this part, a comparison among the most widely used re-refining technologies is performed. For this comparison, advantages and disadvantages of each re-refining technology are given (Islam Sazzad et al., 2024; Jafari & Hassanpour, 2014; Kanokkantapong et al., 2009; Kupareva et al., 2013; Lau et al., 2022; Oladimeji et al., 2018; Pinheiro et al., 2021; Sarkar et al., 2023; Shakirullah et al., 2006; Velasco-Calderón et al., 2020).

1. Acid-clay re-refining technology:
   - **Advantages:** Low capital, simple process, low cost of production, and low potential for global warming.
   - **Disadvantages:** Little environmental safety, low oil recovery, substantial quantities of acidic sludge in the primary by-product, and time-consuming.
2. Solvent extraction re-refining technology:
   - **Advantages:** Recycled solvent, zero pollutants production, good base oils recovery, need for relatively low temperature and pressure, and minimal quantity of generated waste.
   - **Disadvantages:** Need for highly skilled operators, the determination of qualities of the resultant oil only using the quality and composition of the original feedstock, and uneconomical for a small capacity factory.
3. Thin-film evaporation re-refining technology:
   - **Advantages:** Good color, mitigation of carbon residue, mitigation of ash content, suitability for large-capacity plants, and production of high-quality base oil.

- **Disadvantages:** Uneconomical for industrial use, machinery contamination by forming a resinous substance, and the need for modern equipment with highly skilled operators.

4. Membrane filtration re-refining technology:
   - **Advantages:** Reduction of the amount of generated waste, process simplicity, flexibility, preservation of the structural integrity of the oil's base stock, cost-effective especially in the stages of installation and energy use, and improvement of re-refined oil quality.
   - **Disadvantages:** High cost of membrane, slow rates of filtration, membrane durability, and membrane fouling.
5. Hydrogenation re-refining technology:
   - **Advantages:** Improvement of the oil's color stability, oxidation stability improvement, reduction of the neutralization level of rare pollutants like nitrogen and oxygen, good product quality regardless of feedstock features, and efficient removal of chloride.
   - **Disadvantages:** Need for relatively significant investments, need for facilities to supply hydrogen gas, elevated requirements for safety process, high operational costs, low operational effectiveness, requisite for pretreatment and feedstock evaluation, and essential for the catalyst regeneration.

4.6 APPLICATIONS OF TECHNOLOGIES USED FOR RE-REFINING WASTE LUBRICANT OIL

The re-refining of waste oil is an important key process for the environmental and industrial fields. It changes used oils into useful products. In industrial sectors, these technologies are crucial to waste oil recycling, environmental damage mitigation, and the support of strategies associated with the circular economy approach. The applications of these technologies are described in Table 4.3 (Islam Sazzad et al., 2024; Lau et al., 2022; Pinheiro et al., 2021; Sarkar et al., 2023).

TABLE 4.3
The Applications of Re-refining Technologies

Applications	Description
The generation of lubricant base oil	They can produce high-quality base oils, as well as the re-refined oils can be applied to the novel lubricant formulation.
Fuel generation	The oils that have been re-refined can beutilized as a fuel source, especially in industrial sectors where controlling GHG emissions is a concern.
Energy recovery	The processes related to the re-refining can involve the stages of recovering energy, like incineration, with the aim of using the waste oil for generating energy.
Other industrial utilizations	Re-refining by-products can be appliedto other industrial utilizations.

4.7 FUTURE OUTLOOKS

The industry of O&G processing is facing quick growth, leading to an increasing amount of waste lubricant oil, which needs a solution for achieving sustainability targets. Stricter guidelines on GHG emissions are navigating the evaluations of ways to reduce the dependence on generating virgin oil. Re-refining oils that are old is becoming a progressively attractive option. Because researchers strive to develop processes that are both more efficient and less dangerous to public health and the environment. As a consequence of advances in processes in re-refining technologies and the introduction of new materials for treatment, two features, namely (1) the amount and (2) the quality related to recovered base oil from used oil can improve. The increase of the amount of automation in the operations related to the recycling process has the potential with the target of building a process that is both more efficient and cheaper. Partnerships between oil-producing companies and technology developers are likely to accelerate innovation in the sector of recycling industry. Given the concerns related to the environment field and technological advancements continue to converge, the recycling of used oil is expected to play an increasingly important role in the future. Waste lubricant oil recycling is a vital research area with important implications for the environmental and economic fields. The studies related to the future could be performed in relation to improve the efficiency, effectiveness, and sustainability of processes associated with the recycling of waste lubricant oil. Important fields involve research into novel catalytic substances, the use of nanoparticles to improve treatment and re-refining systems, investigating the solvents that are environmentally friendly and exploring the sector of chemical processes, utilizing developed mathematical simulations with the aim of optimizing present re-refining systems, by executing the novel approaches. These approaches can be based on the techniques of artificial intelligence and machine learning in order to optimize the parameters and outcomes of re-refining system, applying expanded systematic and logical approaches. They can be gas chromatography-mass spectrometry (GC-MS) and Fourier transform infrared (FTIR) for achieving the purposes of monitoring the quality of oil recycled, managing inclusive life cycle assessments with the aim of evaluating different systems of recycling in relation to their impacts on the environment, supporting more rigorous standards and regulations for the targets such as used oil recycling and disposal, and boosting knowledge in relation to the advantages related to the recycling of waste lubricant oil. The oil recycled may be inclined to wear and corrosion, which they mitigate the equipment life. The corrosion problem when using oil recycled needs more examination. A feasibility analysis for the use of oil re-refined as a valuable fuel source by considering the reduction of GHG emissions, the compatibility of hardware, and the compliance with regulatory, would be a valuable and interesting area of research to explore. Also, examining the capability of applications related to re-refined oil for diverse industries could be taken into consideration. Strengthening cooperation among the stakeholders of industry, government agencies, and academia could help handle innovation and adoption of the most appropriate practices for recycling waste lubricant oil. By considering these directions for future research, the recycling of re-refined oil could develop into more sustainable and economically viable, considerably mitigating the adverse impacts on the environment and increasing the utilization and quality related to the recycled products.

4.8 CONCLUSIONS

Disposing of waste lubricant oil into the environment may result in an intense risk of harm to ecosystems. Waste lubricant oil is a waste oil product that is generated by consumers in the O&G energy sector. Waste lubricant oil is considered hazardous waste. This type of waste has adverse effects on the environment and human health. The suitable practices of a system for managing waste must be on the basis of a valuable approach such as the circular economy. Policymaking to navigate the sustainability in the O&G processing operations can be planned by implementing the circular economy approach. The circular economy is a resource efficiency approach. It focuses on reducing waste production, resource consumption, and environmental pollutants. The transition for navigating the waste products of the O&G energy operations sector toward a circular economy approach is a useful and valuable strategy in order to resolve the problems that have arisen (including GHG emissions, growth of renewable and sustainable energies, and market headwinds). Incorporating a waste lubricant oil re-refining process into the processing sector related to the value chain of the O&G energy can be a practical action for solving the environment and health challenges. The process of re-refining relates to the system that treats used lubricant oil with the target of recovering a mineral-based oil that is a useful and applicable resource. Re-refining technologies involve multiple alternatives. These alternatives are acid-clay treatment, hydrogenation, thin-film evaporation, solvent extraction, and membrane filtration. Hence, integrating the technology used for re-refining into the processing operations related to the O&G energy sector as a circular economy strategy to help the sustainability of the supply chain of this sector could be a worthwhile investment. This could be practical for developing the part related to waste lubricant oil management and achieving the goals of carbon emissions mitigation, water pollution reduction, and renewable energy growth for the industry.

REFERENCES

Agamuthu, P., Mehran, S.B. (2020). Circular Economy in Malaysia. In: Ghosh, S. (ed.) *Circular Economy: Global Perspective*. Springer, Singapore. https://doi.org/10.1007/978-981-15-1052-6_13

Ali, S. M. W., Ripin, A., & Ahmad, A. (2010). Adsorption of heavy metal from recovered base oil using zeolite. *Journal of Applied Sciences*, *10*(21), 2688–2692. https://doi.org/10.3923/jas.2010.2688.2692

Ani, I., Okafor, J., Olutoye, M., & Akpan, U. (2015). Effects of process variables and a comparative study of methods for transfer oil production from spent engine oil. *British Journal of Applied Science & Technology*, *9*(1), 65–76. https://doi.org/10.9734/bjast/2015/17578

Anisuzzaman, S. M., Jumaidi, M. H., & Nasir, N. N. M. (2021). Used lubricating oil recovery process and treatment methods: A review. *IOP Conference Series: Materials Science and Engineering*, *1195*(1), 12031. https://doi.org/10.1088/1757-899x/1195/1/012031

Botas, J. A., Moreno, J., Espada, J. J., Serrano, D. P., & Dufour, J. (2017). Recycling of used lubricating oil: Evaluation of environmental and energy performance by LCA. *Resources, Conservation and Recycling*, *125*, 315–323. https://doi.org/10.1016/j.resconrec.2017.07.010

Buchmann-Duck, J., & Beazley, K. F. (2020). An urgent call for circular economy advocates to acknowledge its limitations in conserving biodiversity. *Science of the Total Environment*, *727*, 138602. https://doi.org/10.1016/j.scitotenv.2020.138602

Campitelli, A., Cristóbal, J., Fischer, J., Becker, B., & Schebek, L. (2019). Resource efficiency analysis of lubricating strategies for machining processes using life cycle assessment methodology. *Journal of Cleaner Production*, *222*, 464–475. https://doi.org/10.1016/j.jclepro.2019.03.073

Chen, J., Mulgundmath, V., & Wang, N. (2011). Accounting for vapor-liquid equilibrium in the modeling and simulation of a commercial hydrotreating reactor. *Industrial and Engineering Chemistry Research*, *50*(3), 1571–1579. https://doi.org/10.1021/ie101550g

de Oliveira, J. A., Lopes Silva, D. A., Puglieri, F. N., & Barrera Saavedra, Y. M. (2021). *Life Cycle Engineering and Management of Products: Theory and Practice*. https://doi.org/10.1007/978-3-030-78044-9

Di Maio, F., & Rem, P. C. (2015). A robust indicator for promoting circular economy through recycling. *Journal of Environmental Protection*, *6*(10), 1095–1104. https://doi.org/10.4236/jep.2015.610096

Ellen MacArthur Foundation. (2015).*Circularity indicators: An approach to measuring circularity*. Ellen MacArthur Foundation, 12.

Elmehbad, D. N. (2017). Enhancement adsorption of heavy metal from recovered base oil using zeolite and kaolinite by phase transfer catalyst and anionic surfactants-Part 2. *International Journal of Advanced Engineering Research and Science*, *4*(2), 124–128.

Fetanat, A., & Tayebi, M. (2022). A picture fuzzy set-based decision support system for treatment technologies prioritization of petroleum refinery effluents: A circular water economy transition towards oil & gas industry. *Separation and Purification Technology*, 122220. https://doi.org/10.1016/j.seppur.2022.122220

Fetanat, A., & Tayebi, M. (2023). Sustainability and resilience-oriented prioritization of oil and gas produced water treatment technologies: A novel decision support system under circular intuitionistic fuzzy set. *Geoenergy Science and Engineering*, *221*, 211379. https://doi.org/10.1016/j.geoen.2022.211379

Fetanat, A., Tayebi, M., & Mofid, H. (2023). Water-energy-carbon nexus and sustainability-oriented prioritization of negative emissions technologies for the oil & gas industry: A decision support system under Fermatean fuzzy environment. *Process Safety and Environmental Protection*, *179*, 462–483. https://doi.org/10.1016/j.psep.2023.09.037

Fetanat, A., Tayebi, M., & Shafipour, G. (2021). Management of waste electrical and electronic equipment based on circular economy strategies: Navigating a sustainability transition toward waste management sector. *Clean Technologies and Environmental Policy*, *23*, 343–369. https://doi.org/10.1007/s10098-020-02006-7

Gonzaga, N. da C., Silva, R. N. da, & Andrade, L. P. de. (2021). Gerenciamento de Resíduos do Óleo Lubrificante: Uma Revisão Sistemática da Literatura. Revista de Gestão Social e Ambiental, *15*, e02812. https://doi.org/10.24857/rgsa.v15.2812

Grafström, J., & Aasma, S. (2021). Breaking circular economy barriers. *Journal of Cleaner Production*, *292*, 126002. https://doi.org/10.1016/j.jclepro.2021.126002

Hassanain, E. M., Yacout, D. M. M., Metwally, M. A., & Hassouna, M. S. (2017). Life cycle assessment of waste strategies for used lubricating oil. *International Journal of Life Cycle Assessment*, *22*(8), 1232–1240. https://doi.org/10.1007/s11367-016-1255-x

Hishammuddin, M. A. H., Ling, G. H. T., Chau, L. W., Ho, C. S., Ho, W. S., & Idris, A. M. (2018). Circular economy (CE): A framework towards sustainable low carbon development in Pengerang, Johor, Malaysia. *Chemical Engineering Transactions*, *63*, 481–486. https://doi.org/10.3303/CET1863081

Huang, K., & Zhang, J. (2011). Circular economy strategies of oil and gas exploitation in China. *Energy Procedia*, *5*, 2189–2194. https://doi.org/10.1016/j.egypro.2011.03.378

Islam Sazzad, M. R., Rahman, M. M., Hassan, T., Al Rifat, A., Al Mamun, A., & Adib, A. R. et al (2024). Advancing sustainable lubricating oil management: Re-refining techniques, market insights, innovative enhancements, and conversion to fuel. *Heliyon, 10*(20), e39248. https://doi.org/10.1016/j.heliyon.2024.e39248

Jafari, A. J., & Hassanpour, M. (2014). Analysis and comparison of used lubricants, regenerative technologies in the world. *Resources, Conservation and Recycling, 103*, 179–191. https://doi.org/10.1016/j.resconrec.2015.07.026

Kanokkantapong, V., Kiatkittipong, W., Panyapinyopol, B., Wongsuchoto, P., & Pavasant, P. (2009). Used lubricating oil management options based on life cycle thinking. *Resources, Conservation and Recycling, 53*(5), 294–299. https://doi.org/10.1016/j.resconrec.2009.01.002

Kirchherr, J., Reike, D., & Hekkert, M. (2017). Conceptualizing the circular economy: An analysis of 114 definitions. *Resources, Conservation and Recycling, 127*, 221–232. https://doi.org/10.1016/j.resconrec.2017.09.005

Kupareva, A., Mäki-Arvela, P., & Murzin, D. Y. (2013). Technology for rerefining used lube oils applied in Europe: A review. *Journal of Chemical Technology and Biotechnology, 88*(10), 1780–1793. https://doi.org/10.1002/jctb.4137

Lau, P. J., Ng, W. P. Q., How, B. S., Lim, C. H., & Lam, H. L. (2022). Paving a way toward circular economy for oil and gas industry: A conceptual modelling of re-refining process through solvent extraction and hydrofinishing pathway. *Journal of Cleaner Production, 380*, 134839. https://doi.org/10.1016/j.jclepro.2022.134839

Lonca, G., Muggéo, R., Imbeault-Tétreault, H., Bernard, S., & Margni, M. (2018). Does material circularity rhyme with environmental efficiency? Case studies on used tires. *Journal of Cleaner Production, 183*, 424–435. https://doi.org/10.1016/j.jclepro.2018.02.108

Michael, A. (2015). Used oil storage and disposal practices in automobile repair garages in Ghana. *International Journal of Science, Technology and Society, 3*(4), 221. https://doi.org/10.11648/j.ijsts.20150304.23

Mikhail, S., & Riad, M. (2014). Hydro-treatment of middle lubricating base oil on sulfided promoted molybdenum catalysts. *Fuel Processing Technology, 128*, 482–489. https://doi.org/10.1016/j.fuproc.2014.07.044

Mohammed, R. R., Ibrahim, I. A. R., Taha, A. H., & McKay, G. (2013). Waste lubricating oil treatment by extraction and adsorption. *Chemical Engineering Journal, 220*, 343–351. https://doi.org/10.1016/j.cej.2012.12.076

Nour, A. H., Elamin, E. O., Nour, A. H., & Alara, O. R. (2021). Dataset on the recycling of used engine oil through solvent extraction. *Chemical Data Collections, 31*, 1–11. https://doi.org/10.1016/j.cdc.2020.100598

Nowak, P., Kucharska, K., & Kamiński, M. (2019). Ecological and health effects of lubricant oils emitted into the environment. *International Journal of Environmental Research and Public Health, 16*, 3002. https://doi.org/10.3390/ijerph16163002

Oladimeji, T. E., Sonibare, J. A., Omoleye, J. A., Emetere, M. E., & Elehinafe, F. B. (2018). A review on treatment methods of used lubricating oil. *International Journal of Civil Engineering and Technology, 9*(12), 506–514.

Park, J. Y., & Chertow, M. R. (2014). Establishing and testing the "reuse potential" indicator for managing wastes as resources. *Journal of Environmental Management, 137*, 45–53. https://doi.org/10.1016/j.jenvman.2013.11.053

Pichler, J., Maria Eder, R., Besser, C., Pisarova, L., Dörr, N., Marchetti-Deschmann, M., & Frauscher, M. (2023). A comprehensive review of sustainable approaches for synthetic lubricant components. *Green Chemistry Letters and Reviews, 16*(1), 997–1016. https://doi.org/10.1080/17518253.2023.2185547

Pinheiro, C. T., Quina, M. J., & Gando-Ferreira, L. M. (2021). Management of waste lubricant oil in Europe: A circular economy approach. *Critical Reviews in Environmental Science and Technology, 51*(18), 2015–2050. https://doi.org/10.1080/10643389.2020.1771887

Pires, A., &Martinho, G. (2013). Life cycle assessment of a waste lubricant oil management system. *International Journal of Life Cycle Assessment, 18*(1), 102–112. https://doi.org/10.1007/s11367-012-0455-2

Rahman, M., Siddique, T., Samdani, S., & Kabir, K. (2008). Effect of operating variables on regeneration of base-oil from waste oil by conventional acid-clay method. *Chemical Engineering Research Bulletin, 12*, 24–27. https://doi.org/10.3329/cerb.v12i0.1494

Richard, C. N., Zabala, G. S., Saitta, M. F., Díaz, C., & Abalos, R. (2020). Circular economy in oil & gas industry. The current situation in Argentina, possibilities, and future applications. *In SPE/AAPG/SEG Latin America Unconventional Resources Technology Conference 2020, LURT 2020.* https://doi.org/10.15530/urtec-2020-1331

Rigamonti, L., & Mancini, E. (2021). Life cycle assessment and circularity indicators. *International Journal of Life Cycle Assessment, 26*(10), 1937–1942. https://doi.org/10.1007/s11367-021-01966-2

Saleem, H. J., & Karim, A. R. (2020). Re-refining of used lubricating oil by vacuum distillation/thin wiped film evaporation technique. *Petroleum Science and Technology, 38*(4), 323–330. https://doi.org/10.1080/10916466.2019.1704782

Sánchez-Alvarracín, C., Criollo-Bravo, J., Albuja-Arias, D., García-Ávila, F., & Raúl Pelaez-Samaniego, M. (2021). Characterization of used lubricant oil in a Latin-American medium-size city and analysis of options for its regeneration. *Recycling, 6*(1), 1–22. https://doi.org/10.3390/recycling6010010

Sarkar, S., Datta, D., Deepak, K. S., Mondal, B. K., & Das, B. (2023). Comprehensive investigation of various re-refining technologies of used lubricating oil: A review. *Journal of Material Cycles and Waste Management, 25*(4), 1935–1965. https://doi.org/10.1007/s10163-023-01685-w

Shaikh, R., & Mahanwar, P. (2018). Reclamation of used engine oil using polymeric flocculants. *International Journal of Chemical Sciences, 16*(2), 1–14. https://doi.org/10.21767/0972-768x.1000254

Shakirullah, M., Ahmad, I., Saeed, M., Khan, M. A., Habib-Ur-Rehman, Ishaq, M., & Shah, A. A. (2006). Environmentally friendly recovery and characterization of oil from used engine lubricants. *Journal of the Chinese Chemical Society, 53*(2), 335–342. https://doi.org/10.1002/jccs.200600042

Sharma, N. K., Govindan, K., Lai, K. K., Chen, W. K., & Kumar, V. (2021). The transition from linear economy to circular economy for sustainability among SMEs: A study on prospects, impediments, and prerequisites. *Business Strategy and the Environment, 30*(4), 1803–1822. https://doi.org/10.1002/bse.2717

Sillanpää, M., & Ncibi, C. (2019). *The Circular Economy: Case Studies about the Transition from the Linear Economy.* https://doi.org/10.1016/C2017-0-02916-6

Silva, N. R., Rodrigues, T. O., Braga, T. E. N., Christoforo, A. L., & Silva, D. A. L. (2022). Re-refining of lubricant oil used and contaminated (LOUC) or its combustion in cement plants? An exploratory study in Brazil based on the life cycle assessment and circularity indicators. *Sustainable Production and Consumption, 33*, 360–371. https://doi.org/10.1016/j.spc.2022.07.017

Syahir, A. Z., Zulkifli, N. W. M., Masjuki, H. H., Kalam, M. A., Alabdulkarem, A., & Gulzar, M.et al. (2017). A review on bio-based lubricants and their applications. *Journal of Cleaner Production, 168*, 997–1016. https://doi.org/10.1016/j.jclepro.2017.09.106

Tan, E. C. D., & Lamers, P. (2021). Circular bioeconomy concepts—A perspective. *Frontiers in Sustainability, 2*, 701509. https://doi.org/10.3389/frsus.2021.701509

Tsambe, M. Z. A., de Almeida, Ć. F., Ugaya, Ć. M. L., & de Abreu Cybis, L. F. (2019). Environmental assessment of used lubricant oil in southern Brazilian region. *Journal of Environmental Accounting and Management, 7*(4), 279–290. https://doi.org/10.5890/JEAM.2019.09.003

Udonne, J. D., & Bakare, O. A. (2013). Recycling of used lubricating oil using three samples of acids and clay as a method of treatment. *International Archive of Applied Sciences and Technology*, *4*(2), 8–14.

Udonne, J. D., Efeovbokhan, V. E., Ayoola, A. A., & Babatunde, D. E. (2016). Recycling used lubricating oil using untreated, activated and calcined clay methods. *Journal of Engineering and Applied Sciences*, *11*(6), 1396–1401.

Velasco-Calderón, J. C., Garcia-Figueroa, A. A., Lopez Cervantes, J. L., & Gracia-Fadrique, J. (2020). Regeneration of used lubricating oil by solvent extraction and phase diagram analysis. *Current Research in Green and Sustainable Chemistry*, *3*, 100010. https://doi.org/10.1016/j.crgsc.2020.06.003

Wong, P. K., & Wang, J. (2001). The accumulation of polycyclic aromatic hydrocarbons in lubricating oil over time - a comparison of supercritical fluid and liquid-liquid extraction methods. *Environmental Pollution*, *112*(3), 407–415. https://doi.org/10.1016/S0269-7491(00)00142-1

Zgheib, N., & Takache, H. (2021). Recycling of used lubricating oil by solvent extraction: Experimental results, aspen plus simulation and feasibility study. *Clean Technologies and Environmental Policy*, *23*(1), 65–76. https://doi.org/10.1007/s10098-020-01893-0

5 From Formation to Remediation
Maneuvering the Regulatory Framework for Waste Chemicals

Abeera Gul, Mushtaq Ahmad, Mamoona Munir, Shazia Sultana, and Nigar Mursal

5.1 INTRODUCTION

5.1.1 Extent and Impact of Chemical Waste Crisis

The rapidly increasing chemical waste is resulting in public health and environmental challenges, and the possible driving forces include industrialization along with the escalating use of synthesized compounds in pharmaceuticals, agriculture, and consumer goods. Every year, the world produces 400 million tons of hazardous waste (UNEP, 2019), a substantial portion of which remains untreated specially in the developing regions (WHO, 2018). The eco-system not only is threatened by this crisis but it's also resulting in serious health concerns, that is, reproductive harm, endocrinal disruption, and cancer (Landrigan et al., 2018).

5.1.2 Objectives and Structure of the Chapter

This chapter focuses on exploring the governance of chemical waste through intersection of green chemistry principles, international compliance framework, and waste management hierarchies. Policy frameworks like **REACH, RCRA,** and **Basel Convention** will be discussed systematically along with real-world case studies and use of advanced technology for smart governance in future (Wang et al., 2020). It also discusses low and middle-income countries which are lacking legislative and enforcement mechanisms for chemical waste management.

5.1.3 Relevance to Sustainable Developmental Goals (SDGs) and Green Chemistry

Green chemistry offers a design-based approach in order to remove hazards at the source which perfectly aligns with SDG 12 (responsible production and consumption) and SDG 6 (clean water and sanitation) (Anastas & Warner, 2000). The

DOI: 10.1201/9781003642589-5

incorporation of green design (which includes closed-loop synthesis, waste valorization, and benign-by-design molecules) into regulatory systems is pivotal for transition toward sustainable chemical economy (Clark & Deswarte, 2015). This chapter explores how a zero waste, circular future can be attained through integration of the above-mentioned approaches into global policy.

5.2 ORIGIN AND CLASSIFICATION OF CHEMICAL WASTE

5.2.1 Sources of Chemical Waste

A wide range of sectors produce chemical waste which exhibits the excessive use of synthetic compounds in the modern world. The major sources for its production include healthcare organizations, agriculture and farming, industrial activities, academic and research activities and commercial or household operations. In industries, the production of chemical waste as a by-product includes intermediates of reactions, catalysts, solvents or sludge from wastewater treatment plants. Research and academic institutes contribute less but still harmful material by means of reactions, expired reactants, and contaminated scientific glassware (Nascimento & Tenuta Filho, 2010). Healthcare organizations produce chemical waste in the form of cytostatic drugs, disinfectants, and chemicals which are toxic, destructive, reactive, or combustible (WHO, 2018). In low- and middle-income countries (LMICs), the use of chemicals occurs in an unregulated manner which often leads to unmanaged dumping and open burning resulting in environmental and health hazards (Cox, 2020).

5.2.2 Chemical Waste Types

Chemical waste is basically divided into two main types, **Hazardous and Non-Hazardous waste** on the basis of its physical, chemical, and toxicological properties. In accordance with the criteria set by United Nations and U.S. Environmental Protection Agency(EPA), the hazardous type of chemical waste possess the following characteristics: **Ignitability** (e.g., combustible solvents), **Corrosivity** (e.g., strong acids and bases), **Reactivity** (e.g., unstable compounds), **Toxicity** (e.g., carcinogens) (EPA., 2022 a-b).

Non-hazardous chemical waste on the other hand is not very dangerous to health or environment under normal conditions, that is, weak buffer solutions or benign salts. However, they can also contribute to environmental degradation if not managed properly or mixed with hazardous waste (Lipnick et al., 2002).

Few more categories for division of chemical waste include **organic and inorganic waste**, **acute vs chronic toxicity,** and the **Persistent, Bioaccumulative and Toxic Substances (PBT)**, as mentioned in the REACH regulations (ECHA, 2023a-d).

5.2.3 A Complete Lifecycle: From Generation to Final Disposal

There is a sequence of stages in chemical waste lifecycles from generation of waste during chemical production to its handling, storage, transportation, treatment, and finally disposal or recycling, that is, the complete lifecycle involves the following steps:

1. **Generation:** It's the initial point where waste is produced.
2. **Characterization:** In this step, the composition and associated hazards of waste are determined.
3. **Storage and Labeling:** The waste is segregated and stored following regulatory standards, for example, usage of secondary containment and label hazardous contents.
4. **Transportation:** Transport of waste by means of secure packaging and trained handlers occurs under regulated conditions.
5. **Treatment:** The chemical waste is treated either by physical (Filtration), chemical (neutralization), biological (Degradation), or thermal (Incineration) methods.
6. **Disposal or Recycle:** The treated waste is further disposed of in landfills or recovered by means of resource recycling or energy recovery.

This "Cradle-to-Grave" approach assures accountability at each step and aligns with international frameworks like Basel Convention and international acts like RCRA (UNEP, 2020). However, systematic monitoring and documentation is required at each step to ensure safe handling in application of effective waste management strategies. Innovative digital management software, that is, waste tracking software and blockchain-based manifest systems are a prime solution in this particular domain (Wang et al., 2020). Figure 5.1 shows lifecycle of chemical waste.

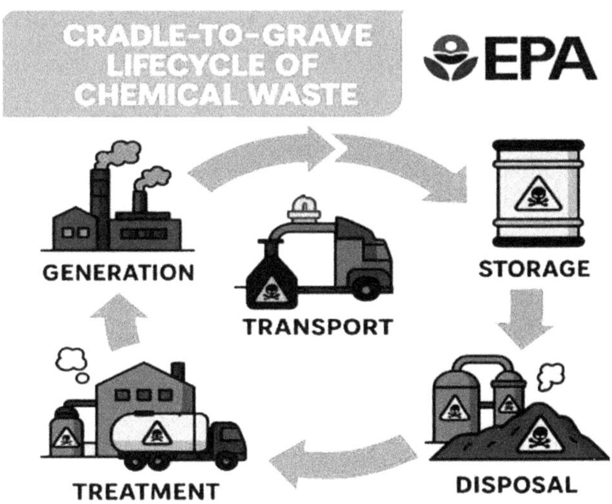

FIGURE 5.1 Visual illustration of waste management lifecycle.

5.3 PREVENTION OF WASTE THROUGH GREEN CHEMISTRY PERSPECTIVE

5.3.1 Waste Minimization and Green Chemistry

The transformative approach of green chemistry focuses on designing safe and sustainable chemical process for handling pollution from the outset, instead of controlling and remediating it later (Anastas and Warner, 2000). It prioritizes the reduction or elimination of hazardous material at the time of its synthesis, utilization, and disposal of chemical products, that is, green chemistry highlights waste prevention at the source rather than treating it at later stages after formation, which aligns perfectly with the first principle: **"Prevention of waste is better than treating or cleaning it up later after its formation."**

This concept represents a groundbreaking development in how the manufacturers; policymakers and researchers see environmental responsibility. By articulating green chemistry principles in the production systems, the economic and environmental costs in chemical waste management can be reduced significantly (Clark & Macquarrie, 2008).

5.3.2 Design for Degradation and Non-toxicity

One of the major goal of green chemistry is to ensure the conversion of a product into harmless and biodegradable entity by the end of its lifecycle and this is known as **"Design for Degradation"** (Wilson & Schwarzman, 2009). For example, the development of surfactants and biodegradable polymers by researchers which do not remain in the aquatic environment, reduces risk of chemical exposure.

In the same manner, green chemistry promotes non-toxic design– the conscious efforts to avoid use of substances which are mutagenic, carcinogenic, or bioaccumulative. Before production, safer alternatives are created by using tools like green toxicology, structure-activity relationship (SAR) modeling, and benign-by-design molecular engineering (Maertens et al., 2021).

5.3.3 Process Intensification and Cleaner Synthesis

Process intensification includes technologies which make the chemical process more energy efficient, compact, and waste minimizing. Some of the examples of intensification which lower energy consumption and generation of waste while enhancing reaction efficiency, include membrane separation process, reactive distillation, and micro-reactors (Stankiewicz & Moulijn, 2000).

Green synthesis strategies involve:

- Utilizing catalysts rather than using stoichiometric reagents to avoid by-product synthesis.
- Conducting reaction under suitable environmental conditions to lessen energy input.
- Using water-based chemistry wherever possible reduces the need for organic solvents and minimizes environmental and health risks.

The above-mentioned techniques result in minimal chemical waste generation and directly reduces the cost of compliance with hazardous waste regulations (Trost, 1991).

5.3.4 Solvent Substitution and Greener Reaction Media

A significant amount of chemical waste generates from using solvents specially in chemical manufacturing and pharmaceutical industries. Green chemistry focuses on solvent substitutions, that is, it encourages use of safe alternatives like water, ionic liquids, supercritical CO_2, and bio-based solvents rather than using traditional volatile organic compounds (VOCs). These green solvents do not require post-reaction treatment and reduce emissions (Jessop, 2011). CHEM21 solvent selection guide is the best tool to rank solvents on the basis of their environmental impact (Prat et al., 2016).

5.3.5 Closed-Loop Recycling and Material Circularity

Green chemistry also overlaps with principles of circular economy by means of closed-loop system where materials gets recycled within the same system without being wasted. Examples include recycling catalysts, solvent recovery units in chemical plants, and chemical leasing models where producers maintain ownership of chemicals and assuring their recovery. These strategies promote raw material preservation and lead to reduced waste disposal need, converting linear processes into regenerative cycles (OECD, 2018).

5.4 THE ARRE FRAMEWORK AND WASTE MANAGEMENT HIERARCHY

5.4.1 ARRE: Avoid, Reduce, Recycle, Eliminate

The waste management hierarchy provides the foundation for sustainable handling of waste, highlighting a proper order of actions to reduce its impact on the environment. A well-known formula for the management of waste is the ARRE framework–Avoid, Reduce, Recycle, and Eliminate. This model is a macro-level intervention strategy for preventing waste generation before the need for its mitigation or disposal (EPA, 2022a-b). The practical framework of ARRE works as follows:

- Avoid: Preventing the use of hazardous substances at the initial stages by following green chemistry principles, that is, choosing non-toxic feedstock.
- Reduce: Optimizing reactions in a way that fewer by-products are generated, or reaction efficiency is improved (e.g., atom-economical synthesis).
- Recycle: Recovering solvents, catalysts, and raw materials from chemical reactions, to be re-used.
- Eliminate: Destruction of hazardous materials mainly through high-temperature incineration with controlled emissions, if prevention or reusability is not possible.

This practical framework is in accordance with global zero-waste objectives and principles of circular economy (UNEP, 2020).

5.4.2 Integration with Zero-Waste and Circular Economy

The zero-waste philosophy is an extension to the ARRE principle with an objective to transform the generated waste into resources by running chemical products through systematic redesign processes. In case of circular economy, materials run in closed loops, reducing input and waste while increasing utility. Green chemistry provides biodegradable materials and technologies like process intensification which support these objectives (European Commission, 2020). For example, industrial symbiosis makes it possible that the generated waste from one source could be used as raw material for the other. Similarly, solvent recovery through closed-loop recycling is practiced in pharmaceutical manufacturing. Moreover, chemical leasing models extend their focus to performance-based services, that is, reduction of waste from over-consumption (UNIDO,2016).

5.4.3 Design for Prevention Over Disposal

Traditional waste management policies focus more on the treatment and disposal of waste, that is, preferring end-of-pipe solutions. However, modern strategies promote source-reduction as it is cost-effective and more environment friendly. This shift has been noticeable in legislative frameworks. Examples include the **Pollution Prevention Act (USA)** which directs industries to find alternatives before employing to disposal option (EPA, 2021 a-c).

Designing for prevention includes opting synthetic pathways which enhance atom economy (Trost, 1991) using enzymatic or catalytic reagents instead of toxic reagents and applying in-process evaluation to figure out inefficiencies at the earlier stage.

By making waste prevention a part of the design and regulation, industries can conform to environmental standards while also improving productivity and safety.

5.5 REGULATORY FRAMEWORKS GOVERNING CHEMICAL WASTE

5.5.1 Importance of Regulatory Chain of Command

Chemical waste is hazardous in nature. It is often poisonous and insistent and results in long-term health and environmental effects. Thus, it is compulsory to manage it via strong regulatory chain of command which regulates the waste through all stages, that is, from synthesis and categorization to transportation and then disposal after treatment while also prioritizing prevention and innovation (UNEP, 2020).

These frameworks also help to systematize stakeholder roles – manufacturers, waste managers, regulators, and transporters – and ensure accountability throughout the lifecycle. The application of these frameworks is essential in international trade to ensure illegal transboundary movement and hence environmental justice.

5.5.2 United States: Resource Conservation and Recovery Act (RCRA)

The Resource Conservation and Recovery Act (RCRA) was constituted by U.S. Congress in 1976 after facing several public health disasters in which the most notable one was the Love Canal contamination which made the government realize the outcomes of unmanaged chemical waste. Under the administration of EPA, RCRA entrenched a diversified system for the cradle-to-grave management of hazardous waste.

Key features of RCRA include:

- Categorization of waste into F-, K-, P- and U-lists.
- Development of hazardous waste manifest system for tracking waste from generation to disposal.
- The RCRA Permit System for treatment, storage, and disposal facilities (TSDFs).
- Strict tracking and reporting obligations.

RCRA is considered to be the pioneer in regulatory frameworks for the integration of waste minimization hierarchy in federal law, encouraging pollution prevention by adopting green chemistry principles, resulting in cleaner production (EPA, 2021a-c).

5.5.3 European Union: REACH Regulation

The Registration, Evaluation, Authorization, and Restriction of Chemicals (REACH) regulation was introduced by the European Union in 2007 and later on, it was updated in 2018. It's the first ever detailed chemical safety framework proposed by the European Union, developed in response to insistent organic pollutants and increasing endocrine disruptions. REACH makes it mandatory for the manufacturers to authenticate their products as safe before they reach the market.

REACH influences waste management in the following ways:

- By promoting safer substitutes for hazardous substances.
- Limiting the use of substances of very high concern (SVHCs).
- Authorizes manufacturers to collaborate and share data.
- Promotes advancement of green chemistry and chemical design.

Practicing REACH framework has resulted in discontinuing use of lead-based pigments and nonylphenol ethoxylates, promoting consumption of safe materials in consumer goods (ECHA, 2023a-d).

5.5.4 Basel Convention: Controlling Transboundary Waste

The Basel Convention is an international treaty adopted in 1989 and implemented in 1992 to address the problem of transporting hazardous waste from developed to developing countries. It was largely prompted by incidents such as the Koko toxic waste scandal, in which Italian companies were found to have illegally dumped about 18,000 drums of hazardous waste in Nigeria.. The Convention's goal is to reduce waste generation, assure environmentally sound management (ESM) and application of Prior Informed Consent (PIC) for the regulation of transboundary movement.

According to 2024 data, 190 countries are members to this convention. However, not all of them are implementing the rules and there are loopholes in following the guidelines and their enforcement, resulting in developed nations transporting their hazardous shipments as "recyclable materials," eluding regulations (Secretariat of the Basel Convention, 2023).

To tackle this problem, some initiatives like Plastic Waste Amendments (2021) and Digital Tracking Systems are in process for improving transparency.

5.5.5 Globally Harmonized System (GHS) of Classification and Labeling

The United Nations developed Globally Harmonized System (GHS) which is now adopted in different forms by around 70 countries. A standard protocol is provided by GHS to classify chemical hazards. GHS stresses on the significance of hazard labeling before its further handling. Central aspects of GHS include:

- Application of standards pictograms and signals words like "Danger" or "Warning."
- Utilization of scientifically derived data to issue hazard statements.
- Providing Safety Data Sheets (SDSs) in a particular 16-section format.

GHS is not a regulation itself, so its implementation occurs through national laws like EU's CLP regulation and the OSHA Hazard Communication Standard (USA), making it a foundation for global chemical safety (UNECE, 2021). Table 5.1 shows the international regulatory frameworks for chemical waste.

5.5.6 Regulatory Framework in Developing Economies

The LMICs adopt basic regulatory frameworks based on international models but fail to implement because of sources limitations, technical capacity, and political involvement. For example:

- India has approved the Management and Transboundary Movement Rules, 2016, which is in accordance with the Basel Convention Principles but unable to practice properly due to fragile pollution control boards (MoEFCC, 2016).

TABLE 5.1
International Regulatory Frameworks for Chemical Waste

Framework	Region	Scope	Key Provisions
RCRA	United States	Hazardous Waste Generation to Disposal	General Classification, Cradle-to-Grave Tracking
REACH	European Union	Chemical Registration and Authorization	Industry-Led Safety Data, Substance Restrictions
Basel Convention	Global (UN Treaty)	Transboundary Hazardous Waste	Prior Informed Consent, Illegal Trafficking Provisions
GHS	Global (UN System)	Classification and Labeling	Harmonized Hazard Symbols, SDS Standardizations

- Nigeria follows the NESREA regulations but struggles to combat e-waste and chemical containers imported from Europe.
- Brazil applies the waste classification standards under CONAMA Resolution 313/2002, but illegal dumping and burning are still practiced, particularly by informal sectors.

There is a growing advocacy for "South-South cooperation," Regional Regulatory Conformity and capacity building such as the efforts made under African Union's Bamako Convention.

5.6 IMPLEMENTATION: COMPLIANCE AND BEST PRACTICES

5.6.1 Bridging Regulation with Practice

After the establishment of chemical waste regulations under different policies, there are multiple factors in the way of their successful implementation. While the standard protocols for the successful management of chemical waste are defined by legal frameworks like RCRA, REACH, The Basel Convention and GHS, it is the practical application which determines their true efficiency. Well-structured protocols, technology utilization, institutional responsiveness and individual accountability are the key factors for successful operationalization of these legal mandates. Their compliance ensures that the waste is not only handled safely but its generation is also minimized which aligns with the core principles of green chemistry (UNEP, 2020).

In this section, we discuss in detail the complete process, that is, from formation to final disposal and obstacles in implementing chemical waste regulations through international examples and real-world case studies.

5.6.2 Waste Identification and Classification

Waste is accurately identified and classified in the first step of regulatory compliance. It is a crucial step to determine the method for further storage, transport, treatment, and disposal of identified waste and failure in correct identification leads to violating regulation and ultimately environmental damage. Waste generators in United States are bound to determine if the waste is ignitable, corrosive, toxic, or reactive, under the criteria provided by **EPA, 40 CFR part 261** (EPA, 2021a-c). Furthermore, the waste is enlisted in **F-, K-, P- and U-lists** which classify it into three types: Industrial process waste, specific source waste, and acute hazardous waste.

The European Union adopts a similar strategy under the REACH and CLP regulations, where the manufacturers properly evaluate the physical, chemical and toxicological properties for proper classification of substances (ECHA, 2023a-d). The GHS and National Legislation design **SDSs** which play pivotal role in passing on hazard information. Organizations which fail to accurately classify hazardous materials are bound to pay fines due to the disintegration of the regulatory chain. For instance, an electronic manufacturer in the United States was fined with $300,000 in 2017 due to illegal classifying of flammable solvent-waste stream which later on resulted in on-site fire and environmental pollution (EPA, 2018).

5.6.3 Labeling and Documentation

This step holds importance in waste management process for transparent and safe regulation. Labels display the nature of hazard in standard formats which include pictograms, code words, hazard signs, and precautions according to GHS guidelines. Besides, each container shows a particular waste identification code and accumulation start dates to adhere to the time-based storage limits, given under RCRA regulations (EPA, 2021a-c).

Documentation goes beyond labeling and encompasses maintaining shipment manifests, internal waste tracking logs, and regulatory reports. While transporting the hazardous waste, the E-manifest system under RCRA tracks hazardous waste activity to assure a verified chain of custody. The European Union utilizes electronic systems under waste shipment regulations to facilitate on-time exchange of data between generators, transporters, and treatment centers (UNEP, 2020). Poor documentation leads to enforcement measures, waste-processing delays, and loss of certification to compliance.

5.6.4 Safe Storage and Segregation

The safety principles, proper separation, and secondary containment must be prioritized in storing chemical waste. Flammable and oxidizing materials, acids, and bases should not be stored together to avoid hazardous reactions. Physical barriers, chemical-resistant shelving and ventilated storage units should be used by the facilities to reduce the risk of cross-contamination or gas buildup. Secondary containments like bunds or spill trays are designed to manage liquid hazardous waste in case of leaks or spills (EPA, 2021a-c).

Storage facilities should also comply with space planning and fire safety standards such as access control, labeling, and fire extinguishing systems. In China, **The Tianjin Port Blast** in 2015 occurred due to co-storage of contradictory substances which is a violation of national safety codes (Zhang et al., 2017). The tragic death of 170 people focused on the criticality of acceding to the regulatory storage frameworks and resulted in major revamp in China's chemical safety laws.

5.6.5 Waste Transport and Tracking System

The transport of hazardous waste might result in new compliance challenges like spilling, accidents, and cross-border legal inconsistencies. Therefore, the RCRA manifest system in the United States makes it mandatory for the waste to be transported by licensed carriers and checked at every transfer point until final disposal. Accountability is ensured by the **end-to-end** approach in tracking waste, facilitating audits and inspections (EPA, 2021a-c).

The Basel Convention governs the transboundary movement of hazardous waste for which the receiving countries provide **PIC.** This process guarantees that receiving countries are accepting waste with free will. However, the implementation remains challenged in many regions due to inadequate scrutiny and false reports.

Various countries have adopted digital solutions for waste flow analysis. For example, **Finland and Germany** have turned to blockchain-based systems which manage each transaction in a secure and transparent manner (Wang et al., 2020). These technologies not only minimize burden on administration but also ensures fraud and illegal dumping.

5.6.6 Treatment and Final Disposal Methods

After the safe transportation of waste to an authorized facility, the next step involves treatment or disposal according to the regulatory standards. Waste is treated by the following methods: Incineration, chemical neutralization, and stabilization.

Incineration is an effective method but pollutants like dioxins, furans, and heavy metals must not exceed the emission limits. To ensure this, **continuous emission monitoring systems (CEMS)** must be installed in facilities and stick to the **Best Available Techniques (BAT)** highlighted in international guidelines (UNEP, 2020).

Acidic or basic waste substances are often treated with chemical neutralization methods while the heavy metals in sludge or ash require solidification/stabilization techniques. Safe land-filling is the last option and is only allowed for residue which passes the **leachability test**. Plasma gasification and super-critical water oxidation are some invitational approaches which promise energy recovery and minimized emissions while treating highly toxic waste (Aggeri et al., 2021).

5.6.7 Monitoring, Reporting, and Continuous Compliance

Compliance is a multi-step procedure which involves performance tracking through monitoring and regular updates by means of documentation and reports. The submission of annual hazardous waste reports (EPA) or performance audits (EU environmental agencies) is compulsory for the governing bodies. The preservation of previous records is mandatory for ensuring timely inspections.

Permit renewal alerts are automatically generated after data collection through digital compliance platforms, enabling up-to-the-minute reporting. The likelihood of non-compliance is reduced when these systems ensure transparency in process. **Environmental, health, and safety (EHS)** metrics are consolidated into one platform by tech giants, facilitating audits, and strengthening regulatory compliance.

5.6.8 Organizational Culture and Behavioral Compliance

The adherence to environmental standards in an organization plays key role in successful implementation of regulations. Workshops, implementation of modern systems and compliance mechanisms basically lead to strong EHS cultures in companies. Empowered employees, strong leadership engagement, and effective incentive programs are all outcomes of an active compliance culture.

Many countries reward efficient performances by several accreditation programs and joint public-private ventures. **The GreenCo Rating System in India and Clean Corporate citizen program of the United States** recognize enterprises which not only meet the standards but outperform and exceed expectations (UNEP,

2020). These incentives increase investors' trust and amplify public opinion, beyond improving compliance.

Thus, an efficient strategy to successfully implement chemical waste regulation involves combining partnerships, standard operating procedures (SOPs), and cultural transformation. Each step in management of waste from synthesis to final disposal needs particular compliance protocols which must be regularly monitored. With the evolving technology, management of environmental risks also gets complex. Hence, dynamic and adjustable compliance frameworks with legal directives and ethical imperatives are obligatory for sustainable management of chemical waste.

5.7 CASE STUDIES OF REGULATORY INTEGRATION IN CHEMICAL WASTE MANAGEMENT

5.7.1 Role of Case Studies in Policy Evaluation

Real-world evidence helps explore the extent to which regulatory frameworks are implemented. They help in linking theory and practice, emphasizing both the existing challenges and the best practical measures. Various case studies are discussed below, from geostrategic to commercial spheres, representing the obstacles, innovations, and expertise from the measures taken for consolidation of the regulatory structure in management of chemical waste. These examples not only display international compliance like REACH, RCRA, Basel Convention but also explains how the governments and organizations design strategies for local needs to meet global standards.

5.7.2 United States: The RCRA Manifest System in Industrial Compliance

The EPA's RCRA has been used as a model from a long time for organized and cradle-to-grave handling of waste. **Dow Chemical's waste program** in Texas is one such example. It created a completely digitized tracking system for hazardous waste containers including bar codes, scanning manifests, and centralized compliance dashboards. This program reduced administrative lag-times and removed errors in documentation by more than 80% in a time-period of three years after being operational.

In addition, the **E-Manifest Program under EPA**, initiated in 2018 nationally has ensured better data transparency, minimized paper usage, and enabled real-time monitoring in many states. In the first two years after implementation, more than 1.5 million manifests were managed digitally, showcasing the potential of electronic transformation in regulatory compliance.

5.7.3 European Union REACH Implementation in the Paints and Coating Industry

A wonderful initiative was taken by **AkzoNobel**, which is a major actor in the paint and coating industry in Europe. It worked on the application of REACH guidelines in its production. A risk-based assessment of over 4000 chemical substances was conducted and the product portfolio was streamlined, which resulted in the elimination of 600 hazardous substances. Its investment in green chemistry R&D led to the development of solvent-free coatings and biocides which are not toxic.

This case is a best example of how REACH Compliance and regulatory practices can lead to innovations. The sustainable report of AkzoNobel which follows REACH guidelines has resulted in its enhancement of public view and investor confidence, setting a perfect example for the industries.

5.7.4 China Post-Tianjin Explosion Reforms and Capacity Building

The Tianjin port incident in 2015,in which over 700 tons of hazardous chemicals which were not properly stored resulted in major regulatory reforms in China. After this, the Chinese Ministry of Ecology and Environment initiated an inspection program nationwide to target warehouses storing chemicals and logistics hubs. IoT sensors were used to establish centralized hazardous chemical registry and real-time inventory management systems.

The government introduced AI-driven monitoring devices, which can interpret anomalies in storage protocols and issue alerts for regulatory inspections, by collaborating with private firms like **Alibaba Cloud**. China adopted preemptive measures to line up with GHS and Basel Framework.

5.7.5 India– Challenges and Reforms in Hazardous Waste Rules

India presents a compound example of a developing country struggling to enact and enforce chemical waste legislation. Due to large number of chemical plants in industrial zones like **Vapi (Gujarat)** and **Kanpur (Uttar Pradesh),** soil and groundwater contamination had been increased. An Indian Comptroller and Auditor General (CAG, 2019) audit showed frequent recurrences of waste treatment norms violation and falsification of records, highlighting policy execution difficulties (CAG, 2019).

To encounter these challenges, initiatives like **Green Rating Project of the Center for Science and Environment (CSE)** have encouraged cooperative conformity through fairness and bench-marking. The design of hazardous waste treatment measures by a large number of countries is a laudable development. The establishment of common hazardous waste treatment, storage, and disposal facilities in several states is also a positive step toward centralized waste management.

This case study indicates the need for skill development, enhanced auditing processes, and public engagement to support legal framework. It also highlights regulatory discrepancies between advanced and developing countries to assure safe and sustainable chemical waste management.

5.7.6 Nigeria: Informal Sector Risk and GHS Integration

Weak institutional framework and preeminence of irregular recycling activities, (particularly medical and e-waste) in Nigeria along with many other sub-Saharan countries, poses major challenges in regulation of chemical waste. The country is signatory to both the **Basel Convention and Bamako Convention** (prohibiting the importation of hazardous waste), but poor enforcement (Santos et al., 2022).

One especially concerning issue is the **Agbogbloshie e-waste site**, often referred to as one of the most contaminated places in the world, where millions of imported electronics are manually dismantled and incinerated by informal workers in order to harvest metals. The process exposes the workers to heavy metals, dioxins and acid residues with zero protection regulation or waste containment (Akese & PLittle, 2018).

Nonetheless, there have been significant attempts to meet worldwide regulatory requirements. Nigeria implemented a national adaptation of the GHS in 2019, mandating hazard classification and labeling for imported and domestically manufactured chemicals. UN- and UNEP-funded pilot schemes are educating custom authorities, port staff, and regulators about GHS use and tracking waste manifests.

The Nigerian experience mirrors the asymmetry in regulatory capability, the risk of unregulated waste activities and the imperative for global cooperation and technology transfer to enable safer waste management system in LMICs.

5.7.7 Japan – Closed-loop Waste and Zero Emission Policy

Japan is generally considered to be at the forefront of worldwide circular chemical waste management, combining legal requirements, business practice, and social custom in an overarching system. After the Basic Act for Establishing a **Sound-Material Cycle Society (2000)**, Japan enacted successive supportive legislation aimed at industrial waste, such as the waste management and public cleansing act and the law for promotion of effective utilization of resources.

The nation has adopted a **Zero Emissions Policy** for industrial parks like **Eco-Town Kawasaki**, where by-products of one industry are used as inputs for another– an use of industrial symbiosis drawn from nature (METI, 2021).

Japanese chemical producers are mandated to submit periodic reports on waste reduction goals, and **extended producer responsibility (EPR)** legislation make sure producers assume responsibility for post-consumer chemicals such as paints, solvents, and pharmaceuticals. Japan's success can be attributed to:

- Substantial local government engagement.
- Public awareness campaigns.
- Company Compliance Culture.
- Enduring regulatory refinement.

Hence, the long-term regulation is achieved only by means of multi-level governance, involving transfer of authority and acts of putting law into practice.

Case histories give convincing evidence of the dynamics of hazardous waste management in vast socio-economic and industrial sectors. They demonstrate that successful implementation is as dependent on adaptive governance, innovation and culture as on legal text. Although high-income countries generally do better because of institutional capacity and technological capability, LMICs are not bereft of agency or success. Pilot projects, local cooperation, and convergence of international standards such as GHS and Basel Convention show a path forward.

Ultimately, regulation of chemical waste must transform itself from a punitive regime into a cooperative one that encourages innovation, rewards sustainability, and guarantees that no community – whether by geography or income – is left to absorb the cost of chemical contamination.

5.8 CHALLENGES IN LOW- AND MIDDLE-INCOME COUNTRIES (LMICs)

Though international conventions and regulatory frameworks haven been progressed considerably in recent decades, their effective implementation in LMICs is still riddled with systemic, institutional, and socio-economic issues. LMICs mostly become recipients of chemical waste hazards – be it from the importation of hazardous chemicals, unorganized recycling mechanisms, or weak domestic regulations – but lack the means and facilities to respond to such hazards in meaningful way (Santos et al., 2022).

The international chemical regulatory regime – implicitly crafted by and for the developed world – tends to assume minimum level of institutional capacity, economic investment, and technological infrastructure that is beyond the reach of many LMICs. This inequality creates climate crisis, where communities with lesser regulatory defenses get affected the most by the harmful effects of chemical pollution.

This portion presents **South Asia, sub-Saharan Africa,** and **Latin America** as practical examples of LMICs which face challenges in adopting efficient chemical waste regulation. It ends with recommendations for global harmonization, capacity building, and inclusive policy frameworks. Table 5.2 shows the challenges and policy recommendations in LMICs.

5.8.1 Weak Institutional and Legal Enforcement

Perhaps the most ubiquitous of the LMIC challenges is the weakness of regulatory institutions. In too many nations, the environmental ministry and pollution boards are short of staff, less funds, and politicized. While there may be legislation on paper – some even based on templates such as RCRA or REACH – there is neither effective follow-through nor enforcement capability.

For example, in **Bangladesh**, regulation of Hazardous waste is carried out by the department of Environment. Yet, according to the 2021 report, the country only had

TABLE 5.2
Challenges vs Policy Recommendations in LMICs

Challenges	Policy Recommendations
Weak Enforcement Capacity	Capacity Building Programs and Inspector Training
Inadequate Treatment Infrastructure	International Investment and Technology Transfer
Informal Waste Sector Dominance	Legal Inclusion and Regulation of Informal Recyclers.
Lack Of Digital Tracking Systems	Mobile-Based Waste Manifests and Cloud Dashboards
Fragmented Legal Frameworks	National Policy Harmonization With Basel/Reach Standards

34 trained inspectors who were in charge of waste activities in over 20 large industrial estates (UNIDO, 2021).

Consequently, regulation is reactive, not proactive, with inspections usually taking place only after accidents or upon complaints of citizens. In addition, overlapping mandates across various agencies – environment, labor, customs – cause regulatory fragmentation, under which responsibility is scattered, and accountability is low.

Environmental legislation in **Indonesia and Pakistan**, for example, is available but is consistently evaded by roundabout informal negotiations or bribery, resulting in systemic non-compliance (Bhagat et al., 2024).

5.8.2 Lack of Infrastructure and Technical Capacity

Chemical waste management is dependent on the availability of centralized treatment facilities, secure storage infrastructure, and trained technical staff – all of which are frequently unavailable in LMICs. For example, most African countries lack certified hazardous waste disposal facilities, leading to local industries to engage in open dumping, burning, or unauthorized burial.

In **Kenya,** an estimated 70% or more of industrial chemical waste is dealt with informally or illegally because of the absence of infrastructure (UNEP, 2020). In **Nepal**, research laboratories and hospitals tend to store expired chemicals indefinitely, hoping for non-existent waste contractors to collect them – presenting a cumulative threat to safety.

Where treatment is available, it is often found in urban areas, with much of the rural and peri-urban countries having no access. Carrying hazardous waste long distances without proper vehicles or containment raises the risk of spills, theft, or unauthorized diversion.

5.8.3 Informal Waste Economies and Unregulated Disposal

LMICs usually have big informal economies that overlap with waste management, particularly in industries such as e-waste, tanneries, textiles, and agriculture. These informal activities are rarely regulated, but they expose tens of millions of individuals to carcinogens, heavy metals, and toxic solvents without environmental controls.

An extreme case is **Kano, Nigeria**, where used batteries and circuit boards are recycled by informal recyclers through burning or acid leaching, typically in residential areas. Samples of soil and groundwater in such communities often becomes hundreds of times higher than WHO safety levels for lead and cadmium.

In the leather industry of Kanpur in India, tanneries release chromium-rich waste into local drains even as national regulations require treatment. The Central Pollution Control Board has consistently identified Kanpur as a critically polluted area, but non-compliance persists as a result of gaps in enforcement and economic constraints on small-scale operators (CSE, 2022).

Informal economies also bypass regulatory monitoring systems such as manifests, which makes it virtually impossible to track the transportation of toxic waste – thus undermining the cradle-to-grave model contemplated by international frameworks.

5.8.4 Global Disparities and Waste Export Loopholes

In spite of the Basel Convention restrictions against transboundary movement of hazardous waste to nations not having appropriate infrastructure, waste exportation goes on – hidden in form of **"recyclables" or "Donations."** This is specifically with regard to e-waste, chemical containers, and outdated medicines.

Based on a 2022 report by the Basel Action Network (BAN), developed nations still export thousands of tons of mixed chemicals and e-waste to LMICs annually. Upon arrival, these waste streams are usually handled without protective equipment, emissions controls, or environmental protection.

The **Koko toxic waste scandal in Nigeria (1988)**, where 18,000 Italian chemical waste drums were dumped illegally in a tiny fishing village, causing global outrage and the creation of the Basel Convention. Such methods continue to this day, especially in the case of free-trade zones and cargo mislabeling.

This practice is an example of "toxic colonialism," wherein global north actors transfer chemicals waste to susceptible populations within the South – substantiating inequality while taking advantage of tax enforcement environments (Ismawati et al., 2024).

5.8.5 Limited Adoption of Digital Compliance Tools

The use of chemicals for waste management in LMICs is at the crossroads of environmental protection, community hygiene, and international equity. These countries face extreme barriers – ranging from weak governance and under-developed infrastructure to illegal waste trade and exclusion from digital universe. For example, the manually filled waste tracking dockets in Uganda, in an irregular manner, makes implementation and reporting very challenging. But they also possess great potential for innovation, cooperation, and leadership on sustainable solutions to waste – if aided through fair global policy.

Solving the problem of chemical waste in LMICs is not just a technical mix but a moral obligation. The future of planetary health requires inclusive systems that leave no country behind, no community behind, and no worker behind.

5.8.6 Sociopolitical Instability and Environmental Neglect

Environmental protection in most LMICs, particularly fragile states or post-conflict countries, is a low political concern. Ministries of the environment are frequently under-financed, overshadowed by sectors such as defense, energy, or health. In some instances, chemical controls are deliberately eliminated to pursue foreign investments.

Civil society organizations and environmental activists in those environments tend to experience threats, such that it becomes risky to criticize chemical contamination – increasing the predicament. Figure 5.2 shows chemical waste generation vs regulatory capacity by income groups.

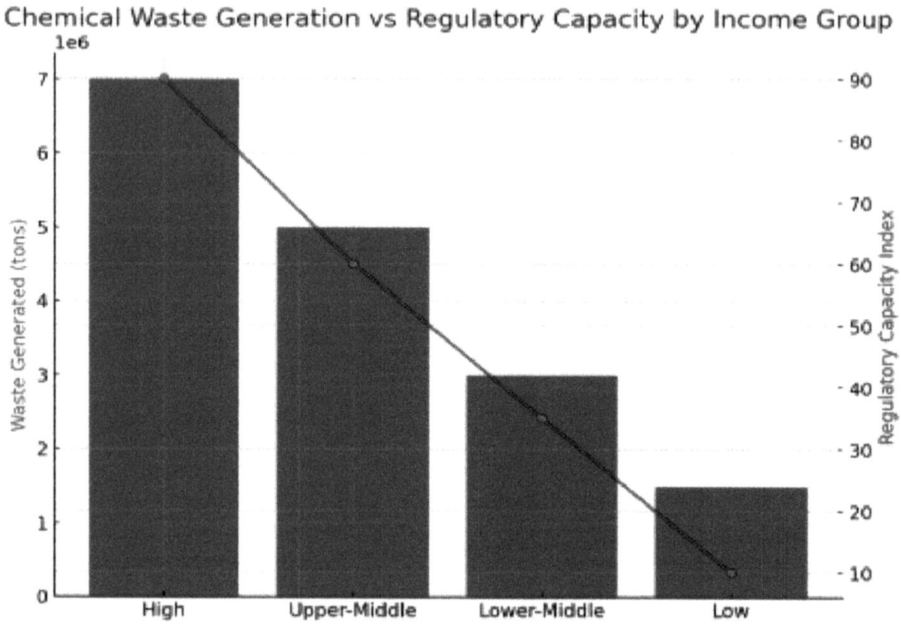

FIGURE 5.2 Visual illustration of chemical waste generation vs regulatory capacity by income group.

5.8.7 Toward Harmonization: Global Support and Policy Equity

In spite of these challenges, progress is attainable – if global systems are realigned to take account of LMIC contexts. A number of solutions have arisen over recent years:

- Capacity-building initiative by UNEP, UNITAR, and the World Bank that instruct inspectors and regulators in LMICs.
- South-South Cooperation, like African regional chemical conventions, that disseminate locally adapted best practices.
- Technology transfers agreements, where developed nations assist in the installation of treatment facilities or computer tracking systems in developing countries.
- Green financing mechanisms including GEF and GCF, which finance waste infrastructure in developing countries.
- Public-private partnerships for pilot GHS labeling, mobile applications for reporting and waste inventory systems at the sub-national level.
- There is also increasing momentum toward a global science-policy panel on chemicals, akin to the IPCC on climate change, which would be able to direct evidence-informed interventions across income groups (UNEP, 2023).

The management of chemical waste in LMICs is at the convergence of global equity, public health, and climate governance. These nations face formidable obstacles – from weak infrastructure and dysfunctional institutions to technology exclusion and informal waste economies. However, they own immense capacity for change, cooperation, and leadership on eco-friendly waste management – if supported by fair global policy.

Resolving the chemical waste issue in LMICs is not only a technical matter but an issue of morals. The fate of planetary health lies in justified systems that leave no country, no population, and no individual behind.

5.9 INNOVATION FRONTIERS IN CHEMICAL WASTE GOVERNANCE

5.9.1 From Reform to Reinvention

Laws, policy regulations, or traditional practices no longer decide the future of chemical waste management. Instead, growing number of examples identify the transformative power of emerging technology– most notably **artificial intelligence (AI), blockchain, and digital systems** – in leading a fundamental change from reactive to preemptive, flexible, and smart waste regulation. These innovations not only improve compliance and transparency but also to anticipate risk, maximize regulation, and implement circular economy principles at the system level.

Developing models in chemical waste management require digitally based solutions able to manage the complexity, magnitude, and hazards of contemporary industrial waste. This part of chapter discusses how digital monitoring, AI, and molecular analysis create base for a more reactive and sustainable regulatory future.

5.9.2 Predictive Waste Analytic and Artificial Intelligence

AI analytic are becoming important forecasting tools to estimate urgent generation of hazardous waste, detect non-compliance and optimize waste logistics. The current literature demonstrates that machine learning-based algorithms can perform an exact estimation of waste quantities, categorize risk types, and even simulate problematic disposal paths for urban waste systems (Ihsanullah et al., 2022).

For example, convolutional neural networks (CNNs) have been applied to automatically sort waste streams based on sensor and visual data, with such systems reducing human error in handling hazardous waste by an order of magnitude (Gulyamov, 2024). Similarly, predictive risk assessment centers around AI, whereby algorithmic models track treatment plant operations in real-time to identify early warning symptoms of containment failure or threshold violations in emissions.

Besides forecasting and classification, other AI technologies such as reinforcement learning are being explored to maximize chemical reuse and selecting least toxic alternatives in process planning (Idrissi et al., 2025). Such methods have a particularly favorable application in the treatment of high-risk chemical waste in sensitive environments or large cities.

5.9.3 Blockchain-Based Digital Waste Manifest

Blockchain provides a tamper-proof, decentralized ledger, well-suited for tracking dangerous substances throughout the waste lifecycle, from synthesis and transportation to treatment and ultimate disposal. Because it records every step irretrievably and reliably, blockchain facilitates cradle-to-grave accountability – a key need in the regulation of chemical waste.

Blockchain systems have recently proven effective in enhancing traceability, stakeholder confidence, and regulatory transparency. An example is that a pilot blockchain system in Asia enhanced e-waste recovery by 27% and enhanced stakeholder participation using token-based incentive schemes (Ping et al., 2024).

The coupling of blockchain with **IoT sensors** and smart contracts also supports real-time collection of data on storage conditions, location, and container integrity. These data are essential for regulatory inspections as well as for self-monitoring by the industry. Blockchain can also minimize compliance fraud, particularly in LMICs, where manual documentation is usually forged or lost (Jiang et al., 2023).

Even with such benefits, scalability in blockchain is an issue. Costs of energy, absence of technical infrastructure, and integrating data issues remain continuing impediments to mass adoption. However, hybrid models that integrate cloud analytic, blockchain, and artificial intelligence are being piloted in smart cities for high-value waste streams.

5.9.4 Bio-Based Materials and Molecular Redesign

Another revolutionary space for innovation involves molecular level chemical and material redesign to avoid waste production entirely. Directed by the tenets of green chemistry, this process employs bio-based synthesis, catalytic enzymes, and computational molecular modeling to create waste free chemical processes de novo.

Current research indicates that bio-inspired synthetic routes have the potential to make the use of harmful solvents obsolete and facilitate energy efficient pollution breakdown (Said et al., 2023). Enzyme-catalyzed reactions are also being employed as alternatives to conventional oxidation-reduction reactions that yield carcinogenic by-products.

One of the bright prospect is the elevation of chemical waste into value co-products like – reprocessing spent solvents as bio-fuel feed-stocks, or utilizing heavy metals for building materials. These techniques are not only waste saver but also create economic value, advancing circular economy models.

AI-retro syntheses are advanced tools which seek green alternatives for toxic reagents and modify reaction conditions to minimize waste output (Nesmachnow et al., 2025). With increase in computing power, these designed resources will be more accessible to industries and regulatory firms as well.

The smart, flexible, and anticipatory systems now shape the future of chemical waste management. Innovations like blockchain, AI, and bio-based molecular design are already transforming the accuracy, safety, and efficiency of tracking, treating, and destroying waste. Hence, these discoveries are offering routes for real-time regulation, where governments and industries are becoming capable to adopt proactive risk management.

With the growth of digital ecosystems, adoption barriers are reduced, enabling their integration to achieve sustainability targets and to protect human health and ecosystems from chemical harm in an increasingly industrialized world.

5.10 POLICY RECOMMENDATIONS FOR ENHANCED CHEMICAL WASTE COMPLIANCE

5.10.1 Bridging Gap between Policy and Practice

The LMICs experience an in-even management of regulatory frameworks like **REACH, RCRA,** and **The Basel Convention,** particularly due to the inability to withstand their developments. One recurring problem is the abandonment of policies which is usually the outcome of fragmented governing bodies, organizational incompetence, and weak science-policy implementation (Kamieniecki & Cohen, 1990).

This chapter proposes an execution strategy for improving policy efficacy and regulatory compliance in chemical waste regulation. The suggestions focus on holistic governance, capacity building, technology advancement, and evidence-driven global harmonization.

5.10.2 Establish a Global Science-Policy Interface Body

One of the key flaw in existing regulatory frameworks is the lacking of a central scientific authority to navigate global policymaking for chemical and hazardous waste. Scientists call for the creation of an international scientific advisory body, similar to the IPCC in climate change, to supervise tracking, risk coordination, and regulatory innovation in jurisdictions (Wang et al., 2021).

The platform could help in:

- Facilitating risk assessments and data standardization.
- Provide early warning systems for new chemical threats.
- Encourage best practices for waste tracking and labeling.

This method is particularly useful for LMICs which may lack autonomous regulatory research capability and instead depend on external standards for application.

5.10.3 Link Compliance Incentives to Circular Economy Goals

Circular economic principles involving compliance incentives should be incorporated in contemporary waste regulation, rather than relying merely on disciplinary actions. Research shows that rewarding industries for waste minimization, recycling and remodeling activities results in regulatory effectiveness (Caldart& Ryan, 1985).

Policy incentives may include:

- Tax breaks or longer permits for companies adopting cleaner technologies.
- Subsidies for waste valorization and closed-loop manufacturing.
- Performance-linked certification schemes to foster public confidence

This paradigm not only fosters compliance but a culture of innovation, especially in industries generating continuous or high-toxicity waste streams.

5.10.4 Build National and Local Capacity in LMICs

Systemic barriers are the main reason most LMICs could not adhere to regulatory compliance. Study shows that 56% of the individuals in Nigeria, despite being aware about waste policies, describe themselves as non-compliant due to poor infrastructure, neglectful supervision, and weak penalties (Nduka, 2021).

Some important capacity-building measures should involve:

- Training for environmental inspectors and facility operators.
- Decentralized compliance centers with digital reporting systems.
- Financing waste treatment infrastructure in high-risk areas.

They make compliance possible rather than merely enforceable. Effective chemical waste supervision is better than strong regulations – it requires systems that have efficient stakeholder engagement, enhanced policy, and global collaboration. These suggestions underscore the transformation from compliance by sanctions to compliance by capacity, coordination, and innovations.

5.10.5 Harmonize Global Chemical Classification and Transport Protocols

Chemical waste management is trans-national in nature, but disparities between national systems often cause regulatory arbitrage, illicit dumping, or negligence of risk factors. System harmonization such as the GHS and Stockholm and Basel Conventions is important for collaborative international measures (Wagner et al., 2013).

Policymakers should:

- Encourage bilateral waste arrangements.
- Enforce common digital documentation for transboundary waste flows.
- Fund regional centers to train customs and environmental officers in compliance with treaties.

This is particularly relevant to nations that are serving as waste transit points or rogue recycling facilities.

5.11 CONCLUSION

A systemic revolution is required to increase the imperative of chemical waste regulation in the 21st century, that is, the world needs to adopt more coordinated, anticipatory, and circular systems. As discussed in this chapter, the regulatory hierarchy dealing with hazardous waste – from global treaties such as the Basel Convention and REACH to local frameworks such as RCRA – come up with the structural framework for managing hazardous chemicals. Yet the structure is not adequate, the actual efficacy of these systems lies in their execution, which remains inconsistent globally, specifically in LMICs.

Initiatives toward green chemistry approaches and sustainable lifecycle regulation are need of the time. ARRE-based waste regulation strategies that incorporate waste negligence at its origin, minimization, recovery, and reintegration are being illustrated in studies to reduce environmental harm while amplifying economic feasibility (Singh, 2024). Such models not only increase resources but lead to chemical synthesis which align with climate goals and circular economic aspirations.

Technological advancement plays a significant role in digital tracking and monitoring of waste. Technologies like AI, blockchain, and intelligent sensor networks give latest data analysis and predict capabilities which traditional regulatory frameworks cannot. These advancements enable regulators and industries to prevent incidents before time, manage compliance at scale, and inculcate transparency into cross-boundary waste trade (Ambaye et al., 2023).

However, in order to adopt these innovations in an equitable manner, policy frameworks should be transformed. This requires enhancing institutional capabilities, exhibiting incentive frameworks for sustainable innovation, and involving international communication. Recent studies prove that intergovernmental, industrial, universal, and civil society collaboration is vital to upgrade integrated waste management solutions (Achillas & Vlachokostas, 2025).

At last, this chapter emphasizes stresses that the future of chemical waste regulation has to embrace the zero-waste vision – not as some long-term objective, but as a scientific and moral prerequisite. This change relies on:

- Institute lifecycle thinking in policy.
- Re-engineering chemicals and processes from the molecular scale.
- Making innovations available worldwide.

By tough regulation, innovative thinking, and international cooperation, the promise of sustainable, zero-waste chemical management is possible.

In this international struggle, capability development in LMICs should be focused. Research repeatedly shows the unfair burden on developing countries, where regulatory compliance is weak, infrastructure is poor, and chemical literacy among actors is extremely low. Conquering these hurdles requires sustained interest in training, community engagement, and trans-national technical cooperation. International funding agencies, including green climate funds and grants from development banks, must be utilized to help develop localized waste treatment capacity, digital compliance systems, and environmental monitoring. Similarly, international chemical classification, labeling, and transport standards should be modified to facilitate efficient cross-border regulation. Inconsistent regulatory standards usually cause inadequacies in waste trading, illegal dumping, or improper risk labeling. Transforming global systems – like REACH, GHS, and the Basel Convention – with standard digital tracking mechanisms could heavily reduce irregularities and enhance traceability (Wang et al., 2021). Policy harmonization, facilitated by common databases and treaty-based inspection, will enhance regulatory coherence globally.

Lastly, repairing the chemical waste crisis needs multi-disciplinary collaboration. Chemists, engineers, decision-makers, data analysts, and civil society members should work together to co-develop solutions that are not only scientifically rational but also socially justified and economically viable. As discussed in this chapter, sustainable chemical waste governance is a multi-layered task accomplished by integrating innovation, regulation, and equity. The chemical industry, if dealt with systemically, can modify from being among the largest pollution causing agents to being one of the leaders in the global shift toward sustainability.

REFERENCES

Achillas, C., & Vlachokostas, C. (2025). Integrated waste management in the circular economy era: Insights from research and practice. *Energies*, *18*(3), 728.

Aggeri, F., Micheaux, H., & Ntsondé, J. Special Issue 23| 2021 Industry and Waste: Toward the Circular Economy.

Ambaye, T. G., Djellabi, R., Vaccari, M., Prasad, S., Aminabhavi, T. M., & Rtimi, S. (2023). Emerging technologies and sustainable strategies for municipal solid waste valorization: Challenges of circular economy implementation. *Journal of Cleaner Production*, *423*, 138708.

Anastas, P. T., & Warner, J. C. (2000). *Green chemistry: Theory and practice*. Oxford University Press.

Bhagat, J., Singh, N., & Shimada, Y. (2024). Southeast Asia's environmental challenges: Emergence of new contaminants and advancements in testing methods. *Frontiers in Toxicology*, *6*, 1322386.

CAG. (2019). *Performance audit of hazardous waste management*. Comptroller and Auditor General of India.

Caldart, C. C., & Ryan, C. W. (1985). Waste generation reduction: A first step toward developing a regulatory policy to encourage hazardous substance management through production process change. *Hazardous Waste and Hazardous Materials*, *2*(3), 309–331.

Clark, J. H., & Deswarte, F. (Eds.). (2015). *Introduction to chemicals from biomass*. John Wiley & Sons.

Clark, J. H., & Macquarrie, D. J. (Eds.). (2008). *Handbook of green chemistry and technology*. John Wiley & Sons.

Cox, P. (2020). Environmental justice, waste management, and the circular economy: Global perspectives. In *The Routledge handbook of waste, resources and the circular economy* (pp. 42–49). Routledge. Edited By Terry Tudor and Cleber JC. Dutra.

CSE. (2022). *Assessment of tannery pollution in Kanpur*. Centre for Science and Environment.

ECHA. (2023a). Persistent, bioaccumulative and toxic (PBT) and very persistent and very bioaccumulative (vPvB) substances. European Chemicals Agency.

ECHA. (2023b). *REACH Regulation*. European Chemicals Agency.

ECHA. (2023c). *Safety Data Sheets and Waste Classification*. European Chemicals Agency.

ECHA. (2023d). *REACH regulation overview*. European Chemicals Agency.

EPA. (2018). *RCRA Enforcement Actions Database*.U.S. Environmental Protection Agency.

EPA. (2021a). *Pollution Prevention Act of 1990*. U.S. Environmental Protection Agency.

EPA. (2021b). *Resource Conservation and Recovery Act (RCRA)*. U.S. Environmental Protection Agency.

EPA. (2021c). *Resource Conservation and Recovery Act (RCRA)*. U.S. Environmental Protection Agency.

EPA. (2022a). *Hazardous waste characteristics*. U.S. Environmental Protection Agency.

EPA. (2022b). *Sustainable materials management: Non-hazardous materials and waste management hierarchy*. U.S. Environmental Protection Agency

European Commission. (2020). *Circular economy action plan*. European Commission.

Gulyamov, S. (2024). Intelligent waste management using IoT, blockchain technology and data analytics. In *E3S web of conferences* (Vol. 501, p. 01010). EDP Sciences. https://doi.org/10.1051/e3sconf/202450101010

Idrissi, A., Benabbou, R., Benhra, J., & El Haji, M. (2025). Solid waste management through the application of AI and ICT: A systematic literature review. *Journal of Environmental Engineering and Science*, *20*(2), 88–121.

Ihsanullah, I., Alam, G., Jamal, A., & Shaik, F. (2022). Recent advances in applications of artificial intelligence in solid waste management: A review. *Chemosphere*, *309*, 136631.

Ismawati, Y., Septiono, M. A., Proboretno, N., & Zaki, K. (2024). Plastic waste trade in Indonesia and Country's response to waste trade challenges. In *Plastic waste trade: A new colonialist means of pollution transfer* (pp. 155–189). Springer Nature Switzerland.

Jessop, P. G. (2011). Searching for green solvents. *Green Chemistry*, *13*(6), 1391–1398.

Jiang, P., Zhang, L., You, S., Van Fan, Y., Tan, R. R., Klemeš, J. J., & You, F. (2023). Blockchain technology applications in waste management: Overview, challenges and opportunities. *Journal of Cleaner Production*, *421*, 138466.

Kamieniecki, S., & Cohen, S. A. (1990). Strategic regulatory planning in the management of hazardous materials. *Policy & Politics*, *18*(3), 207–216.

Landrigan, P. J., Fuller, R., Acosta, N. J., Adeyi, O., Arnold, R., Basu, N. N., Baldé, A. B., Bertollini, R., Bose-O'Reilly, S., Boufford, J. I., Breysse, P. N., Chiles, T., Mahidol, C., Coll-Seck, A. M., Cropper, M. L., Fobil, J., Fuster, V., Greenstone, M., Haines, A.,…& Zhong, M. (2018). The Lancet Commission on pollution and health. *The Lancet*, *391*(10119), 462–512.

Lipnick, R. L., Mason, R. P., Phillips, M. L., & Pittman, C. U. (Eds.). (2002). *Chemicals in the environment: Fate, impacts, and remediation.* American Chemical Society.

Maertens, A., Golden, E., & Hartung, T. (2021). Avoiding regrettable substitutions: Green toxicology for sustainable chemistry. *ACS Sustainable Chemistry & Engineering*, *9*(23), 7749–7758.

METI. (2021). *Japan's eco-town program and zero emissions strategy.* Ministry of Economy, Trade and Industry.

MoEFCC. (2016). *Hazardous and other wastes rules.* Government of India.

Nascimento, E. D. S., & Tenuta Filho, A. (2010). Chemical waste risk reduction and environmental impact generated by laboratory activities in research and teaching institutions. *Brazilian Journal of Pharmaceutical Sciences*, *46*, 187–198.

Nduka, I. (2021). Compliance to waste management policy guidelines in Abia State, South-East, Nigeria. *International Journal of Innovative Research and Development*, *10*(8), 41–47.

Nesmachnow, S., Rossit, D. G., & Moreno-Bernal, P. (2025). A literature review of recent advances on innovative computational tools for waste management in smart cities. *Urban Science*,*9*(1), 16.

OECD. (2018). *Green chemistry for sustainable development.* Organisation for Economic Co-operation and Development.

Ping, G., Wang, S. X., Zhao, F., Wang, Z., & Zhang, X. (2024). Blockchain based reverse logistics data tracking: An innovative approach to enhance e-waste recycling efficiency. *World Journal of Innovation and Modern Technology*, *7*(4):11–31.

Prat, D., Wells, A., Hayler, J., Sneddon, H., McElroy, C. R., Abou-Shehada, S., &Dunn, P. J. (2016). CHEM21 selection guide of classical-and less classical-solvents. *Green Chemistry*, *18*(1), 288–296.

Said, Z., Sharma, P., Nhuong, Q. T. B., Bora, B. J., Lichtfouse, E., Khalid, H. M., Luque, R., Nguyen, X. P.,& Hoang, A. T. (2023). Intelligent approaches for sustainable management and valorisation of food waste. *Bioresource Technology*, *377*, 128952.

Secretariat of the Basel Convention. (2023). *Basel Convention Overview.*

Singh, S. B. (2024). Circular chemistry: An enabler of circular economy to achieve the zero-waste goal. *Sustainability & Circularity NOW*, *1*(a24514726.

Stankiewicz, A. I., & Moulijn, J. A. (2000). Process intensification: Transforming chemical engineering. *Chemical Engineering Progress*, *96*(1), 22–34.

Trost, B. M. (1991). The atom economy—A search for synthetic efficiency. *Science, 254*(5037), 1471–1477.

UNECE. (2021). *Globally Harmonized System of Classification and Labelling of Chemicals (GHS), Rev. 9.* United Nations Economic Commission for Europe.

UNEP (2019). *Key messages for beyond 2020: Global Chemicals Outlook II - From legacies to innovative solutions: Implementing the 2030 agenda for sustainable development.* United Nations Environment Programme.

UNIDO. (2016). *Chemical leasing: Towards sustainable chemicals management.* United Nations Industrial Development Organization.

UNIDO. (2021). *Capacity gaps in South Asia's chemical waste governance.* United Nations Industrial Development Organization.

Wagner, B. O., Aziz, E. R. A., Schwetje, A., Shouk, F. A., Koch-Jugl, J., Braedt, M., Choudhury, K. A., &Weber, R. (2013). Recommendations on chemicals management policy and legislation in the framework of the Egyptian–German twinning project on hazardous substances and waste management. *Environmental Science and Pollution Research, 20*, 2087–2097.

Wang, Z., Altenburger, R., Backhaus, T., Covaci, A., Diamond, M. L., Grimalt, J. O.,Lohmann, R., Schäffer, A., Scheringer, M., Selin, H., Soehl, A., & Suzuki, N. (2021). We need a global science-policy body on chemicals and waste. *Science, 371*(6531), 774–776.

Wang, Z., Walker, G. W., Muir, D. C., & Nagatani-Yoshida, K. (2020). Toward A global understanding of chemical pollution: A first comprehensive analysis of national and regional chemical inventories. *Environmental Science & Technology, 54*(5), 2575–2584.

WHO. (2018). Health-care waste. World Health Organization.

Wilson, M. P., & Schwarzman, M. R. (2009). Toward a new US chemicals policy: Rebuilding the foundation to advance new science, green chemistry, and environmental health. *Environmental Health Perspectives, 117*(8), 1202–1209.

Zhang, H., Duan, H., Zuo, J., Song, M., Zhang, Y., Yang, B., & Niu, Y. (2017). Characterization of post-disaster environmental management for hazardous materials incidents: Lessons learnt from the Tianjin warehouse explosion, China. *Journal of Environmental Management, 199*, 21–30.

6 Production of Chemicals Using Microbial Consortia

Bisma Javed, Mamoona Munir, Sadia Abdullah, Mushtaq Ahmed, Shazia Sultana, Nigar Mursal, and Maliha Asma

6.1 INTRODUCTION

A microbial consortium is a collection of two or more symbiotic microorganisms that survive in various environments by forming synergistic population structures such as stromatolites, microbial mats, and biofilm (Gautam et al., 2022). The beneficial influence and promising outcomes have significantly increased the acceptance and recruiting of microbial consortiums. Microbial consortiums live in the bulk interacting closely with each other instead of living as separate cells, they adapt to environmental stress as a distinct organism. In the positive internal interaction, microbial consortiums embrace and sustain than single strain. When bacterial consortiums form, two or more bacteria of different species interact additively or synergistically. In certain circumstances, a mixture of various strains of the same species are combined together to increase activity and are also recognized as a consortium (Singh & Das, 2024). Inoculant strains in microbial consortia must possess desired traits including long shelf life, the ability to survive in challenging environments, cost-effectiveness, effective delivery with the host plant, genetic stability, physiological adaptability to the host environment, etc. The bacterial consortia is typically produced in a liquid medium to maximize biomass output (Gautam et al., 2022; Duncker et al., 2021).

Microbial consortia accomplish the process of biotic transformation in multiple steps which cannot be performed by a single organism itself, forcefulness, the functional maintenance by internal and external shocks, important for the effective constancy of the community in a bioprocess. Furthermore, the complex relationships and changing aspects between species within a bacterial community that are essential for any successful application. Unlike diverse affiliates of the bacterial community have access variety of metabolic pathways, variations in substrate composition can be addressed in a mixed population, depending on the type of bacterial community. If major metabolic organism route has a limited capacity, greater availability and consequently metabolism of one substrate lead to decreased absorption of other substrates due to their increased metabolic flexibility and stability of microbial communities has been linked to their variety. Due to thediversity of members and the

DOI: 10.1201/9781003642589-6

redundancy of their functions, natural microbial communities are naturally more resilient to environmental stresses and perturbations such as pH shifts, temperature fluctuations, and the presence of inhibitory compounds (Lahiri et al., 2022; Schmidt, 2021).

The ability of mixed consortia to complement the produced metabolites and operate while unwanted microorganisms are excluded, improving process stability, differentiates them apart from competitors. The bacteria have the ability to produce lactic acid (LA) and other organic acids during the processes of fermentation of food, while the yeast can produce alcohol. The acids can alter the atmosphere from aerobic to anaerobic which helps to keep most harmful bacteria and molds out. Furthermore, the infections of bacteriophage can be minimized in the mixed-culture fermentations, which results in the complete closing of a specific manufacturing process. The wide-ranging capacity of the infection in the culture is not likely to be happen for the reason that the mixed cultures have a robust genetic foundation for resistance of the phage. A second or third resistant strain from the inoculum may take over and carry on the fermentation process if one strain in the culture contracts phage infection. Even when the fermentations are conducted in an open manner, contamination happens less frequently in these processes(Schmidt, 2021).

In microbial consortia, low-value waste streams are converted into high-value chemicals, with minimal impact on the environment. This also helps reduce raw material costs and supports a circular bioeconomy. In engineered pure cultures, a common problem is metabolic bottlenecks, where intermediate compounds build up or enzyme activity is limited, reducing overall production. These bottlenecks can be overcome through metabolic cross-feeding, where one microbe uses intermediate compounds produced by another, including those that may be inhibitory or toxic. This interaction helps remove bottlenecks and improves overall efficiency. Moreover, the corporal distance between metabolic steps among innumerable organisms can prevent feedback by inhibition which enables each segment of pathway at optimal rate without impacted adversely by upstream or downstream products (Mittermeier et al., 2023, Jiang et al., 2017, Paul et al., 2023; Li et al., 2021).

The adherents in cross-feed nutrients for microbial consortia, allows the system to perform a number of processes which includes the synthesis of biochemical and bioremediation. The genetically identical cells have different sub-populations which can supply harmful intermediates or split the exertion of biosynthesis. When these concepts are applied on different bioprocesses, new possibilities for regulating cell populations and building routes to preserve homeostasis rise. The metabolic load of partners can be divided by microbial consortia. Microbial consortia have previously been employed in industries such as waste treatment, anaerobic digestion, the food and nutrition sector, and medicine. Recently, co-cultures have been effectively produced using synthetic biology to produce pharmaceuticals and commodity chemicals. It is possible to create synthetic consortia from countless strains from the similar species (Roell et al., 2019).

The design and operation of bioreactors for bioremediation demand advanced stages of breakdown of pollutants and further unchanging performance of the system, which can be easily achieved by increasing the number of microbial species in a system. Artificial microbial consortia systems provide a number of benefits over natural microbial communities, which include the simpler composition, stronger division of work, and the capacity to be tailored to a wide range of targeted products. Furthermore, by utilizing microbial consortia comprising just contributing species, competition with substrates will be reduced(Bhatia et al., 2018). A microbial consortium allows bacteria to exploit a variety of carbon sources. It gives bacteria the ability to respond robustly to environmental stress conditions. Microbes in a consortium may accomplish sophisticated activities that a single organism cannot ensure (Mishra et al., 2021).

Microbial consortia are also adaptable exceedingly and robust in complex settings. When different varieties of the cells with distinct activities are mixed with each other, the equilibrium is sustained through complex associations between different cells which can enhance the overall flexibility and stability of the system in the face of external shocks. The cooperative development of synthetic biology and systems will make it possible to understand these intricate consortia thoroughly and logically and then create them for new uses (Cao et al., 2022). These developments in the biotechnology contribute toward the expansion of microbial consortia with anticipated functionalities of bioremediation. Furthermore, ideas like preserving multifunctionality, reducing stress, and distributing effort in metabolic pathways must be taken into deliberation as engineering advances. In order to properly govern microbial consortia and accomplish desired functions and effects, further study on metabolic interactions among species is required in the future. Although engineered microbial consortia have significant drawbacks, they remain a promising method for bioremediation (Li et al., 2021). The cooperative development of synthetic biology and systems will yield comprehensive understanding and logical engineering of intricate groups for novel uses (Cao et al., 2022; Jia et al., 2016).

Chemical production involves the use of complex and unpredictable microbial consortia. The genetic stability in microbial consortia can result in functional drift, and they can be challenging to predict and manage over extended periods of time. Real-time monitoring and scaling these processes to industrial volumes are both difficult. Economic feasibility is being assessed, and regulatory frameworks are changing. Microbial consortia present chances for robustness, modular design, improved product spectrum, complex substrate utilization, and integration with the circular bio economy. They can be applied to environmental applications and bioremediation as well. The complete potential of microbial consortia for sustainable chemical production will be unlocked by addressing these issues with advanced engineering, "omics" technologies, and computational modeling (García-Solares et al., 2023; Li et al., 2021; Mittermeier et al., 2023).

6.2 THE INTERACTIONS AMONG CELLS IN MICROBIAL CONSORTIA

6.2.1 Cross-Feeding

Numerous studies have been conducted on the environmental and developmental aspects that affects transport mechanisms of the metabolite. There ways in which the metabolites can be exchanged consist of contact-dependent mechanisms which include vesicle chains, nanotubes, flagella-like filaments, and cell-cell contact, and contact-independent processes like passive diffusion, active transport, and vesicle-mediated transport. The stability and effectiveness of cross-feeding are greatly influenced by the makeup of microbial consortia as well as environmental factors including nutrient availability, pH, and temperature. The cross-feeding connection was changed when cross-fed nutrients were added because one bacterial strain was no longer dependent on the other. Additionally, cross-feeding connections are impacted by geographical factors. Diffusion of resources and dispersion of microbes control the dynamics of a microbial consortium of bacteria. If the distribution of microbes is restricted but nutrients are widely distributed, cross-feeders may outcompete cheaters. The colony may eventually collapse if microbe dispersion is high and nutrients are widely distributed, even when cheaters replaced cross-feeders. It offered strategies for maintaining the stability of microbial consortia throughout bioremediation (Li et al., 2021).

The members of microbial consortia cross-feed nutrients, enabling the system to perform a variety of tasks. Even genetically identical cells can have different subpopulations that store harmful intermediates split the work of biosynthesis. When these ideas are applied to bioprocesses, new possibilities for regulating cell populations and building routes to preserve homeostasis arise. The metabolic load of partners can be divided by microbial consortia. Microbial consortia have previously been employed in industries such as waste treatment, anaerobic digestion, the food and nutrition sector (such as dairy products, soy sauce, wines, and vitamins), and medicine (such as gut microbiome). Recently, co-cultures have been effectively produced using synthetic biology to produce pharmaceuticals and commodity chemicals. It is possible to create synthetic consortia from many strains of the same species. For example, modified strains of *Escherichia coli* may generate high-value compounds like anthocyanins, curcuminoids, and flavan-3-ols. It is still difficult to regulate population dynamics and maximize output in the industrial DoL applications which include the utilization of microbial consortia (Roell et al., 2019; Carlson et al., 2018).

6.2.2 Quorum Sensing

In a microbial consortium, interaction among different species of microorganisms is characterized by both mutualism and competition for resources within the same environment. These relationships are not only observed between different species but also occur among similar microbial cells of the same species, highlighting a complex web of interactions. A key process that enables these dynamics is quorum sensing (QS), where microbial cells communicate with one another and adjust their behavior based on the density of the population. The fermentation process serves as a prime example, illustrating how these mutual exchanges, particularly through metabolic interactions, can substantially influence metabolic pathways and enhance the yield of desired bioproducts. Overall, the balance of competition and cooperation among various microbial entities underscores the intricacies of microbial ecosystems and their significant role in bioproduct synthesis, reflecting a sophisticated network of interactions that drive microbial community dynamics (Jiang et al., 2017).

Bacterial density-dependent communication information characterizes QS, which allows the understanding of corresponding activities through responsive gene regulation. When there is a sufficient quorum, the microbial cells can coordinate their activities by releasing certain signal molecules and sensing changes in their concentrations on their own. The common QS signals that are utilized by Gram-negative bacteria include N-acyl-homoserine lactones (AHLs). In contrast to Gram-negative bacteria, auto-inducing peptides (AIPs) are synthesized and transmitted by Gram-positive bacteria as communication signals. Each kind of bacterium, whether Gram-positive or Gram-negative, produces a distinct AIP or AHL. Therefore, only individuals belonging to the same species are able to identify and react to it. In a mixed population, the species-specific QS appears to enable self-recognition and encourages intraspecies communication. Bacteria develop the capacity to recognize various species in their environment in addition to using signals from the AI-2 (auto-inducer-2) family for interspecific communication. Research on *Vibrio harveyi*, a Gram-negative bioluminescent pathogen, demonstrated the existence of AI-2. The enzyme LuxS is responsible for producing AI-2.Nonetheless, the genomes of numerous Gram-positive and Gram-negative bacteria contain the luxS gene. Consequently, any bacterium with a functioning luxS gene can produce an activity that can be detected by a *V. harveyi* reporter strain unique to AI-2 (Jiang et al., 2017).

A more general signal called AI-2 may facilitate communication between different types of bacteria. In natural microbial interactions, QS is an important function that regulates the synthesis of virulence factors, forms biofilms, enhances microbial stress tolerance, and more. In biofilm the collection of microorganisms consist of cells that attach to surface or one another. The extracellular polymeric substance (EPS) matrix of these adhering cells is often self-produced. Microbial resistance to oxygen, substrate, and toxic/inhibitory substances can be greatly increased by biofilm development. Anaerobic bacteria can be given suitable microenvironments by creating an oxygen gradient, for example, when one community member in a biofilm consumes the dissolved oxygen (Jiang et al., 2017; Stephens & Bentley, 2020).

6.2.2.1 Mutualism and Synergism

The term "mutualism" describes how two or more species gain from coexisting, but when they are separated, their existence will be negatively impacted and may even end. The microbial consortia can be found in the processes of fermentation. For example, bacteria and archaea have a mutualistic relationship during the anaerobic fermentation process, which produces methane. Using stoichiometric models and flow balance analysis, the mutualistic metabolite exchange between the methanogen *Methanococcus maripaludis* and the sulfate reducer *Desulfovibrio vulgaris* was first studied. In a co-culture experiment, the proportions of *D. vulgaris* with *M. maripaludis* are predictable. Each strain supplies the other strain with a required metabolite. However, the emergence of "cheater" means the individuals who profit from the facilitation without contributing can easily cause this reciprocal connection to break down. Depending on the spatial organization of the surroundings, the possible breakdown by the cheater may be avoided or postponed. The spread of genotypes that cheat can be limited by the physical composition of the environment (Jiang et al., 2017).

Microbial mutualism can take many forms, and synergy is one in which the development of one species or genotype is influenced by the metabolites generated by another. Synergy interactions include the scavenging of inhibitory by-products from a producer population by a community member who eats organic acid and the generation of biofuel through cellulose bioprocessing. In the compost extract two strains of bacteria, aerobic non-cellulolytic bacterium and *Clostridium straminisolvens* CSK1 was capable of breaking down cellulose in anaerobic environments. By combining these two strains, they were able to create a bacterial community that breaks cellulose effectively. By producing an anaerobic environment, absorbing metabolites, and neutralizing pH, the non-cellulolytic bacteria basically support in the breakdown of cellulose, rendering to the mixed culture (Jiang et al., 2017).

6.2.2.2 Competition and Antagonism

When one species adversely affects another without suffering any negative consequences in return, this is known as antagonistic relationships. It stops spoiling organisms from growing and is commonly found in food fermentations. The selection that encourages the manufacturing of antimicrobial chemicals occurs within a microbial community as competition for scarce natural resources. Recent research has demonstrated that these naturally occurring antimicrobial compounds serve as communication tools that control the homeostasis of microbial communities rather than being predominantly employed as weapons for competition. For example, co-culturing *Lactobacillus acidophilus* with yoghurt starter species *Lactobacillus delbrueckii* subsp. *Bulgaricus* and *Streptococcus thermophilus* increased the strain's production of lactacin B (Jiang et al., 2017).

The interactions between the cells in microbial consortia are shown in Figure 6.1. The stability of the bacterial community is significantly impacted by the interactions

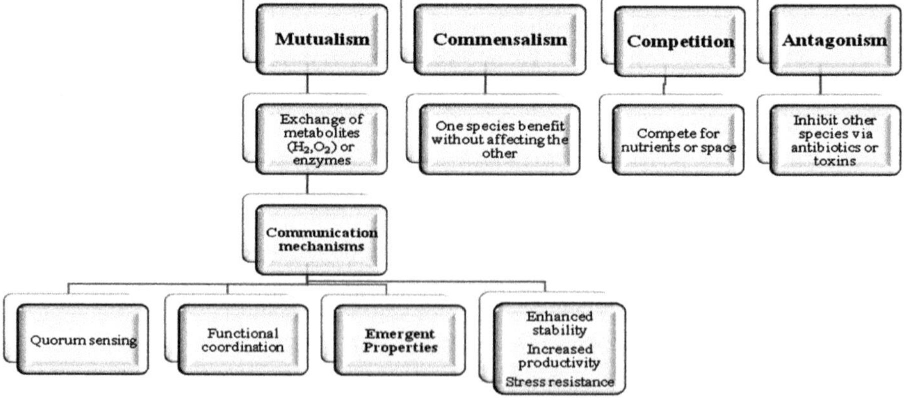

FIGURE 6.1 Diagram showing the interactions among cells in microbial consortia.

between the cells in a microbial consortium. The environmental stability of a microbial consortium may be disrupted by the collaboration of several types of microbial cells. Furthermore, by utilizing negative feedback to offset the instability brought on by microbial diversity, the competitive connection between probiotics would preserve the stability of the gut ecology. The mutualism interaction became more sensitive than competition with the advent of cheaters, leading to several claims that are backed by the environment. In order to maintain the stability of the microbial consortium, the competitive interaction seems to be more beneficial (Jiang et al., 2017).

6.2.3 Strategies for the Assembly of Microbial Consortia

The assembly of microbial consortia with preferred functionalities involves the strategies, ranging from different methods such as natural selection to cultured synthetic biology. These selected methods can be influenced by the intricacy of the desired function, the prevailing knowledge about the microbial consortia, and the intentional application (Cui et al., 2023).

6.2.3.1 Top-Down Approach

The top-down method is a strategy for transforming waste streams into value-added products, including soil, rumen, compost, and wastewater treatment systems. This approach involves selecting a natural microbial community and modifying environmental conditions to shape its structure and functions. Microbial communities are found in environments such as soil, wetlands, compost, herbivore digestive tracts, activated sludge, and anaerobic digesters. Strategies are employed to encourage the growth of microbial populations with specific functional abilities. Enrichment is characterized as a strategic methodology that utilizes a top-down approach, which involves executing multiple cycles of dilution. Each of these cycles takes place in tightly regulated environments that are continuously monitored to ensure precision and accuracy in the process. This particular control is essential for managing the various factors that could influence the results, ultimately improving the reliability and effectiveness of the overarching strategy. By maintaining microbial groups that perform specific tasks, this method can simplify the system without requiring detailed knowledge of each microbial population and their cellular interactions. Due to their metabolic abilities, mixed microbial communities play a vital role in converting complex and varied biomass waste into valuable products. When compared to a monoculture system, functional redundancy in mixed microbial communities increases stability by offering resistance against external disruptions (Lyu et al., 2024).

The top-down approach to microbial characterization has multiple limitations, including reduced control over the preliminary microbial makeup, unpredictable results due to unknown microbial roles, adverse competitive interactions, and the risk of overlooking rare yet functionally significant populations. Mixed-culture research has made significant advancements by employing innovative methods to address the complexities inherent in studying diverse microbial populations. A key approach is the integration of amplicon or shotgun sequencing with stable isotope probing techniques for DNA, RNA, and protein. This method allows for the identification of in situ connections and interactions among various microbial communities that are involved in specific metabolic processes. Furthermore, a groundbreaking technique has emerged for detecting functionally active populations within these intricate microbial consortiums, utilizing fluorescence-activated cell sorting alongside bio-orthogonal, non-canonical amino acid tagging. These methodological innovations notably enhance researchers' understanding of microbial dynamics and the functional roles these organisms play within different ecosystems (Lyu et al., 2024).

6.2.3.2 Bottom-Up Design Approach

The top-down approach in microbiome engineering focuses on the development of synthetic microbial communities, which can consist of both naturally occurring and engineered microorganisms, including strains that are not typically found together in natural environments. A crucial element of this method is the comprehensive characterization of the genetic, metabolic, and physiological properties of bacterial species, alongside an understanding of the communication mechanisms between strains within the consortium. Progress in synthetic and systems biology has notably enhanced these biological methodologies, enabling the replication and alteration of distinct cellular functions. To bolster the bottom-up approach, strategies include the design and construction of microbial consortia, their optimization through ecological engineering, and the adoption of metabolic engineering techniques, all aimed at improving the functionality and performance of these microbial communities (Lyu et al., 2024).

6.2.3.3 Approaches for Scheming and Collecting Synthetic Microbial Consortia

The purpose of synthetic microbial consortia is vital in order to perform tasks that single strains cannot achieve independently. They are assembled using bottom-up approaches, such as computational model-driven design, combinatorial screening and community enrichment or reduction. Community reduction focuses on removing microbial species while preferentially promoting beneficial ones to enhance the consortia's performance and precision. Combinatorial screening is a method that integrates well-characterized microbial species to enhance the efficiency of biomass bioconversion. This process involves the simultaneous construction and evaluation of paired microbial strains, which can be achieved through advanced tools such as the k-Chip, along with other innovative high-throughput culturing and phenotypic screening methods. By employing genetic and metabolic data for modeling bacterial interactions, this approach allows researchers to design models that significantly diminish reliance on traditional trial-and-error experimental methods. Additionally, synthetic microbial consortia utilize the principle of metabolic division of labor (DOL) to efficiently convert various biomass substrates and syngas waste into valuable products. This strategy alleviates metabolic stress on individual strains by promoting cooperation among them, thus reducing competition and enabling efficient task distribution among the consortium members. In the realm of bottom-up approaches, the development of biofilm-based systems improves the division of labor, while well-characterized microbial strains are assembled into various co-cultures. Strategies in metabolic engineering are also employed to enhance both strain performance and metabolic capabilities. The initial design of a microbial consortium is specifically crafted to allow controlled manipulation, which leads to improved optimization and performance in bioconversion processes (Lyu et al., 2024).

Understanding microbial communications and their impact on community dynamics is important for predicting community functions. The collective activity of microbial consortia members is not merely the sum of their individual contributions due to the presence of complex interactions among them, which include mutualism, competition, commensalism, amensalism, predation, neutralism, and facilitation. These dynamics lead to community-level functions that can significantly diverge from the actions and impacts of individual microbes. This highlights the intricate nature of microbial ecology, underscoring the necessity to comprehend these relationships for an accurate assessment of microbial community behavior and function. The flexibility of synthetic microbial consortia can increase the spectrum of products and enhance substrate use. Interspecies interactions can be made simpler to avoid unintended consequences. A synthetic microbial consortium can be stabilized over the long run by comprehending community dynamics and establishing a link between composition and overall activity. Computational methods have seen a significant rise in application within the field of genomics, particularly through the use of artificial intelligence (AI) and machine learning (ML) techniques. These methodologies are employed to effectively estimate the optimal composition and metabolic pathways of microbial consortia, which enhance the design and assembly processes of these communities. By leveraging large datasets and sophisticated algorithms, AI and ML provide deep insights into the functionalities and interactions among microorganisms. This advancement not only facilitates a more efficient analysis but also presents new possibilities for applications that might not be easily discernible through conventional approaches. Overall, the integration of computational techniques is proving to be revolutionary within microbiology and biotechnology, highlighting their transformative potential for understanding complex microbial ecosystems (Lyu et al., 2024).

6.2.4 Synthetic Biology Approaches for De Novo Consortia

For microbial consortia, synthetic biology entails both the creation of new biological entities and the redesign of preexisting ones. Although the media frequently uses the term "radical novelty" to create buzz and elicit public responses, it is unclear what novelty actually addresses. De novo and "designed and produced from scratch" are sometimes used interchangeably, and it is anticipated that functional innovation will arise when systems are built from scratch. It has been disputed in the field whether synthetic biology provides a new strategy utilizing already-existing methods or employs innovative assembly procedures that are different from those employed in molecular biology and biotechnology. Some claim that synthetic biology is a paradigm shift, while others contend that it is just a catchphrase used to create excitement. For instance, the well-known synthetic biology structures known as "Bio Bricks" are DNA sequences with particular functionalities that are anticipated to be ubiquitous, interchangeable parts in a range of biological systems. However, because Bio Bricks are technically and structurally traditional creations rather than the improvisational design of genetic constructs in biotechnology, it is difficult to determine how they achieve originality by design (Delgado & Porcar, 2013).

6.2.5 "Omics" Technologies and Computational Modeling in Microbial Consortia

Computer modeling in conjunction with omics technologies like transcriptomics, metagenomics, and metabolomics, which are currently helpful tools for engineering microbial consortia—a group of two or more genetically modified strains that contribute to a higher objective by carrying out specific tasks on the system—allows for a deeper understanding of microbial interactions and the programming of complex features beneficial for biotechnological applications. For the construction of an engineered consortium, design concepts including biofilm development, detoxification, and cross-feeding are essential. Additionally, the optimization and simple assembly of the engineered strains to construct a synthetic community are made possible by the modularity that is frequently offered by the establishment of engineered microbial consortia. From a biotechnology standpoint, engineered microbial consortia have been applied in wastewater treatment, agriculture, fisheries, medicine, and biogas production, among other industries (González et al., 2021).

Meta-transcriptomics involves the sequencing of entire genetic material from an environmental sample or a microbial consortium without the need to cultivate individual species previously. It identifies the genes which are actively expressed under particular conditions, offering insights into metabolic activities, gene regulation, and stress responses in real time. Metaproteomics involves the identification and quantification of proteins at large-scale expressed by all members of a microbial consortium, revealing changes at the protein level in response to environmental conditions, important enzymes involved in product formation or substrate utilization, and post-translational modifications that control protein activity. Metabolomics is the study of all small-molecule metabolites found in a biological system; it provides information on cellular regulation and the end products of metabolic pathways. Computational modeling is essential for understanding, predicting, and optimizing the behavior of complex microbial consortia and improves experimental "omics" data. Metabolic fluxes within a metabolic network are modeled using constraints in a mathematical method called Flux Balance Analysis (FBA). Different agents interact with one another and their environment in computer simulations using agent-based models (ABMs) in compliance with preset rules (Nagarajan et al., 2022).

6.3 PRODUCTION OF PLATFORM CHEMICALS

6.3.1 Organic Acids

6.3.1.1 Lactic Acid

A basic organic acid, LA is used in food, medicine, and cosmetics. It is utilized as the monomer in the production of polylactic acid (PLA) plastic, which is biodegradable and bioabsorbable and helps reduce "white pollution" brought on by petroleum-based plastics. It is anticipated that LA would reach 19.601 million tons by 2025, with a market value of around USD 9.8 billion.LA is neither suitable for use in food and medicine, nor does it satisfy the specifications needed to produce PLA plastic. As a result, the pure isomer has more utility. Although LA may be produced chemically or biologically, microbial fermentation is the most used method because of its high efficiency, low cost, outstanding optical purity, and environmental friendliness. *Saccharomyces cerevisiae*, *Lactobacillus cruzi*, *E. coli*, *Bacillus* sp., *Rhizopus oryzae,* and *Aspergillus niger* can be utilized in the fermentation of LA. The fermentation initially used more expensive feedstocks, such as refined sugars or starches. Straw, rice husk, and bran are examples of agricultural by-products that have been used as substrates to promote sustainability. On the other hand, mixed sugars and inhibitory substances created during fermentation can reduce the amount of LA generated. High early sugar concentrations, toxic buildup in the late stage, and changes in the cultural context and circumstances are all key variables in LA fermentation. Researchers must overcome these obstacles and create more affordable and effective methods to synthesize LA with high optical purity in order to satisfy the expanding demand for LA (Huang et al., 2023).

Feedstock costs typically account for 40–70% of the total expenses in lactose (LA) production. To reduce these costs, three types of carbon feedstocks have been studied for their use in fermentation to produce LA. The first-generation feedstock made from refined sugar or starch crops; second-generation feedstock made from lignocellulose biomass or other waste resources from everyday life and food production; and third-generation feedstock made from easily culturable ocean microalgae or macroalgae in captivity. First-generation feedstocks are not appropriate for production at a larger scale because of their high cost and the competitive relationship with human diet. Second-generation feedstocks include food waste from living things like fruit peels and leftover bread; food processing wastes like molasses and cheese whey; and agricultural wastes like straw, bagasse, and maize cobs. For microbial fermentation into LA, these materials may often be hydrolyzed to yield significant quantities of sugar. Because of their cellulose and hemicellulose content, sugarcane bagasse, corn cobs, and molasses are good substrates for LA synthesis. Bagasse was fermented by *Bacillus coagulans* DSM ID 14-300 to produce optically pure LA. Compared to MRS medium, molasses and powdered maize pulp are more economical. Since food waste contains a lot of nutrients, solid waste management is an important study topic. Despite the effective use of a variety of lignocellulose-based biomasses, LA fermentation is influenced by season, climate, and geography. Third-generation feedstocks that are easy to cultivate, have short growth cycles, and lack lignin include microalgae and macroalgae. Future research will focus on increasing the effectiveness of LA synthesis from algae (Huang et al., 2023).

Non-sterilized fermentation has been investigated by researchers for the synthesis of (LA), which can lower production costs and energy usage. The high D-LA concentration in fed-batch fermentation from cheese whey at 42°C is evidence that anaerobic fermentation prevents the development of certain bacteria. By preventing biomass saccharification and releasing enough sugar for increased output, simultaneous saccharification and fermentation (SSF) can reduce energy expenses. However, SSF is less efficient at mesophilic temperatures because saccharifying enzymes are less active. The fermentation process makes extensive use of microorganisms that can ferment at thermophilic temperatures. *Pediococcus acidilactici* DQ2, *B. coagulans* LA204, and *Enterococcus faecium* QU 50 have demonstrated promise for industrial use. According to a research, *Enterococcus faecalis* DUT1805 produced the greatest amount of LA with starting starch while simultaneously achieving liquefaction, saccharification, and fermentation at 50°C or above. Manufacturing time and energy might be decreased if these techniques are successfully applied in industry. Finding strains that can directly make LA from starchy feedstock can help reduce the energy needed for the saccharification process. The goal of the research is to reduce energy expenses by streamlining and improving the fermentation process (Huang et al., 2023).

6.3.1.2 Succinic Acid

Lignocellulose has to be processed using a number of physical or chemical techniques, including acid hydrolysis, ultrasonic treatment, and ionic liquid pretreatment, in order to effectively create succinic acid from it. Both the microbial fermentation of cellulose and hemicellulose hydrolysates and the enzymatic hydrolysis of lignocellulosic biomass are involved in the pre-treated lignocellulose fermentation process. There are three main types of traditional techniques for making succinic acid from lignocellulose. The two-step process known as SHF (Succinic Hydrolysis of Lignocellulose) entails the hydrolysis of lignocellulose by hydrolytic enzymes and the fermentation of the hydrolyzed sugars into a variety of target products. This process has been thoroughly investigated for the manufacturing of succinic acid from various lignocellulosic biomass. The current research has shown the encouraging outcomes. For example, pretreatment *Arundodonax hydrolysate* was converted to succinic acid by *Basfiasucciniciproducens* BPP7, producing around 6.06 g/L of succinic acid (Lu et al., 2021).

Whenever microbial consortia combine the saccharification and fermentation processes to produce succinic acid, SSF (saccharification of lignocellulose and bioconversion of hydrolyzed sugars into succinic acid) offers a number of advantages over SHF. By removing the partial inhibitory impact of hydrolytic products on enzymes, it lowers hydrolytic enzyme supplementation quantities and increasesthe product output. Additionally, SSF lowers equipment investment expenses. Research has demonstrated that the maximum concentration of succinic acid may be produced by SSF from diluted alkaline-pretreated maize stover. However, optimal enzymatic hydrolysis is compromised by the fermentation temperature in SSF. For SSF to be optimized, strains with high temperature tolerance and succinic acid production yield are essential. SSCF (sucrose-sucrose condensed), which is based on SSF, allows pentose and hexose to ferment concurrently in the same reactor. Genetic engineering is used to produce microbial strains that can consume pentose and hexose at the same time (Lu et al., 2021).

6.3.1.2.1 Production of Succinic Acid from Lignocellulose by CBP

The idea behind biomass bioprocessing (CBP) is to lower biological processing costs by lowering equipment purchases and operational expenses.The ultimate setup for inexpensive hydrolysis and fermentation of cellulosic biomass is acknowledged as the integrated bioprocessing process. Hydrolyzing lignocellulosic biomass and generating high titer, yield, and productivity are essential for an effective CBP microbial strain. Few natural strains, however, are capable of producing succinic acid and hydrolyzing lignocellulose. Microbial co-cultivation systems and metabolic engineering are methods for directly generating succinic acid from lignocellulose via CBP (Lu et al., 2021).

6.3.1.2.2 *Metabolic Engineering-Based CBP*

CBP strains for the direct production of succinic acid from lignocellulose can be developed using either natural methods or recombinant strategies. It involves giving naturally occurring lignocellulose breakdown microbes the capacity to produce succinic acid, which is now restricted in research. The majority of research focuses on developing CBP strains that produce biofuels like butanol and ethanol from lignocellulose. The formation of genetically modified instruments and the examination of strain metabolic networks are the primary obstacles. The recombinant approach gives microorganisms that produce succinic acid the capacity to break down lignocellulose. Through the heterologous expression and secretion of hydrolytic enzyme systems in succinic acid-producing microorganisms, the genetically modified hosts may directly transform lignocellulosic materials into succinic acid. Microbial strains modified by this recombination approach mostly target model microorganisms like *Bacillus subtilis*, *S. cerevisiae*, and *E. coli*. These microbes may be readily altered through metabolic engineering to allow the conversion of lignocellulose into desired products since they have reasonably developed genetic engineering techniques and genetic modification tools (Lu et al., 2021).

The bioconversion of lignocellulose to succinic acid has been accomplished in recent years by a number of researchers using CBP. For example, direct succinic acid synthesis from xylan without the usage of external xylan hydrolase and any organic components was accomplished by combining the expression and secretion of three hemicellulase enzymes with the establishment of a succinate metabolic pathway in *E. coli*. However, metabolic engineering-based CBP has not yet been effectively used to produce succinic acid from lignocellulose. Multiple enzymes and stages are involved in the breakdown of lignocellulose; when these enzymes are combined in a single strain, the rate is often low, resulting in low productivity and yield (Lu et al., 2021).

6.3.1.2.3 *Microbial Co-cultivation-Based CBP*

The utilization of single microbial strains, which might result in an excessive metabolic burden and poor product titer and yield, presents difficulties intended for the manufacturing of succinic acid from lignocellulose. Synthetic pathways should be distributed among several microbial hosts in order to solve this problem. Using a solid-state pre-fermentation stage with *A. niger* and *Trichoderma reesei* and a two-stage slurry fermentation stage with *Phanerochaete chrysosporium*, a mixed fungal consortium has been successfully developed to directly produce succinic acid from minimally pretreated lignocellulose through CBP. Succinic acid was roughly extracted from birch wood chips and soybean hulls. The consortium is a viable substitute for producing succinic acid from lignocellulose and can lower production costs. However, the generation of succinic acid and hydrolases is divided into two phases, which lengthens the fermentation process. Additionally, to directly produce succinic acid from hemicellulose-enriched lignocellulose, a microbial consortium including thermosanaerobacterium, thermossaccharolyticum M5, and *A. succinogenes* 130Z was established. After optimization, succinic acid concentrations from xylan reached 32.50 g/L, resulting in 0.39 g/g. However, the high xylose consumption requirement of strain 130Z is still not met by the hemicellulose hydrolysis efficiency of the M5 strain, which results in a prolonged fermentation duration. To effectively bio convert lignocellulose to succinic acid, additional research is needed for the development of greater capacity microbial consortia (Lu et al., 2021).

6.3.1.3 Acetic Acids

In addition to having the best selectivity and the lowest necessary expenses, biological acetic acid synthesis from bacteria has the strongest resilience to poisoning and allows for the least amount of pollution emission. Acetic acid from microbial consortia involves the treatment of food waste been investigated in recent years using aerobic fermentation technology. The creation of high-value products from food waste is important for the environment since it can reduce food waste and improve acetic acid yield annually and lowers manufacturing costs (Li et al., 2021). In the industrial sector, acetic acid plays a crucial role; it is mostly used to make vinegar. These days, acetic acid is used extensively to make vinyl acetate, which is used in latex paints, vinyl plastic, adhesives, textile finishes, and a market that is expanding quickly because of the need for synthetic fibers (Merli et al., 2021).

A non-corrosive substitute for chloride salts in the defrosting of roadways was discovered in 1979: calcium magnesium acetate (CMA). In the United States, between 10 and 12 million tons of street salt are used annually; solid CMA, which is more sustainable, would be a good substitute. Liquid potassium acetate partially replaces ethylene glycol as a heat exchange fluid and as a deicer for airport runways. Furthermore, it has been claimed that calcium acetate, or CMA, can be added to boilers and other coal-fired combustion units. The need for acetic acid would skyrocket if these replacements occurred on a worldwide scale. Anaerobic and aerobic biological methods are two alternatives to chemical procedures for producing acetic acid. Numerous reviews exist about the synthesis of acetic acid such as process description, bacterial physiology and metabolism, or biotechnological application using microbial consortia (Merli et al., 2021).

6.3.2 Alcohols

6.3.2.1 Ethanol

A notable alternative to petrol, ethanol is characterized by its affordability, cleanliness, safety, renewability, and environmental friendliness. The primary objective of ethanol production focuses on converting non-food raw materials, particularly lignocellulose, into ethanol. Lignocellulosic hydrolysates contain several hazardous by-products such as phenol, acid, aldehyde, and sugars including hexoses and pentoses. Generally, *S. cerevisiae*, a common yeast used in ethanol fermentation, struggles to efficiently convert the combination of these sugars. In contrast, a natural bacterial consortium known as the HP consortium demonstrates a significantly higher ethanol yield. Notably, the composition of the microbial community plays a crucial role in the efficiency of cellulose conversion to ethanol. Moreover, biomass-degrading enzymes produced by anaerobic gut fungi found in herbivores exhibit a remarkable ability to deconstruct lignocellulose. This microbial consortium approach to cellulosic ethanol production is promising, particularly given that the enzyme activity from *Piromyces finnis*, which hydrolyzes xylan, is reported to be three times more effective than the best commercial enzyme preparations derived from *Aspergillus*. Additionally, natural microbial consortia are capable of synthesizing various cellulases tailored to different types of lignocellulosic materials (Jiang et al., 2017).

6.3.2.2 Butanol

Butanol is a four-carbon primary alcohol that serves as a significant bulk chemical feedstock and is poised to become a next-generation liquid fuel. It presents numerous advantages over ethanol, including a higher energy content, reduced hygroscopicity, better blending capabilities, and an energy density similar to that of gasoline. Despite these benefits, the production of butanol via bio-generation currently faces economic challenges compared to petrochemical processes. This economic uncompetitiveness is largely due to the low concentration of butanol in fermentation broths, high feedstock costs, and the production of low-value by-products such as ethanol and acetone. Recently, there has been growing interest in biosynthesizing butanol from lignocellulosic materials, which could help mitigate feedstock costs. This microbial conversion process involves several biological steps: the enzymatic hydrolysis of lignocellulose to yield soluble saccharides, the production of saccharifying enzymes, and the subsequent metabolism of these saccharides into butanol. This emerging approach highlights the potential for more sustainable and cost-effective production methods for butanol as a desirable biofuel option. By simultaneously fermenting and hydrolyzing cellulolytic microorganisms and solventogenic bacteria in the same bioreactor, consolidated bioprocessing (CBP) has been proposed as an effective and cost-effective way to produce butanol from lignocellulose.The cooperating microorganisms could learn to use interspecies metabolic pathways in a synergistic way. Butanol synthesis from lignocellulose in a direct, pure culture proved to be fairly difficult (Jiang et al., 2017).

6.3.2.3 Biogas

Anaerobic microbial consortia are used in anaerobic digestion to convert organic waste into biogas, a mixed gas. Methane, H_2, CO_2, and other gases are present. The process of turning organic waste into biogas is usually believed to entail four stages. During the hydrolysis phase (Stage I), biopolymers disintegrate into monomers or oligomers; in the acidogenesis phase (Stage II), that they transform into volatile organic acids, alcohols, CO_2, and H_2. The molecules created in Stage II are converted into acetic acid, a little amount of CO_2, and H_2 during the acetogenesis phase (Stage III). Methanogens employ acetate or CO_2 and H_2 to create CH_4 during the methanogenesis phase (Stage IV). Certain bacteria are challenging to produce using traditional techniques like pure culture because of their unique development requirements inside microbial consortia, including a low hydrogen partial pressure. The most often used type of bioreactor is the up flow anaerobic sludge bed (UASB). Methanogen-producing microbial consortia granules play a crucial role in a two-stage reactor process specifically designed to treat wastewater with high sugar content, such as that generated from bread waste. The initial stage involves the fermentation of the bread waste at a controlled temperature of 55°C using thermophilic anaerobic sludge. This fermentation process leads to the production of hydrogen along with volatile fatty acids, particularly acetate and butyrate. In the second phase, these intermediate products are converted into methane, yielding a biogas that can be harnessed for energy. The thermophilic species introduced into the microbial community demonstrate a remarkable ability to thrive during the hydrogenotrophic phase, showcasing their resilience in a non-sterile environment, which is vital for maintaining effective microbial operations. This thermophilic method not only enhances microbial activity but also significantly reduces the potential for pathogen contamination throughout the wastewater treatment process, thereby improving overall safety and efficiency (Jiang et al., 2017).

6.3.2.4 Hydrogen

In recent years, fermenting organic waste to produce hydrogen has drawn a lot of interest as a potential clean fuel. Hydrogen generation is a subject of significant economic interest due to its diverse applications in various chemical industries. It plays a vital role in the synthesis of essential chemical compounds, including amines, alcohols, and aldehydes. This underscores hydrogen's importance not only in specific industrial processes but also within the broader context of chemical manufacturing, where it contributes substantially to the production of numerous chemicals that are crucial for various applications. Not only is hydrogen a perfect fuel, but as it burns, it also releases water. The main obstacle to manufacturing hydrogen by microbial anaerobic fermentation at the moment is the low hydrogen yield. Methanogens and homoacetogens are two examples of bacteria in microbial consortia that utilize hydrogen throughout the mixed culture. Alkali, acid, or heat treatment can be used to render the hydrogen-consuming bacteria inactive; heat treatment yields the best results, save for homoacetogens. The internal metabolic pathway of hydrogen is made more complicated by the complexity of the microbial community. At a pH of 3.5, hydrogen production was exclusively facilitated by a microbial consortium using the pyruvate decarboxylation-ferredoxin-hydrogenase pathway. Research conducted with clone libraries validated an initial model designed to predict the organizational structure of communities involved in mixed-culture fermentative bio-hydrogen generation. This model utilized assessments of electron fluxes and $NADH_2$ balances to gain insights into metabolic processes. Under optimal conditions, a significantly expanded microbial population derived from activated sludge demonstrated the ability to bioconvert crude glycerol into hydrogen, reaching a conversion rate of about 1.1 moles of hydrogen for every mole of glycerol consumed (Jiang et al., 2017).

6.3.2.5 1,3-Propanediol

In recent years, the market demand for 1,3-propanediol (1,3-PD), an essential monomer in the production of polytrimethylene terephthalate (PTT), has been experiencing consistent growth, despite its industrial production levels. The conventional method of producing 1,3-PD is through pure culture microbial fermentation, utilizing both wild-type and genetically modified bacteria to convert glucose and glycerol directly into 1,3-PD. An increasing focus has been placed on the conversion of crude glycerol into 1,3-PD due to the surplus of crude glycerol generated from large-scale biodiesel production. However, for crude glycerol to be viable for pure culture fermentation, it must undergo filtration to remove various impurities, including alcohol, salts, esters, lipids, and colors, which can inflate production costs significantly. Most research efforts have concentrated on optimizing the fermentation process for 1,3-PD through techniques such as strain screening, genetically engineered microorganisms, and methods involving pure cultures. Unfortunately, achieving biological production of 1,3-PD incurs higher costs, largely due to the requirement for strict aseptic conditions and the use of purified substrates in the fermentation process associated with pure culture methods (Jiang et al., 2017).

During the microbial production of 1,3-PD, a variety of by-products such as alcohols, LA, succinic acid, and acetic acid are generated alongside the primary product. These by-products play a vital role by contributing to the synthesis of adenosine triphosphate (ATP), which is essential for providing cellular energy. Furthermore, these compounds are instrumental in maintaining intracellular redox balance, an important aspect that influences the overall efficacy of the biosynthetic process. The simultaneous generation of these metabolites is critical for optimizing microbial metabolism, creating a stable and balanced intracellular environment. Achieving this balance is crucial for enhancing the production yields of 1,3-PD during microbial fermentation, thereby improving the efficiency of the entire process. These by-products often interrupt the NADH 1,3-PD pathway, hinder cell growth, reduce the quantity of 1,3-PD that can be recovered from glycerol, and make it more difficult to separate and purify the desired product. Some benefits of microbial consortia fermentation over pure culture include the following: the potential to use mixed or less expensive substrates; the cooperation of various enzyme systems and combinations of the metabolic pathways of various microbes, which can lead to a more limited production spectrum and less costly product purification. The vast microbial variety will lower manufacturing costs since it does not require sterility when dealing with microbial consortia (Jiang et al., 2017).

Microbial consortia are a desired addition to or replacement for pure culture-based traditional biotechnology in industrial biotechnology for chemical synthesis. In an effort to reduce the cost of biological synthesis of 1,3-PD and overcome the limitations of pure culture, fermentation utilizing microbial consortia has been the subject of much recent research. Pure cultures of a single strain and mixed cultures of microbial consortia were compared for the biological production of 1,3-PD.A microbial consortium mixed culture often showed greater productivity and substrate tolerance than a pure culture of a single strain. Although the exact processes are seldom known, it is clearly caused by the interactions between the cells in the microbial consortium. However, the metabolites or inter-metabolites (such as nucleotides and amino acids), coenzymes (such as NADH/NADPH), or cofactors (such as ATP) produced by one strain may control the growth and metabolism of another strain. The durability of the microbial consortia structure during fermentation is another critical concern in industrial operations, in addition to the mixed-culture approach (Jiang et al., 2017).

6.3.3 Dicarboxylic Acids

Microbial consortia are necessary for the synthesis of dicarboxylic acids. In the polymer sector, fumaric acid, a strong but non-toxic acidulant and antibacterial agent, has great promise as a practical replacement for petroleum-based monomers. Additionally, it is a nutritional supplement for cow feed that can cut methane emissions by up to 70%, with a global decrease of 10% attributable to human activity. One of the compounds with added value that may be extracted from biomass is fumaric acid. There are major economic and environmental benefits to producing fumaric acid using microbial consortia through a biotechnological method. Maleic acid isomerization employing metallic catalysts under high pressure and temperature conditions makes up the actual manufacturing process. Enzymes and fermentative processes are alternatives that might be more appealing since they allow waste materials to be valued under moderate process settings (Martin-Dominguez et al., 2020).

Fumaric acid hydration employing entire *Rhyzopusoryzae* cells under anaerobic circumstances can provide malic acid. It is promising for the recovery of important metals from batteries, the synthesis of copolymers and polymalic acid for controlled medication release, and the synthesis of deep eutectic solvents for malic acid due to its chelating characteristics.The majority of research on biotechnologically derived fumaric acid focuses on creating an efficient fermentative method for the industrial manufacture of fumaric acid utilizing microbial consortia. However, kinetic models for fumaric acid production—especially for free or suspended fungi—have not received much attention (Martin-Dominguez et al., 2020).

6.3.4 Other Platform Chemicals (2,3-Butanediol)

In the manufacture of 2,3-BDO, microbial manufacturing using microbial consortia has been established as an alternative, economical, and eco-friendly technique. The *Enterobacter*, *Klebsiella*, and *Bacillus* species were identified in the majority of research on microbial consortia for the synthesis of 2,3-BDO. In order to control the NADH/NAD+ balance, stop intracellular acidification, and store energy, certain bacterial species are capable of producing 2,3-BDO on their own. The use of several indigenous 2,3-BDO producers in industrial 2,3-BDO production is restricted because they are categorized as Risk Group 2 diseases (pathogenic to humans).Although *Bacillus* species and *Klebsiella oxytoca* are generally considered harmless (GRAS) bacteria, they may create 2,3-BDO from a variety of substrates. Because of their safe traits and innate capacity to produce 2,3-BDO, *Bacillus* species and *K. oxytoca* are perfect hosts for industrial 2,3-BDO production (Martin-Dominguez et al., 2020).

Using industrially relevant host strains such *S. cerevisiae*, *Lactococcus lactis*, and *E. coli*, efficient 2,3-BDO production has been investigated. Due to its well-defined physiology, genetic tractability, and tolerance to environmental stresses, *S. cerevisiae* has been extensively manipulated to manufacture 2,3-BDO. It is also suitable for commercial fermentation of bio-based products because it is a GRAS bacterium. Microbial synthesis can result in the production of 2,3-BDO stereoisomers, such as meso-BDO, (2R,3R)-BDO, and (2S,3S)-BDO. Enantiopure stereoisomers have unique physicochemical properties that make them very useful in the cosmetic and agricultural industries. Microorganisms manufacture 2,3-BDO utilizing cheap and plentiful feedstocks such lignocellulosic biomass, food processing by-products, industrial wastes, and agricultural wastes through the use of recent metabolic engineering (Martin-Dominguez et al., 2020).

6.4 PRODUCTION OF VALUE-ADDED CHEMICALS AND BIOPOLYMERS

6.4.1 Amino Acids (e.g., Lysine, Tryptophan)

Amino acids, the main metabolites that make up all proteins, were formerly believed to be impossible for microbial organisms to produce. However, in 1956, Japanese researchers discovered a bacterium that produced a significant quantity of glutamic acid: *Corynebacterium glutamicum*. Since *C. glutamicum* was discovered, it has been known that microbial cells produce amino acids. Microbial cells are producing a wide range of amino acids, and consumer demand for them has expanded significantly. Numerous studies on the logical structure and metabolic engineering of microbial cells for the synthesis of amino acids have been reported. The design of microbial cells for the synthesis of amino acids has lately made use of systems metabolic engineering, which applies the ideas and methods of systems biology to metabolic engineering. Furthermore, utilizing state-of-the-art techniques like evolutionary engineering, synthetic biology, and single cell analysis, microbial cells that produce amino acids have been produced (Hirasawa & Shimizu, 2016).

6.4.2 Vitamins and Cofactors

The fermentation of VB2 and VB12 is used in commercial production, the fermentation of additional B-group vitamins necessitates a considerable increase in output. Although there is a sizable market for vitamin C production, the current state of affairs shows overcapacity, challenging downstream processing, and concentrated demand in food and pharmaceuticals. Vitamin C fermentation technology can employ isotope technology to identify and track particular metabolites, create a database of the accompanying bacteria's anabolism, or investigate the mechanism of those microorganisms. It is also possible to develop heterologous assembly modules for the synthesis of 2-KGA and study microbial cell adaptation mechanisms in order to improve the productivity. The primary focus of vitamin A biosynthesis was β-carotene, which has been successfully produced on a massive scale using realistic and traditional microbial metabolic engineering. However, because intermediate industries has high barriers and synthesis and metabolism are intricate processes (Wang et al., 2021).

The fermentation of vitamin E has reduced overall industry costs, encouraging further expansion.. Though light-driven fermentation is expensive, photosynthetic bacteria have a great deal of potential for producing tocopherols, which makes commercialization challenging. For the synthesis of tocopherol, building a specially built photo-bioreactor might be a viable research avenue. Another important control technology for the synthesis of vitamin E might be controllable temperature-sensitive control systems. *B. subtilis* is a suitable choice for vitamin K production, and many studies have improved the design of fermentation modes, medium components, and growth conditions. MK output is being increased by genetic engineering and other methods. With the use of CRISPR/Cas9 genome editing technology, automated gene assembly, and high-throughput screening of high-yield strains, synthetic biotechnology opens up new possibilities for the development of vitamin cell factories. Vitamin cell factories are becoming more and more supported by advanced engineering technologies including computer-aided design, cell-free systems, dynamic regulation of gene expression systems, cold-shock-triggered temperature control systems, and various biosensors(Wang et al., 2021).

6.4.3 Bioplastics and Biopolymers (Polyhydroxyalkanoates)

A large number of microorganisms accumulate polyesters, known as polyhydroxyalkanoates (PHAs), which are naturally occurring macromolecular biopolymers. PHA is presented as a substitute for biodegradable plastics because of its full biodegradability and comparable characteristics to traditional polymers. The industrial manufacturing processes rely on cultures or genetically engineered strains of microorganisms in their natural form. Even though the production technique is sterile and clean substrates are employed in PHA synthesis, cultures remain prohibitively expensive to compete as a commercial plastic product. Because of this, using mixed cultures to create PHA has grown in popularity recently as a way to lower PHA manufacturing costs. By using renewable carbon sources from industrial or agricultural waste, mixed cultures of microbial consortia may manufacture PHA in non-sterile conditions, thereby lowering substrate and equipment costs (Jiang et al., 2017).

6.4.4 Biosurfactants

The low-molecular-weight, surface-active compounds that come from microorganisms like bacteria, yeasts, and fungi. Because of their biodegradability, non-toxicity, structural variability, and durability under harsh circumstances, they provide environmentally friendly substitutes for chemically produced surfactants. The main sources of biosurfactants include microbes, plants, and animal fats. Oil recovery, home and food businesses, agriculture, and personal care materials all make extensive use of them. In order to enhance the human health and environmental sustainability, more research is being done on the commercial production of biosurfactants. A variety of biosurfactants are produced by yeast, fungus, and bacteria; the majority of microbial biosurfactant makers are bacteria. Approximately 19% of all microbial biosurfactants are produced by fungi. The three most common microorganisms that create surfactants are *Pseudomonas aeruginosa*, *B. subtilis*, and *Acinetobacter calcoaceticus*. These chemically varied microbial surfactants include alasan (*Acinetobacter radioresistens*), emulsan (*A. calcoaceticus*), liposan (*C. lipolytica*), and biodispersan (*A. calcoaceticus*). Surfactin and rhamnolipid, which are made from the bacteria *Bacillus* and *Pseudomonas*, respectively, are illustrations of biosurfactants that have been used extensively for a range of applications. Few research concentrate on the biosurfactants that yeasts create, particularly for *Pseudozima* sp., *Candida* sp., and *Yarrowia* sp. (Datta et al., 2024).

6.4.5 Industrial Enzymes and Biocatalysts

A wide range of microbes create most commercially useful enzymes; bacteria including *Pseudomonas*, *Bacillus*, and *Clostridium* produce proteases; however, some fungi have also been discovered to produce these enzymes. The xylanases that are frequently utilized in bioindustries are produced by fungi belonging to the genera *Trichoderma*, *Penicillium*, and *Aspergillus*. It has been shown that xylanases in these bacteria are highly active throughout a broad temperature range (40–60°C). Proteases are essential enzymes in many industries, including the manufacturing of animal feed, textiles, detergents, leather goods, and pharmaceuticals. Research is being done on the stability and efficacy of microbial proteases in alkaline pH settings. Keratinases, often referred to as serine or metallo-proteases, are essential for the manufacturing of animal feed, textiles, detergents, leather, and pharmaceuticals. Amylases are used in the starch conversion, liquefaction, paper, food, sugar, and pharmaceutical industries. *Geobacillus* sp. produces a thermostable alpha-amylase Which was recently discovered, and an isolated and characterized thermotolerant iso-amylase (Nigam, 2013).

Xylanase is a new kind of alpha-amylase that exhibits alkali resistance and thermotolerance. Pentose manufacturing, fruit juice clarifying, rumen digestion, and the bioconversion of agricultural leftovers are just a few of the businesses that depend on xylanase enzymes. They are stable at high reaction temperatures and active at alkaline pH. For cellulases, thermostability is an essential technical feature as saccharification happens faster at higher temperatures. For industrial applications focused on the economic potential of glucose feedstock synthesis from agricultural cellulosic resources, cellulase enzymes rank third. To develop cellulases of the necessary quality, researchers have cultivated microorganisms in both solid and submerged habitats. The characteristics and uses of microbial enzymes, such as lipases and pectinases, have been thoroughly investigated in several fields. They are especially crucial for analytical testing and diagnostic kits in the pharmaceutical sector (Nigam, 2013).

A number of results have been produced by recent biocatalysis research, creating bio-industries that are profitable. Current bioprocess technology is capable of fulfilling the demands and problems of both contemporary and traditional industries. Trinconehas examined methods for special enzymatic glycoside preparation. Enzymes can be modified using contemporary methods to satisfy process specifications. Long-term biocatalysis research is required to solve fundamental issues with the transition from traditional to bio-based processes for commercial applications (Nigam, 2013).

6.4.6 Pharmaceutical Precursors and Natural Products

The antimalarial drug artemisinin is synthesized using amorphadiene. Secondary metabolite extraction from plants may be replaced by the synthesis of artemisinin and amorphadiene from metabolically modified microorganisms. The DOL is usually made possible by microbial consortia, and the microbial co-culture system may be used to produce natural commodities at a reasonable cost. Table 6.1 displays the chemical production capability of microorganisms. Co-utilization of various carbon sources is essential for enhancing amorphadiene production, which is achieved by strategically increasing the surface area of the endoplasmic reticulum (ER). Additionally, the targeted localization of the ADS gene, responsible for coding amorphadiene synthase, into the ER plays a vital role in this process. Implementing

TABLE 6.1
Production of Chemicals in Microbial Consortia

Production of Chemicals	Microbial Consortia	Applications	References
Ethanol	*Saccharomyces cerevisiae*	Used in the production of biofuels	Ruiz-Marín et al. (2016)
Lactic acid	*Lactobacillus* sp.	Used in the production of bioplastics and preservation of food	Abedi and Hashemi (2020)
Butanol	*Clostridium*sp.	Production of advanced biofuels and chemicals feedstocks	Yu et al. (1985)
Methane	Anaerobic bacteria and archaea	Used in the production of biogas and treatment of wastewater	Wang et al. (2022)
Succinic acid	Escherichia coli	Used in the production of bio refinery platform chemicals	Ahn et al. (2016)
Polyhydroxyalkanoates (PHA)	Mixed microbial cultures from wastewater	Production of biodegradable chemicals	Coats et al. (2016)
Hydrogen	*Rhodobactersphaeroides, Clostridium butyricum*	Renewable energy source	Fang et al. (2006)
Biosurfactants	*Pseudomonas* and *Rhodococcus*sp.	Used to enhance oil recovery and bioremediation	Pacheco et al. (2010)

a co-culture system with *Yarrowia lipolytica* strains Po1f and Po1g further amplifies the synthesis of amorphadiene. These interrelated strategies not only improve the efficiency of the metabolic pathway responsible for amorphadiene production but also broaden its applicability in several biotechnological applications. The main components of microbialites and lithified microbial mats include pseudomonadota, cyanobacteria, sulfate-reducing bacteria, diatoms, and microalgae. Plants can benefit from the additive or synergistic effects of rhizosphere microbial consortia, which are made up of two or more interacting microorganisms, in terms of growth, pathogen defense, nutrient absorption, and resistance to environmental stress. Simplified microbial consortia (SMC) are utilized in the breakdown of keratin (Marsafari et al., 2022).

An effective keratinolytic microbial consortia (KMCG6) was previously cultivated in keratin medium to enhance an environmental sample. In 2020, an SMC with comparable keratinolytic activity but fewer species was extracted. The process involves creating a library, enriching for desired traits, performing repeated dilutions, and selecting the most promising synthetic microbial consortium (SMC). Humans naturally host microbial consortia, such as those in the gut and on the skin, which support nutrition and provide protection. Recent studies have also shown that the brain contains bacteria, and metagenomic data suggest that these microbes may originate from the intestines. It is quite plausible that these species have built a symbiotic consortium within the brain because they seem well-established, have no appreciable effect on human health, and are known to form consortia when discovered in the gut (Marsafari et al., 2022).

6.5 BIODEGRADATION OF COMPLEX COMPOUNDS BY MICROBIAL CONSORTIA

Natural microbes find it difficult to break down complex substances because of their intricate structures, especially environmental contaminants. To break down these substances, including waste plastics, microbial consortiums are frequently employed. For the creation of natural microbial consortia, strain interaction research is crucial. Researchers can construct basic microbial consortiums to gain a better understanding of degradability, gene regulation, or enzymatic activity. As an alternative, they can incorporate microbes to increase the effectiveness of some isolated natural microbial consortia that are capable of breaking down plastics. Almost 90% of the n-alkanes and aromatic hydrocarbons in petroleum hydrocarbons can be broken down by rebuilt microbial populations. To increase the effectiveness of the adsorption and degradation of different chemicals and heavy metals in sewage, microalgae can be added to activated sludge. As an alternative, a microbial consortium, whose substrates are more specialized and can only break down one or more particular kinds of complicated chemicals, may be built based on the interactions between microorganisms. The modulation of the degrading process may be easily studied and clearly represented by artificial microbial consortia. Microbial consortiums are useful for bioremediation of complex substances, particularly contaminants, refractory degradation, and environmental complexity (Cao et al., 2022).

6.5.1 Petroleum Hydrocarbons Degradation

They are known as the essential raw materials and energy sources for a variety of living things. These pollutants, which are both recalcitrant and classified as priority substances, include aromatics, cycloalkanes, and normal paraffins. Certain naturally occurring bacteria are capable of degrading these specific compounds. The benefits of microbial consortia in crude oil bioremediation stem from their genetic mutations and environmental adaptation. Microbial consortiums for the breakdown of petroleum hydrocarbons are frequently built using the top-down method. However, low biodiversity frequently limits these consortiums' ability to perform bioremediation in oil-contaminated environments. Artificial compounds or engineered bacteria can be added to the microbial consortia system to increase its capacity for degradation or substrate extensiveness. When dealing with hydrocarbon pollutants in petroleum crude oil as possible carbon source, microbial consortia outperform individual bacteria (Cao et al., 2022).

Microbial consortia constructed on petroleum hydrocarbon breakdown pathway may likewise be formed via the bottom-up technique. Anaerobic and aerobic degradation are two of the several mechanisms that make up the biodegradation of petroleum hydrocarbons. The early breakdown of alkanes and oxidation of methyl groups, which yields alcohols, and the dehydrogenation of aldehydes, which yields carboxylic acids, are typical processes for linear alkanes. Because hydrolase or isomerase must first progressively open cycloalkanes and aromatic hydrocarbons before they can be broken down by other routes, they are more difficult to break down than linear alkanes. A single strain cannot break down petroleum hydrocarbons and create biosurfactants at the same time to guarantee declining efficiency. In response to this, a number of research employ microbial consortia to degrade petroleum hydrocarbons, with several bacterial species carrying out distinct functions. When several bacterial species cooperate, they create a strong and potent degradation process that not only lowers the growth pressure of individual cells but also increases the resistance of bacterial groups to adverse conditions (Cao et al., 2022).

6.5.2 Biodegradation of Petroleum Hydrocarbons

Modifying engineered bacteria in accordance with the degradation route can improve the efficiency of petroleum hydrocarbon degradation. Complex substances can be broken down or aided in their breakdown by genetically engineered microbes. For example, when *E. coli* was given the alkane hydroxylase (alkB) gene, diesel breakdown rates increased from 31% to 50%. Increasing the production of surfactants also increases the accessibility of petroleum hydrocarbons to these strains, improving the breakdown efficiency. By modifying the interspecific connection, a three-bacterium system may greatly increase the breakdown productivity of phenanthrene through the synergistic effects of many bacterial species in microbial consortia. The synergistic microbial consortia frequently had a far larger repair impact than a single strain. A microbial consortium's ability to avoid interfering with complex chemical breakdown processes is an advantage of including strains with varying activities. Using the microorganism that is capable of using breakdown products as carbon sources or substrates for the biosynthesis of high-value goods is the most economical approach (Cao et al., 2022).

6.5.3 Degradation of Plastics

Only 21% of plastics are recycled or burned, making plastic manufacturing a major worldwide problem. The remainder is buried or thrown away, adding to the degradation of the environment. Plastic can be decomposed through physical, chemical, or biological sources. Microbial degradation has gained a lot of interest because it is cost-effective, rapid, and safe. Microbial consortia, which outperform strains in biodegradation effectiveness, are capable of breaking down multiple polymers. The decomposition process of consortia acting on deteriorated foam blocks retrieved from landfills has been investigated using top-down techniques. It was demonstrated that strains' carbon consumption efficiency in the presence of microbial consortia was more effective in degrading different types of complex polymers due to interspecific interaction. According to the study, there is a significant amount of possibilities for bioremediation in the future for certain degradable enzyme genes in the microbial consortia (Cao et al., 2022).

Microbial consortia based on various plastic degrading pathways may be built using the bottom-up technique. Three phases can be used to describe the process: biodeterioration, bio fragmentation, and biodegradation. Biodeterioration is the process by which biofilms break down the surfaces of plastic polymers and degrade them, while biofragmentation is the process by which extracellular enzymes and free radicals break down plastic polymers into smaller components, whereas biodegradation is the breakdown of plastic polymer surfaces by biofilms. Enzymes that break down plastic can be broadly classified as either intracellular or extracellular. In order to create combinations of oligomers, dimers, and monomers, these enzymes depolymerize the lengthy carbon chains of polymeric polymers. These monomers are broken down into oxidative metabolites by enzymatic processes that are carried out by various strains. Microbial consortiums are a good option for accelerating the breakdown of plastics, particularly when dealing with a blend of different polymers (Cao et al., 2022).

6.5.4 Biodegradation of Plastics

Artificial microbial consortia constructed using bottom-up approaches can indirectly enhance biodegradation by including compounds that promote co-metabolic breakdown or metabolic cross-feeding. The degradation pathways may be divided into discrete steps, each of which can be associated with a certain strain, using key enzymes and intermediates that regulate the step of plastic breakdown. This strain group is the first microbial consortia. This method lessens the metabolic load on each strain while increasing the microbial community's ability to withstand challenging conditions through interactions between strains. The microbial consortium is effective in degrading polyethylene terephthalate (PET) which comprises of *Rhodococcus*, *Pseudomonas putida*, and two genetically modified strains of *B. subtilis*. The modified *B. subtilis* strains play a vital role by producing specific enzymes, notably PET hydrolase and monohydroxyethyl terephthalate hydrolase, which facilitate the initial breakdown of PET polymers. Following this enzymatic action, *Rhodococcus jostii* and *P. putida* collaborate to further degrade the by-products ethylene glycol and terephthalic acid. Remarkably, under ambient temperature conditions, this microbial collaboration can fully degrade up to 23.2% of a PET film. Moreover, based on comprehensive research into the degradation pathways of synthetic plastics, a microbial community degradation platform has been developed that not only facilitates the breakdown of synthetic polymers but also aids in transforming them into valuable high-value products (Cao et al., 2022).

6.5.5 Bioremediation

Many industries release hazardous pollutants into the environment, especially heavy metals such as lead, chromium, nickel, zinc, and copper. To address the contamination caused by these metals, a variety of remediation methods have been developed. These methods include adsorption, which involves the attachment of pollutants to solid materials; chemical oxidation, which transforms harmful substances into less toxic forms; ion exchange, a process that swaps harmful ions in solution with less harmful ones; reduction, which removes oxygen or adds hydrogen to decrease oxidation states of metals; and chemical precipitation, where metals are converted to insoluble salts. Additionally, techniques like solvent extraction, which separates metals from a solution using organic solvents; electrolyzation, which employs electrical currents for metal recovery; biological flocculation, utilizing biological agents to aggregate pollutants; and biosorption, where biological materials absorb heavy metals, are also effective in remediation efforts. These diverse strategies highlight the complexity and necessity of environmental cleanup measures in industrial contexts. Algae, fungus, and bacteria use cost-effective and environmentally benign biosorption techniques. *Aspergillus*, *Bacillus*, *Pseudomonas*, and *Flavobacterium* are among the microorganisms whose capacity to reduce metals has been documented. Microbial activity plays a significant role in the adsorption of heavy metals through the production of various EPSs. These EPSs contain functional groups, such as carboxyl and hydroxyl, which are capable of forming complexes with heavy metals. This biochemical interaction enhances the capacity of microbes to bind and remove heavy metals from the environment, thereby contributing to bioremediation efforts and the maintenance of ecological balance. Microbes create extracellular polymeric compounds called bioflocculants, which include proteins, polysaccharides, and nucleic acids. Flocculants are very helpful for treating industrial effluent and wastewater. Both organic and inorganic flocculants are hazardous and can lead to cancer and other health issues in people. Bioflocculants don't pose any risks and are both ecologically benign and biodegradable. Bioflocculant production is an expensive procedure, however a mixed-culture technique can be cost-effective (Bhatia et al., 2018).

6.5.6 Degradation of Antibiotics and Azo Dyes

Antibiotics are a commonly used family of medications, and their presence in drinking water, surface water, groundwater, bio solids, and sewage treatment plant effluent poses a serious threat to the world. To reduce pollution after treatment, future bioremediation projects will favor biotechnology over chemical treatment and concentrate on enzymatic remediation. Microbial consortia have demonstrated remarkable degradability in research on antibiotic biodegradation because they can break down various antibiotic kinds in various ways. *Bacillus* and *Flavobacterium* are the predominant bacteria in consortia that break down sulfates. Microbial consortia for antibiotic degradation are often developed using the top-down method. Antibiotic degradation in natural microbial consortia is commonly studied using activated sludge. Artificial microbial communities might be created using a bottom-up method. More than 80% of tetracycline may be degraded by artificial microbial consortia. Because of their stability, these consortia may be utilized in environmental restoration to create a co-culture system that may be applied to real sewage for bioremediation (Cao et al., 2022).

The most common synthetic color used in the printing and dying of textiles and clothing is azo dyes, of which 10–15% are released into the environment untreated. For leftover azo dyes, microbial decolorization is thought to be the most efficient and eco-friendly solution. The bacterial breakdown of azo dyes begins with the disruption of the azo bonds inside the dye molecules. Then, chemical connections are broken by the release of extracellular enzymes. Using both intracellular and extracellular azo bond breaking mechanisms, microorganisms can cooperate to degrade azo dyes. Most current microbial consortia are built using the top-down approach since the underlying molecular mechanism of synergistic metabolism in the microbial consortia system has not been determined.Azo dyes may be more extensively biodegraded and mineralized using the microbial consortia approach because of the consortia's DOL and cooperative interaction. Microbial consortia have shown remarkable success in real-world applications of azo dye degradation, such as irrigating inedible plants and lowering water shortages globally. Azo dyes have a major detrimental impact on the environment despite their limited use, hence more study into their ability to degrade azo dyes is necessary to support environmental protection strategies (Cao et al., 2022).

6.5.7 Treatment of Sewage

The rising sewage outflow from industries presents a significant global environmental challenge, compounded by the inadequacies of traditional sewage treatment methods, which are often costly, energy-intensive, and insufficiently effective in managing sewage-related issues. In response, microbial consortia have been identified as a promising and sustainable alternative for sewage treatment, specifically effective in mitigating problems like sewage eutrophication. These consortia often incorporate microalgae, which possess the remarkable capability to alternate between autotrophic and heterotrophic metabolic pathways depending on the availability of carbon sources and nutrients in their environment. Techniques employing both top-down and bottom-up approaches have shown considerable stability in wastewater treatment; notably, the former requires less extensive screening processes compared to the latter, highlighting their potential in contemporary sewage treatment frameworks. This adaptability and efficiency make microbial consortia a critical component in the evolution of sewage management practices aimed at addressing environmental degradation. Microalgae and bacteria have been demonstrated to have a positive growth-promoting impact when combined; both produce carbon dioxide and oxygen. In addition to secreting extracellular chemicals like auxin and vitamin B12 that are essential for microalgae development, bacteria can convert organic resources into mineral forms. The detoxification of organic and inorganic contaminants as well as the removal of nutrients from sewage is enhanced when bacteria and microalgae are combined. The benefits of artificial microbial consortia in treating wastewater eutrophication have been shown in a co-culture system of *Chlorella vulgaris* and *P. putida*, where the simultaneous removal of nutrients and COD occurred. It implies that there are several potential applications for the combination of microalgae and microbial consortia in sewage treatment in the future (Cao et al., 2022).

6.5.8 Consolidated Bioprocessing

CBP) represents an economical approach for generating biofuels and other valuable products. This process combines fermentation lignocellulose breakdown and simultaneous production of lignocellulose-degrading enzymes. Identifying bacteria that can produce all enzymes is a challenging process. A great alternative is microbial consortia, which adopt bottom-up approaches. Experiments indicate that microbial consortia enhance the performance of CBP, with a combination of *C. phytofermentans* and *S. cerevisiae* achieving greater ethanol production from α-cellulose compared to single cultures. The effectiveness of CBP could be greatly enhanced by designing a microbial consortium utilizing task specialization among strains. For example, enzymatic degradation of biomass by *Candida tropicalis* and *Zymomonas mobilis* can yield 97.7% of ethanol. Additional products synthesized by CBP include halomethanes and LA. For example, halomethanes were produced from raw switchgrass, bagasse, maize stover, and poplar by a group of *Actinotalea fermentans* and modified yeast. An artificial microbial consortium consisting of facultative anaerobic LA bacteria and *T. reesei* produced the theoretically maximum amount of LA (Cao et al., 2022).

6.6 BIOREACTOR DESIGN AND OPERATION FOR CONSORTIA-BASED PRODUCTION

6.6.1 Types of Bioreactors Suited for Consortia

Bioreactors have gained significant attention in recent years due to the increasing global demand for biopharmaceuticals and sustainable products. This interest is driven by the escalating use of industrial biotechnological processes and the emergence of concepts such as bio-refineries and the bio-economy. Biotechnological processes can be conducted in varied environments, including liquid, solid, and low water activity conditions, and can involve agitation and aeration or not. The primary types of bioreactors employed for submerged fermentations (SmFs) that involve microbial cells are stirred tank reactors (STR), airlift reactors, and bubble column reactors. The bioreactor market stood at an estimated value of $9.3 billion in 2022 and is expected to experience substantial growth in the coming decade, with a projected compound annual growth rate (CAGR) of 12.58%, potentially reaching a market value of approximately $30.42 billion by 2032. This growth is largely attributed to major corporations engaged in the manufacturing and marketing of bioreactors tailored for both industrial and laboratory applications(Palladino et al., 2024).

6.6.1.1 Stirred Tank Bioreactors

The biotechnology industry favors the aerated STR bioreactor because of its ease of use, adaptability, scalability, control over growing conditions, and lower danger of contamination. Since they are less susceptible to shear stress, it is usually applied to bacterial or yeast microbial cells. The mechanical agitation system facilitates effective mixing of the culture media by using turbines, propellers, or blades. Nutrient homogeneity, even oxygen distribution, and the preservation of optimal conditions throughout the reactor volume are all ensured by the agitation. The most popular agitation technologies in bioprocesses are Rushton turbines and blades on pitched-blade impellers. Pitched-blade impellers are perfect for low-shear cultivation and sensitive cells, whereas Rushton turbines are frequently utilized for microbial cultures resistant to shear stress. Different types of impellers can be used to enhance mixing efficiency and reduce shear force, depending on the application (Palladino et al., 2024).

6.6.1.2 Packed Bed and Fluidized Bed Bioreactors

Modified STR bioreactors, such as those that employ marine-type impellers to reduce shear stress and prevent foam formation and perforated-type sprinklers for aeration, were developed as a result of the development of submerged cell production from animal cells. Despite the fact these bioreactors are essential for several aerated processes in the pharmaceutical industry as they come with limitations, such as high mass transfer, low energy demand, user-friendliness in sterile settings and simple design. In recent years, aerated processes utilizing airlift and bubble column bioreactors have garnered significant attention due to their distinct advantages. These bioreactors are noted for their high mass transfer rates, effective mixing capabilities with minimal shear stress, low energy consumption, simplicity in use within sterile environments, and straightforward design. However, a significant challenge arises when attempting to scale these systems up to larger volumes. As a solution to this limitation, wave bioreactors emerged in the late 1990s, offering an alternative to stirred-tank reactors (STRs) for the submerged cultivation of both plant and animal cells, especially those sensitive to shear stress. Wave bioreactors comprise a sterile plastic cell bag equipped with a hydrophobic filter facilitating gas exchange, which is linked to a rocking table that regulates temperature and provides oscillatory motion. Despite their innovative design and operational advantages, wave bioreactors also face constraints regarding scalability, effectively operating up to a maximum capacity of 500 liters without the need for sterilization. The development of these bioreactors represents a crucial step in biotechnology, particularly for applications involving delicate cell cultures (Palladino et al., 2024).

The pharmaceutical sector relies increasingly on single-use or disposable bioreactors for the cultivation of animal cells in SmF. These bioreactors are crafted from high-quality plastic materials such as low-density polyethylene, polypropylene, or polycarbonate, which eliminates the need for extensive cleaning and sterilization between culture batches. Their designs and sizes vary from bench-scale to industrial applications, allowing for versatility and efficiency in use. Significant advancements in disposable bioreactor technology include the development of non-invasive sensors, the incorporation of process analytical technologies (PAT) aligned with quality by design (QbD) principles, the ability to tailor designs to specific applications, and a range of vessel and impeller geometries optimized for mixing and mass transfer. These developments will enable the development of durable and reproducible procedures as well as more adaptable bioreactors for the culture of different cells under physiologically appropriate circumstances (Palladino et al., 2024).

Supercritical fluids (SSFs) techniques use both mixed bioreactors, including stirred drums, fixed beds, and trays. However, these systems face challenges like clumping during fermentation harm to fungi and difficulties, with medium volume injection, regulation, and sterilization. Custom modifications to the bioprocess-including the choice of microorganisms, substrate type, and operating parameters-are often needed to optimize bioreactors for solid-state fermentation (SSF). Heat transport and humidity management are issues at the industrial scale because of the greater substrate quantities.To reduce these problems and increase production, research has concentrated on creating creative designs and automated technologies. Even though these systems are widely used, they nevertheless require constant assessment and development (Palladino et al., 2024).

6.6.1.3 Membrane Bioreactors

Membrane bioreactors (MBs) are permeable devices that can be utilized for industrial wastewater treatment. They hold back cells or particles while allowing selective transfer of materials of variable sizes. Based on the size of particles to be removed from the culture medium MBs can be classified as micro-, ultra- or nano-filtration devices. Advantages include separation, continuous process, contaminant management, and reduced space requirements. Nevertheless, disadvantages involve cost, challenging maintenance, and limited flow rates. To enhance and streamline bioprocesses for the production of value-added products such as polyols, bio-surfactants, organic acids, and bacterial polyesters, MBs can be utilized effectively. These systems allow for the purification of biomolecules through a membrane filtration process, which subsequently reduces the number of processing stages required. This reduction not only influences the duration of the processing but also leads to lower overall expenses, rendering the entire bioprocess economically viable (Palladino et al., 2024).

The construction of micro- and macro-bioreactors for bench-scale cultivation of cells or enzymes demonstrates a significant application of 3D printing technology, facilitating innovative approaches in biotechnological processes. Furthermore, it can be employed to produce matrices and micromixers that immobilize enzymes and cells. By simplifying the assembly of bioreactors and related components 3D printing offers a means to enhance both SmF and SSF processes facilitating controlled bioprocessing. Macro bioreactors enable the generation of tissues for tissue engineering studies while micro bioreactors are employed in pharmaceutical processes for drug screening optimizing cellular responses and modeling diseases. Substances such, as PDMS (polydimethylsiloxane) play a role in ensuring safety during these operations. Reaction micro-devices greatly gain from the adaptability, clarity, and gas permeability of PDMS, a silicon-based polymer. AI has enhanced process output and energy usage simultaneously reducing waste and improving efficiency. The engineering, safety, and energy efficiency of bioreactors on an industrial scale are challenges. Operations are made safer and more effective by combining AI, automation, modeling, and sophisticated materials. Through the incorporation of Internet of Things (IoT) devices and advanced sensors, wastewater treatment scenarios have the potential to evolve into intelligent and self-adaptive systems. Additionally, the application of neural networks in the management and control of these systems can further enhance their efficiency and responsiveness (Palladino et al., 2024). Pros and cons of microbial consortia are represented in Table 6.2.

TABLE 6.2
Pros and Cons of Microbial Consortia

Pros	Cons	References
The utilization of complicated carbon sources is made possible by mixed culture as different bacteria may develop different kinds of enzymes that can target diverse substrates.	It is challenging to form a biotechnologically relevant consortium since individual capabilities and interactions might impact fermentation, making research challenging.	Bhatia et al., (2015), Bhatia et al. (2018)
Microbes in co-culture encourage one another growth by influencing their surroundings and cross-feeding nutrients.	Microbial contamination can happen sometimes. It becomes challenging to identify the pollutant when this occurs.	Shou et al. (2007), Bhatia et al. (2018)
The ability to complete multistep reactions more quickly than a single organism may lead to increased productivity through mutual interaction.	As fermentation goes on, consortium stability becomes a problem as well since one organism may become dominant over the others, which would reduce process productivity.	Stewart (2012), Bhatia et al. (2018)
In laboratory conditions, many microorganisms that normally coexist are not incompatible with one another. It is possible to employ some unculturable microorganisms in mixed cultures. Co-cultures can prevent contamination by preventing the growth of undesirable microorganisms, which further stabilizes the culture	The formation of a consortium is further hampered by a lack of prior understanding regarding the metabolite profiles, biological functions, and nutritional requirements of different bacteria.	Bhatia et al. (2018)

6.6.2 Control Parameters for Consortia Stability and Productivity

1. The pH levels in an environment significantly influence the survival and functionality of microorganisms. Microbes can be categorized into three distinct groups based on their optimal pH levels: alkaliphiles, which thrive at a pH above 9; neutrophiles, whose optimal pH ranges from 5 to 9; and acidophiles, which flourish at a pH below 5. The designated optimal pH for each group is crucial as it affects the structural integrity of proteins, the stability of cellular membranes, and the activity of enzymes necessary for microbial processes. Variations in pH not only favor certain microbial species but also alter the diversity and composition of microbial communities.
2. Temperature is one of the most significant environmental factors that regulate microbial growth and metabolism. Temperature plays a crucial role in the growth of microbial species, as each species has a specific optimal temperature range that promotes its best growth. Extremely high temperatures can hinder or completely stop the growth of some microbes. The impact of

temperature goes beyond growth rates; it influences the rate of biochemical reactions within the microbial cells, the fluidity of cellular membranes, and the activity levels of enzymes. Understanding these relationships is essential for studying microbial behavior and optimizing conditions for various applications in microbiology. Temperature in microbial consortia may affect that how the different microbial species interact with one another, which could alter the composition and functionality of the community.
3. Microbial growth and activity are based on the availability of necessary nutrients like carbon, nitrogen, phosphorus, and micronutrients. The composition and function of microbial communities can be strongly affected by nutrient availability and stoichiometry. Some species may be more favored over others due to changes in microbial metabolism and community structure triggered by nutrient limitations. For instance, microbial consortia that promote phosphate acquisition can offset the effect of phosphate limitation, which in turn can reduce the diversity in the rhizosphere of the bacterial populations. Excessive carbon concentrations, demonstrated by acetic acid, can impede cellular growth and lead to increased operational costs.

6.6.3 Challenges

1. The interactions within microbial consortia are poorly understood and difficult to engineer in artificial, synthetic, or semi-synthetic systems. Key challenges include predicting the behavior of multiple species within a community and maintaining long-term stability. It is essential to understand how different microbial species interact indirectly, including potential synergistic as well as antagonistic or competitive interactions, under operational conditions.
2. Microbial consortia products face significant challenges due to a continuously changing regulatory landscape. The ambiguity surrounding their legal status, coupled with safety concerns, poses obstacles that could hinder the authorization process for the commercial utilization of artificial, synthetic, or semi-synthetic consortia.
3. Large-scale production of microbial consortia enhances technological transfer and the supply chain globally. However, this process has some logistical and financial challenges. The key aspects involved are ensuring low-cost production with maintenance of consistency in microbial compositions.
4. The stability and performance of microbial consortia should be ensured for the long term under different field conditions. Over time, environmental variables and other uncontrollable factors can influence their performance.
5. The introduction of artificial semi-synthetic organisms into nature raises the concerns about ethics and society. At the same point, public opinion needs to be taken into consideration and a balance has to be achieved between prospected benefits and unforeseen disadvantages. The outcomes of the current chemical treatments will need to be weighed against this indirect protection from diseases and pests.

6. The successful application of microbial consortia hinges on effective collaboration among various disciplines, including microbiologists, geneticists, agronomists, and regulatory agencies. To ensure both development and implementation are effective, it is critical to address and bridge the existing gaps between these fields.

6.7 FUTURE DIRECTIONS AND INNOVATIONS IN MICROBIAL CONSORTIA

1. Upcycling renewable feedstock using microbial consortium-based processes is an innovative approach to sustainable biomanufacturing, it is a concept that fits well within a circular bioeconomy. The rapid advancement of synthetic biology and genetic engineering has led to notable developments in the single-strain approach, which focuses on the manipulation of metabolic pathways within specific strains. This methodology facilitates targeted alterations to enhance the productivity and efficiency of microbial systems, ultimately targeting for the improved outcomes in biotechnology applications. Through these advancements, researchers can optimize the metabolic functions of selected strains, paving the way for innovative solutions in various fields such as pharmaceuticals, agriculture, and renewable energy. Harnessing the synergistic potential of several microbial species, the microbial consortium-based approach now offers a valid and versatile means of upcycling a range of different renewable feedstock streams into valuable products. Further research needs to be focused on further method enhancement toward better scalability and efficiency, diversifying reused feedstocks, and assessing the environmental sustainability and economic viability of the technologies. Legislative environment and public acceptance of renewable feedstock upcycling technologies are also crucial for successful implementation (Hossain et al., 2024).
2. Cooperation and a multidisciplinary approach involving microbiologists, engineers, environmental scientists, and legislators could generate more creative ideas and efficient implementation techniques. Under sterile growing conditions, saccharification, and fermentation of polysaccharides; nonchemical pretreatment; and a zero-waste strategy, microbial consortia are an efficient system for generating biofuels in CBPs and bio refineries. These characteristics contribute to a circular economy model with a minimum generation of greenhouse gases and no landfill requirements. A circular economy promotes the cycle of lignocellulosic materials back into the industrial market for resource recovery. Nevertheless, there are various challenges that have to be faced, such as commercialization, synergic analysis, elimination of raw material pretreatment, and government policies for using biofuels efficiently (García-Solares et al., 2023).
3. The microbial consortia have great potential for improving the yields of products. To successfully engineer a consortium, it is essential to grasp the underlying mechanisms of interactions among microbes. Different screening methodologies, such as systems biology and computer modeling, can

facilitate the creation of co-cultures where microbes collaborate by sharing the metabolic responsibilities of synthetic pathways, establishing mutually beneficial relationships, and creating dependencies between each other. This approach can also provide a way to use more microorganisms within one artificial consortium (Mittermeier et al., 2023).

4. Future strategies for constructing and improving microbial co-cultures include directed evolution, metabolic engineering, and enzyme engineering. Before microbial consortia can be applied in a wide range of industrial bioprocesses, several challenges have yet to be overcome. Successful scale-up will involve increasing the amount of microbial biomass and the development of complex control mechanisms. For proper viability assessment, integrated technical and economic evaluations need to take place. In spite of these challenges, co-cultured bioprocesses remain a promising approach for future applications (Mittermeier et al., 2023).
5. Advances in biotechnology have made it possible to develop microbial consortia for bioremediation.While bottom-up engineering reorganizes social relationships and metabolic processes, microbial engineering modifies environmental factors to carry out the desired objectives.Despite the fact intricate pollutants work well with bottom-up engineering, top-down engineering is more economical for basic pollutants. The division of labor, multifunctionality, and stress reduction are necessary concepts which play an important role in advancing engineering. Metabolic interaction between species needs to be the focus of future research in order to control microbial consortia for desired functions and long-lasting impacts. Limitations notwithstanding, engineered microbial consortia remain a promising approach for bioremediation (Li et al., 2021).

6.8 TRENDS

1. Synthetic biology holds significant potential for transforming the manufacturing sector by utilizing microbial consortia as foundational elements for feedstocks and processes. This advancement presents an opportunity for enhancing bio-based production, which is inherently more efficient and sustainable, aligning with circular economy principles. With the advent of sophisticated DNA manipulation and bio-foundry technologies, there is potential to reconfigure natural biosynthesis pathways, effectively turning microorganisms into efficient producers of valuable chemicals on demand. Additionally, feedstocks derived from renewable biomass offer an environmentally friendly alternative, significantly reducing reliance on fossil fuels. Nonetheless, the long-term sustainability of these innovations hinges on addressing various challenges that may arise (Mao et al., 2021).
2. Synthetic biology has significantly impacted the pharmaceutical industry by enhancing therapeutic manufacturing through the use of microbial consortia. The advent of recombinant DNA technologies has led to the development of an increased variety of biologics, including protein- and RNA-based therapeutic products. Furthermore, synthetic biology has revolutionized the

production of small molecule drugs like artemisinin and cannabinoids, which were traditionally extracted from plants. This innovation has facilitated the creation of a broader range of biologics and improved therapeutic options available in the market (Mao et al., 2021).

3. Synthetic biology facilitates the creation of innovative cell-based and gene therapies through the use of microbial consortia. A significant milestone in this field occurred in 2017 when the FDA approved the first chimeric antigen receptor (CAR)-T cell therapy, followed by the 2018 approval of the first viral vector-based gene therapy. The expansion of engineered immune cells is ongoing, focusing on incorporating natural killer (NK) cells and macrophages to enhance therapeutic effectiveness. Additionally, the engineering of bacteria is progressing, targeting treatment for skin diseases, gastrointestinal disorders, and systemic metabolic disorders. Bacteriophages are being recognized as powerful antimicrobial agents, with engineered phages being utilized as modulating agents to bolster chemotherapy effectiveness in cancer treatment. This combination of advancements demonstrates the potential of synthetic biology in developing cutting-edge therapies for various diseases (Mao et al., 2021).
4. Microbial consortia play a crucial role in advancing new therapeutic developments, particularly in response to the increasing prevalence of non-communicable diseases like diabetes, cancer, and cardiovascular issues. These health problems are exacerbated by urbanization, lifestyle changes, and an aging population. To address these challenges, local governments are fostering medical innovation aimed at alleviating healthcare burdens and fulfilling local healthcare requirements. They provide various forms of support, including capital investments, tax incentives, necessary infrastructure, and regulatory frameworks. In addition, there is a notable focus on developing technologies that tackle environmental and demographic challenges, exemplified by advancements in biofuel production. Complementing these efforts, investments are also directed toward food and agricultural technologies in order to cater to the expanding global population and adapt to climate change impacts (Mao et al., 2021).
5. Synthetic biology research cores have become prevalent across Asia, supported by eight members of the Global Bio-foundry Alliance and the Asian Synthetic Biology Association, which facilitate academic collaborations and the commercialization of technology. In 2019, venture investments in therapeutic biotech firms within the region exceeded US$2.6 billion, rivaling those in Europe and accounting for 40% of investment levels found in the United States. In terms of agricultural biotechnology, five Asian nations attracted approximately US$2.5 billion during the first half of 2020, falling short of the US$4.9 billion raised in the United Statessduring the same period. To foster synthetic biology innovation, governments are urged to transition from merely being supporters to becoming essential enablers of the industry (Mao et al., 2021).
6. The cross-border challenges primarily encompass economic development, the availability of natural resources, regulatory frameworks, and investment

strategies. To address these issues, fostering regional and cross-border collaboration, along with enhancing public-private partnerships, may be effective solutions. Additionally, synthetic biology represents a dual-use technology that promises significant benefits but also poses inherent risks. To facilitate the progress of synthetic biology while minimizing potential dangers, both regional and global collaboration are essential (Mao et al., 2021).

6.9 RECOMMENDATIONS

1. Synthetic biology research on microbial consortia requires investment in and technological advancement of DNA synthesis, computer-assisted design, biological data science, and ML.
2. Practices for the collection, curation, and sharing of biological sequencing data should be streamlined and harmonized globally.
3. Researchers involved in projects at an early stage of R&D should focus on the "end product" and develop economic models, downstream processing, and manufacturing scale-up with industry partners.
4. Integrated research, development, and innovation (R&D&I) ecosystem models are crucial for accelerating the translation and deployment of technology. These models aim to streamline the process through which innovations transition from conceptual stages to practical application, thereby enhancing the overall efficiency of technological advancements. By fostering collaboration among various stakeholders within the ecosystem, such as academic institutions, industry players, and government entities, these integrated models promote a more cohesive approach to innovation. The implementation of such models is expected to significantly contribute to more rapid and effective technology deployment in various sectors.
5. Governments should apply "push-and-pull" policy instruments to support the adoption of bio-based production through microbial consortia.
6. Stakeholder involvement is critical in the development of effective strategies and rules for biosecurity surveillance, risk mitigation, and response mechanisms. Engaging various stakeholders, including government agencies, private sectors, non-governmental organizations, and community members, ensures a collective approach to identifying risks and establishing surveillance protocols. By fostering collaboration among stakeholders, strategies can be tailored to meet specific regional needs and challenges, enhancing the overall resilience to biosecurity threats. Furthermore, stakeholder input in the formulation of response mechanisms is essential as it promotes transparency, accountability, and trust, which are pivotal during crisis situations. Continuous communication and feedback loops among stakeholders allow for adaptive strategies that respond effectively to emerging threats, thereby strengthening biosecurity efforts on a broader scale.
7. Cooperation and standard synchronization play a crucial role in enhancing biosecurity measures, which are essential for the advancement of synthetic biology. By fostering collaborative efforts among stakeholders and

establishing uniform standards, the field can significantly improve its protective strategies and mitigate potential risks associated with synthetic biological research and applications.

8. Strong technology translation and the commercialization of synthetic biology within microbial consortia have demonstrated promising initial advancements in both biomanufacturing and medicinal applications. These developments highlight the potential of engineered microbial systems to produce valuable bioproducts and facilitate innovative therapeutic solutions. The approach harnesses the diverse capabilities of microbial communities, allowing for enhanced efficiency and effectiveness in production processes. As the field progresses, the integration of technology translation practices is essential for maximizing the practical utility of synthetic biology in addressing various industry needs and improving health outcomes.
9. Further development is necessary in various fields such as industrial production, smart medicine, and environmental applications. This call emphasizes the importance of advancing technologies and practices that can enhance efficiency, improve health outcomes, and promote sustainability. The focus on industrial production indicates a need for innovation in manufacturing processes, potentially integrating smart technologies to optimize operations. In smart medicine, there is an opportunity to leverage advancements in data and analytics to personalize healthcare and improve patient care. Moreover, environmental applications are crucial for addressing ecological challenges, necessitating collaboration across sectors to implement sustainable solutions. Overall, fostering growth in these areas is imperative for achieving economic and environmental benefits.

6.10 CONCLUSION

Microbial consortia represent a paradigm shift in state-of-the-art biotechnology for efficient, robust, and sustainable biosynthetic production beyond the limitations imposed by monoculture systems. They are capable of self-coordinating on different metabolic processes in a complementary manner, thus enhancing tolerance against hazardous intermediates or end products, reducing metabolic bottlenecks, and allowing better utilization of substrates. They can adapt to environmental changes, thus embedding resilience that is difficult to mimic using monocultures. Design and optimization of consortia require deep knowledge of microbial interactions, including commensalism, competition, and mutualism, and how these interactions impact resource distribution, exchange of metabolites, and stability of the community. These need to be integrated with metabolic modeling, systems biology, synthetic biology, and bioprocess engineering approaches. Recent progress has been made in the ability to tune microbial behavior and communications through the advancement of QS, modular gene expression, and genetic circuit design, allowing better control over the function of consortia. Different microbial consortia have been successfully applied to produce a wide variety of industrial chemicals, which enhances their prospects in the green and circular bioeconomy. Community stability, population dynamics, cross-talk, or metabolic leakage still presents challenges for ensuring performance stability in an industrial environment. Further efforts should be directed at interdisciplinary research and innovation to fill these gaps.

REFERENCES

Abedi, E., & Hashemi, S. M. B. (2020). Lactic acid production–producing microorganisms and substrates sources-state of art. *Heliyon*, *6*(10), e04974.

Ahn, J. H., Jang, Y.S., & Lee, S. Y. (2016). Production of succinic acid by metabolically engineered microorganisms. *Current Opinion in Biotechnology*, *42*, 54–66.

Bhatia, S. K., Bhatia, R. K., Choi, Y.K., Kan, E., Kim, Y.G., & Yang, Y.H. (2018). Biotechnological potential of microbial consortia and future perspectives. *Critical Reviews in Biotechnology*, *38*(8), 1209–1229. https://doi.org/10.1080/07388551.2018.1471445

Bhatia, S. K., Yi, D.H., Kim, Y.H., Kim, H.J., Seo, H.M., Lee, J.H., Kim, J.H., Jeon, J.M., Jang, K.S., & Kim, Y.G. (2015). Development of semi-synthetic microbial consortia of Streptomyces coelicolor for increased production of biodiesel (fatty acid methyl esters). *Fuel*, *159*, 189–196.

Cao, Z., Yan, W., Ding, M., & Yuan, Y. (2022). Construction of microbial consortia for microbial degradation of complex compounds. *Frontiers in Bioengineering and Biotechnology*, *10*, 1051233. https://doi.org/10.3389/fbioe.2022.1051233

Carlson, R. P., Beck, A. E., Phalak, P., Fields, M. W., Gedeon, T., Hanley, L., Harcombe, W. R., Henson, M. A., & Heys, J. J. (2018). Competitive resource allocation to metabolic pathways contributes to overflow metabolisms and emergent properties in cross-feeding microbial consortia. *Biochemical Society Transactions*, *46*(2), 269–284.

Coats, E. R., Watson, B. S., & Brinkman, C. K. (2016). Polyhydroxyalkanoate synthesis by mixed microbial consortia cultured on fermented dairy manure: Effect of aeration on process rates/yields and the associated microbial ecology. *Water Research*, *106*, 26–40.

Cui, C., Jiang, M., Zhang, C., Zhang, N., Jin, F.J., Li, T., Lee, H.G., & Jin, L. (2023). Assembly strategies for rubber-degrading microbial consortia based on omics tools. *Frontiers in Bioengineering and Biotechnology*, *11*, 1326395.

Datta, D., Ghosh, S., Kumar, S., Gangola, S., Majumdar, B., Saha, R., Mazumdar, S. P., Singh, S. V., & Kar, G. (2024). Microbial biosurfactants: Multifarious applications in sustainable agriculture. *Microbiological Research*, *279*, 127551.

Delgado, A., & Porcar, M. (2013). Designing de novo: Interdisciplinary debates in synthetic biology. *Systems and Synthetic Biology*, *7*, 41–50.

Duncker, K. E., Holmes, Z. A., & You, L. (2021). Engineered microbial consortia: Strategies and applications. *Microbial Cell Factories*, *20*(1), 13.

Fang, H. H., Zhu, H., & Zhang, T. (2006). Phototrophic hydrogen production from glucose by pure and co-cultures of *Clostridium butyricum* and *Rhodobacter sphaeroides*. *International Journal of Hydrogen Energy*, *31*(15), 2223–2230.

García-Solares, S. M., Mena-Cervantes, V. Y., Sosa-Rodríguez, F. S., Hernández-Altamirano, R., & Vazquez-Arenas, J. (2023). Circular economy involving microbial consortia in consolidated bioprocesses to produce biofuels. In *Biofuels in circular economy* (pp. 279–301). Springer. Edited by: Suhaib A. Bandh and Fayaz A. Malla. https://doi.org/10.1007/978-981-19-5837-3_15

Gautam, S. S., Gondwal, M., Soni, R., & Gautam, B. P. S. (2022). Potash biofertilizers: Current development, formulation, and applications. In *Trends of applied microbiology for sustainable economy* (pp. 481–500). Elsevier. Edited by: Ravindra Soni, Deep Chandra Suyal, Ajar Nath Yadav and Reeta Goel. https://doi.org/10.1016/B978-0-323-91595-3.00014-8

González, C., Wu, Y., Zuleta-Correa, A., Jaramillo, G., & Vasco-Correa, J. (2021). Biomass to value-added products using microbial consortia with white-rot fungi. *Bioresource Technology Reports, 16*, 100831.

Hirasawa, T., & Shimizu, H. (2016). Recent advances in amino acid production by microbial cells. *Current Opinion in Biotechnology, 42*, 133–146.

Hossain, G. S., Liang, Y., Foo, J. L., & Chang, M. W. (2024). Engineered microbial consortia for next-generation feedstocks. *Biotechnology Notes, 5*, 23–26.

Huang, Y., Wang, Y., Shang, N., & Li, P. (2023). Microbial fermentation processes of lactic acid: Challenges, solutions, and future prospects. *Foods, 12*(12), 2311.

Jia, X., Liu, C., Song, H., Ding, M., Du, J., Ma, Q., & Yuan, Y. (2016). Design, analysis and application of synthetic microbial consortia. *Synthetic and Systems Biotechnology, 1*(2), 109–117.

Jiang, L.L., Zhou, J.J., Quan, C.S., & Xiu, Z.L. (2017). Advances in industrial microbiome based on microbial consortium for biorefinery. *Bioresources and Bioprocessing, 4*(1), 10.

Lahiri, D., Nag, M., Ghosh, S., Dey, A., & Ray, R. R. (2022). Microbial consortium and crop improvement: Advantages and limitations. In*Trends of applied microbiology for sustainable economy* (pp. 109–123). Elsevier. Edited by: Ravindra Soni, Deep Chandra Suyal, Ajar Nath Yadav and Reeta Goel. https://doi.org/10.1016/B978-0-323-91595-3.00016-1

Li, X., Wu, S., Dong, Y., Fan, H., Bai, Z., & Zhuang, X. (2021). Engineering microbial consortia towards bioremediation. *Water, 13*(20), 2928. https://www.mdpi.com/2073-4441/13/20/2928

Lu, J., Li, J., Gao, H., Zhou, D., Xu, H., Cong, Y., Zhang, W., Xin, F., & Jiang, M. (2021). Recent progress on bio-succinic acid production from lignocellulosic biomass. *World Journal of Microbiology and Biotechnology, 37*(1), 16. https://doi.org/10.1007/s11274-020-02979-z

Lyu, X., Nuhu, M., Candry, P., Wolfanger, J., Betenbaugh, M., Saldivar, A., Zuniga, C., Wang, Y., & Shrestha, S. (2024). Top-down and bottom-up microbiome engineering approaches to enable biomanufacturing from waste biomass. *Journal of Industrial Microbiology and Biotechnology, 51*, kuae025.

Mao, N., Aggarwal, N., Poh, C. L., Cho, B. K., Kondo, A., Liu, C., Yew, W. S., & Chang, M. W. (2021). Future trends in synthetic biology in Asia. *Advanced Genetics, 2*(1), e10038.

Marsafari, M., Azi, F., Dou, S., & Xu, P. (2022). Modular co-culture engineering of *Yarrowia lipolytica* for amorphadiene biosynthesis. *Microbial Cell Factories, 21*(1), 279. https://doi.org/10.1186/s12934-022-02010-0

Martin-Dominguez, V., Bouzas-Santiso, L., Martinez-Peinado, N., Santos, V. E., & Ladero, M. (2020). Kinetic modelling of the coproduction process of fumaric and malic acids by *Rhizopus arrhizus* NRRL 1526. *Processes, 8*(2), 188. https://www.mdpi.com/2227-9717/8/2/188

Merli, G., Becci, A., Amato, A., & Beolchini, F. (2021). Acetic acid bioproduction: The technological innovation change. *Science of the Total Environment, 798*, 149292.

Mishra, M., Singh, S. K., & Kumar, A. (2021). Microbial consortia: Approaches in crop production and yield enhancement. In*Microbiome stimulants for crops* (pp. 293–303). Elsevier. Edited by: James White, Ajay Kumar and Samir Droby. https://doi.org/10.1016/B978-0-12-822122-8.00013-3

Mittermeier, F., Bäumler, M., Arulrajah, P., García Lima, J. J., Hauke, S., Stock, A., & Weuster-Botz, D. (2023). Artificial microbial consortia for bioproduction processes. *Engineering in Life Sciences, 23*(1), e2100152.

Nagarajan, K., Ni, C., & Lu, T. (2022). Agent-based modeling of microbial communities. *ACS Synthetic Biology, 11*(11), 3564–3574.

Nigam, P. S. (2013). Microbial enzymes with special characteristics for biotechnological applications. *Biomolecules, 3*(3), 597–611. https://www.mdpi.com/2218-273X/3/3/597

Pacheco, G. J., Ciapina, E. M. P., Gomes, E. d. B., & Pereira Junior, N. (2010). Biosurfactant production by *Rhodococcus erythropolis* and its application to oil removal. *Brazilian Journal of Microbiology*, *41*, 685–693.

Palladino, F., Marcelino, P. R. F., Schlogl, A. E., José, Á. H. M., Rodrigues, R. d. C. L. B., Fabrino, D. L., Santos, I. J. B., & Rosa, C. A. (2024). Bioreactors: Applications and innovations for a sustainable and healthy future—A critical review. *Applied Sciences*, *14*(20), 9346. https://www.mdpi.com/2076-3417/14/20/9346

Paul, D., Bohacz, J., & Bhatia, S. K. (2023). Biowaste valorization utilizing microbial systems. *Frontiers in Microbiology*, *14*,. 1213598.

Roell, G. W., Zha, J., Carr, R. R., Koffas, M. A., Fong, S. S., & Tang, Y. J. (2019). Engineering microbial consortia by division of labor. *Microbial Cell Factories*, *18*(1), 11.

Ruiz-Marín, A., Canedo-López, Y., Narváez-García, A., & Robles-Heredia, J. C. (2016). Production of ethanol by *Saccharomyces cerevisiae* and *Zymomonas mobilis* coimmobilized: Proposal for the use of organic waste. *Agrociencia*, *50*(5), 551–563.

Schmidt, S. (2021). Chapter 13 Applications of mixed microbial cultures in industrial biotechnology. In *The autotrophic biorefinery: Raw materials from biotechnology*. De Gruyter. Edited by Robert Kourist and Sandy Schmidt. https://doi.org/10.1515/9783110550603-013

Shou, W., Ram, S., & Vilar, J. M. (2007). Synthetic cooperation in engineered yeast populations. *Proceedings of the National Academy of Sciences*, *104*(6), 1877–1882.

Singh, N., & Das, P. P. (2024). Microbial consortium as promising biostimulants for plant health: A future perspective for agriculture. In *Microbiome drivers of ecosystem function* (pp. 123–143). Elsevier. Editors: Javid A. Parray, Nowsheen Shameem and Dilfuza Egamberdieva. https://doi.org/10.1016/B978-0-443-19121-3.00019-3

Stephens, K., & Bentley, W. E. (2020). Synthetic biology for manipulating quorum sensing in microbial consortia. *Trends in Microbiology*, *28*(8), 633–643.

Stewart, E. J. (2012). Growing unculturable bacteria. *Journal of Bacteriology*, *194*(16), 4151–4160.

Wang, Y., Liu, L., Jin, Z., & Zhang, D. (2021). Microbial cell factories for green production of vitamins. *Frontiers in Bioengineering and Biotechnology*, *9*, 661562.

Wang, S., Xu, C., Song, L., & Zhang, J. (2022). Anaerobic digestion of food waste and its microbial consortia: A historical review and future perspectives. *International Journal of Environmental Research and Public Health*, *19*(15), 9519.

Yu, E., Chan, M.H., & Saddler, J. (1985). Butanol production from cellulosic substrates by sequential co-culture of *Clostridium thermocellum* and *C. acetobutylicum*. *Biotechnology Letters*, *7*, 509–514.

7 Sustainable Biohydrogen Production from Waste via Dark Fermentation and Its Efficient Disposal

Ashwini S. Thakre, Diwakar Z. Shende, and Kailas L. Wasewar

7.1 INTRODUCTION

The heavy and extended use of fossil energy has also led to the extensive buildup of greenhouse gases and the degradation of natural ecosystems. International energy is responding to this by moving slowly to renewable energies in the form of wind, solar, hydropower, and bio-energy, which are long-term, sustainable, and environmentally friendly (Dincer, 2000). Advances in grid technologies, storage systems, and distributed energy systems have also contributed to the incorporation of these options into the national power networks (Panwar et al., 2011). Biofuels have a special place among renewable resources as they may be obtained with the help of a great variety of organic substances: agricultural wastes, woody biomass, and algae. Moreover, bioenergy contributes to waste management in numerous areas, as well as providing the local area with energy independence (Liu et al., 2021). Therefore, the current energy crisis is not only an environmental problem, but it is also closely related to the economic, geopolitical, and social welfare (Solomon, 2010).

Hydrogen is entering the picture as part of this cleaner energy transition. It has a high-energy density and involves zero-carbon combustion, which makes it an appealing alternative to industries and sectors that cannot be electrified easily. To make hydrogen a sustainable energy source in the future, however, it must be produced using a low-carbon or renewable source. The area of biological production of hydrogen or biohydrogen is one of the ways to meet this demand and turn organic waste streams into clean fuel with the help of microorganisms (Mokhtarani et al., 2025). With the depletion of traditional fossil fuel reserves and the increase in environmental challenges, renewable solutions like biohydrogen are being considered quite viable in the long term.

Biomass-produced biohydrogen has a number of advantages in comparison to hydrogen produced by fossil-based methods, which involve energy-intensive steam reforming to achieve the final product. It works in mild conditions, has a lesser amount of energy input, and provides an alternative that is environmentally

DOI: 10.1201/9781003642589-7

friendly (Hawkes et al., 2007). Moreover, it promotes waste management through materials of food waste, agricultural residual, and wastewater sludge (Wang & Wan, 2009). It is also consistent with the principles of a circular economy (CE) since waste is sold to transform into useful energy and it will decrease environmental burdens.

The use of hydrogen is not limited to production. It can be stored, transported, and used in fuel cells to produce electricity, or it can be used as a clean industrial fuel. The flexibility of this makes it a good candidate to be widely adopted in the world's decarbonization strategies. Despite the fact that existing world hydrogen production depends mainly on fossil fuels, with a small percentage being electrolyzed or deriving bio-mass (Sinha & Pandey, 2011), developments in the microbial process and reactor construction have given biohydrogen an even greater push to be considered as a viable alternative.

To conclude, the transition to the use of sustainable energy means that people should not just use more renewable resources but also work on the creation of clean energy sources, which can be used to provide energy to a wide range of demands. Biohydrogen can play this purpose by using processes of microbes to transform waste into valuable energy of value. Its versatility, green nature, and the ability to complement current systems make it one of the prospects in providing a contribution toward energy horizons of the future. Further research and development will see biohydrogen become more prominent in global energy solutions as a technology.

7.2 BIOHYDROGEN PRODUCTION METHODS

Various biological pathways may be used to produce biohydrogen, though each of them is based on a particular microbial activity and environmental conditions. Dark fermentation, photofermentation, biophotolysis, and microbial electrolysis cells (MECs) are the four significant routes that have been reported in the literature. Of them, dark fermentation is particularly distinguished due to its comparatively high hydrogen production rates, simplicity of operation, and the ability to work with a variety of organic feedstock, including low-value waste streams (Srivastava et al., 2024).

Photofermentation is a photosynthetic process in which organic acids are changed to hydrogen and carbon dioxide by photosynthetic bacteria with the use of light energy. The key role is taken by enzymes of nitrogenase, which reduce nitrogen and protons simultaneously in the conditions of light (Łukajtis et al., 2018). The available light, pH, and temperature have a strong impact on the performance of this process.

Biophotolysis, a process that depends on algae and cyanobacteria, is a process that depends on sunlight or artificial light to cleave water molecules into hydrogen and oxygen. These reactions involve the enzymes hydrogenase and nitrogenase (Akhlaghi & Najafpour-Darzi, 2020; Javed et al., 2022). Despite its effectiveness under laboratory conditions, biophotolysis is vulnerable to the buildup of oxygen, which inactivates the main enzymes, and thus it is not easy to put biophotolysis into practice (Azwar et al., 2014).

One more technological advancement is MECs. In such systems, the oxidation of organic matter takes place at the anode by microorganisms that release electrons and protons. The electrons move to the cathode with the assistance of a small external voltage to produce hydrogen (Ndayisenga et al., 2021). MECs are highly conversion efficient, but are limited with respect to cost, electrode material stability, and scale-up.

Dark fermentation is the opposite of the latter methods in that it does not need light. Rather, carbohydrates and other biodegradable substances are degraded by anaerobic bacteria in the absence of oxygen. It is done at a low temperature of mesophilic (25–45°C) to thermophilic (50–65°C) or even extreme thermophilic (above 80°C) temperature (Mokhtarani et al., 2025). The dark fermentation is a widely studied process owing to its ability to provide quick growth of microorganisms, constant generation of hydrogen, as well as efficient treatment of organic matter. Sugar-rich crops, agro-industrial residues, or food waste have all been successfully used as feedstocks (Mokhtarani et al., 2025). Nonetheless, the process also generates gases like CO_2, H_2S, and CO, which might need purification.

Photofermentation, biophotolysis, and MECs can potentially achieve high levels of hydrogen in the laboratory, but dark fermentation is the most feasible and scalable. This is because it is relatively high in its hydrogen concentrations, has low-energy input, and can process various waste streams (Balachandar et al., 2020). The developed state of dark fermentation has made it possible to proceed with the advancement of inoculum, reactor, and operational optimization (Ghimire et al., 2015). However, demonstrations are mostly at the laboratory level, and pilot-scale systems are relatively uncommon (Kotay & Das, 2009).

Other feedstocks, including lignocellulosic biomass and agricultural residues, have high potential, yet they can frequently be subjected to pretreatment to extract fermentable sugars (Mokhtarani et al., 2025). In order to prevent competition with food production sectors, waste-based substrates at low costs are favored over crops based on sugar. This has led to the emergence of more research on low-cost biomass sources to ferment in the dark (Hallenbeck & Benemann, 2002).

Research shows that process efficacy is highly affected by such operational conditions as substrate concentration, temperature, and hydraulic retention time (Łukajtis et al., 2018). Mixed cultures are stronger and do not need sterilization as compared to the pure ones, whereas pure cultures can be used in large processes with higher process control (Mohanakrishna & Pengadeth, 2024).

In the world, issues pertaining to exhaustion of fossil fuels, emission of greenhouse gases, and increasing energy demands have been escalating. The International Energy Agency estimates that fossil fuels continue to contribute to about three-quarters of CO_2 emissions in the world (Mokhtarani et al., 2025). The combination of these environmental effects and geopolitical instability, and population growth has prompted the development of sustainable energy sources.

Biohydrogen derived from organic wastes is a good source of reducing such environmental stresses, besides solving the waste disposal problems. Figure 7.1 demonstrates the variety of available microbial processes by presenting various biological hydrogen production pathways.

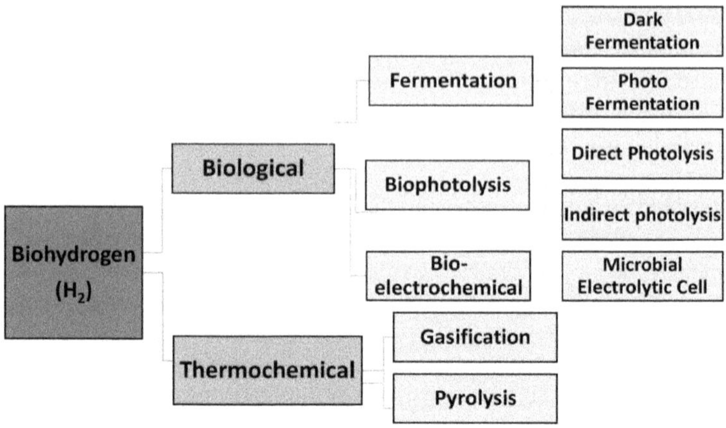

FIGURE 7.1 Biological methods used for the production of biohydrogen.

7.3 DARK FERMENTATION

Dark fermentation refers to a biological process, in which microorganisms break down organic materials through fermentation, to produce hydrogen gas, under anaerobic and light-independent conditions. The process involves the metabolic activity of anaerobic bacteria, mainly the species of the genus *Clostridium*, which utilize carbohydrates and other biodegradable substances to produce hydrogen and organic acids (Kumar et al., 2015). In contrast to dark fermentation, photo fermentation and biophotolysis are light-dependent processes and therefore, require the presence of light. Dark fermentation can take place without any light and can thus be carried out indefinitely, making it more suitable for large-scale or industrial applications. The method finds its use as the most efficient way to generate hydrogen in a sustainable manner, when agricultural residue, food waste, or other low-value organic by-products are the starting materials (Mokhtarani et al., 2025).

7.3.1 Mechanism of Dark Fermentation

The dark fermentation process is initiated by the microbial breakdown of sugars, mainly glucose, through the glycolytic metabolism. In the process of glycolysis, glucose molecules are dismantled to yield pyruvate, which becomes a central pathway in other fermentation pathways. The total yield of hydrogen depends on these pathways because various reactions of microbes yield varying amounts of hydrogen, carbon dioxide (CO_2), and organic acids like acetate, butyrate, or lactate (Rangel et al., 2021).

One of the principal reactions associated with hydrogen generation follows the acetate pathway:

$$\text{Glucose} - 2\,\text{Acetate} + 2\,CO_2 + 4\,H_2$$

This particular hydrogen-production process is the most efficient method in terms of the amount of hydrogen produced. Other processes, including the butyrate pathway, generate less hydrogen per mole of substrate because they require more reducing equivalents. The dominance of the different pathways is, to a large extent, determined by various environmental factors such as substrate concentration, pH, inoculum type, hydrogen partial pressure, and temperature. The accumulation of volatile fatty acids (VFAs), like that, in an uncontrolled manner can not only inhibit the production of hydrogen along with bacterial growth, but the latter process may also be inhibited until management is properly carried out (Ghimire et al., 2015; Wang & Wan, 2009).

Different metabolic pathways for the production of hydrogen are represented in Figure 7.2. After the biomass pretreatment that is supposed to liberate the fermentable sugars such as glucose, the sugars follow the glycolysis pathway to yield pyruvate, which can undergo various metabolic pathways (Palanivel et al., 2025). One of the major hydrogen-producing pathways is pyruvate: ferredoxin oxidoreductase (PFOR), where, with the aid of hydrogenase activity, the electron carrier reduction leads to the liberation of hydrogen together with the formation of acetyl-CoA and CO_2 (Mokhtarani et al., 2025). Moreover, pyruvate can be changed to formate via the enzyme pyruvate formate lyase (PFL). Then formate can be changed into CO_2 and H_2 by formate-hydrogen lyase, thus making the hydrogen production quantity higher (Ghimire et al., 2015). If things were to get worse, pyruvate might be converted into the lactate pathway, which would consume reducing equivalents and thus lead to less hydrogen being produced (Stabel et al., 2022). Besides, acetyl-CoA can be changed into one of the four: acetate, ethanol, butyrate, or butanol. Indeed, the generation of acetate is related to the maximum hydrogen production, whereas the production

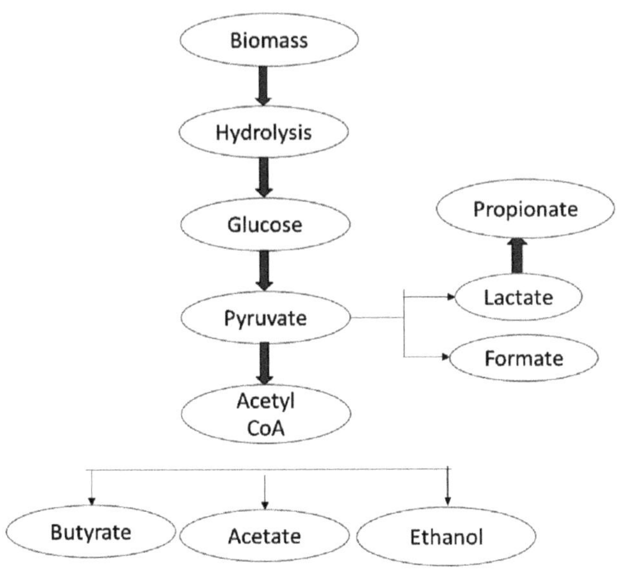

FIGURE 7.2 Metabolic pathways involved in biohydrogen production through dark fermentation.

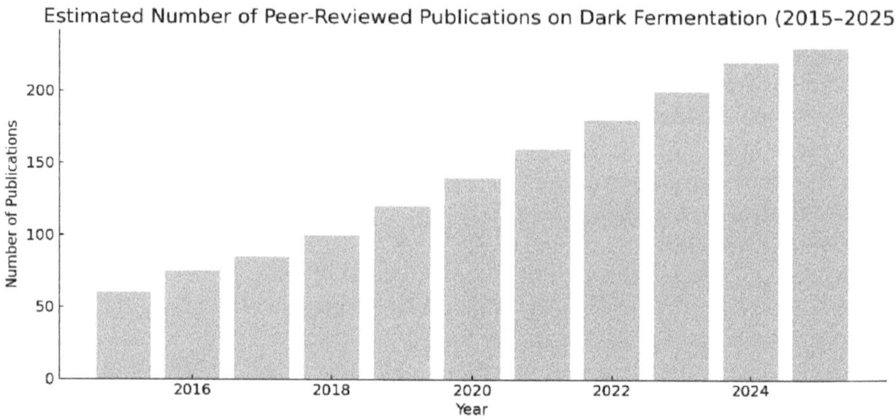

FIGURE 7.3 Estimated number of peer-reviewed publications on dark fermentation (2015–2025).

of solvents such as butanol or ethanol leads to almost no hydrogen being released (Akhlaghi & Najafpour-Darzi, 2020).

Essentially, the question of dark fermentation turning out to be successful is mainly related to the ability to control microbial metabolism in such a way as to favor the hydrogen-generating pathways. When hydrogen production is desired along with the necessity to keep the system stable, then appropriate measures in regard to pH, temperature, microbial structure, and substrate loading should be taken. The graph in Figure 7.3 showing current research trends in this field, also indicates that the number of publications has been steadily increasing over the years from 2015 to 2025, which is due to both growing scientific interest and technological changes. Besides these, the main reasons behind this increase are the clean energy and waste management demands and the necessity for more efficient bioreactor systems (Das and Veziroğlu, 2001; Lee et al., 2011; Vidal et al., 2025).

7.4 PROCESS OPTIMIZATION OF DARK FERMENTATION

A combination of interconnected operation parameters is critical to the performance of dark fermentation systems and thus the quantity of hydrogen generated. These encompass the nature of the inoculum, the pH and the temperature of the medium, the type and concentration of substrate, and system loading as a whole (Ayodele et al., 2023). These variables should be optimized since they affect the structure of microbial communities, fermentation metabolism, fermentation stability, and the allocation of by-products. The microbial activity is mainly shaped by the inoculum. The commonly used pretreatment methods include heat-based conditioning and acid-based conditioning, which reduce undesirable hydrogen consumers, especially methanogens, and enrich hydrogen producers, respectively (Tiegam Tagne et al., 2024). Similarly, the material composing the substrate and the rate at which the substrate is loaded with organic matter are heavily influential aspects that affect the yield of hydrogen through the nutrient availability and fermentability. The pH and

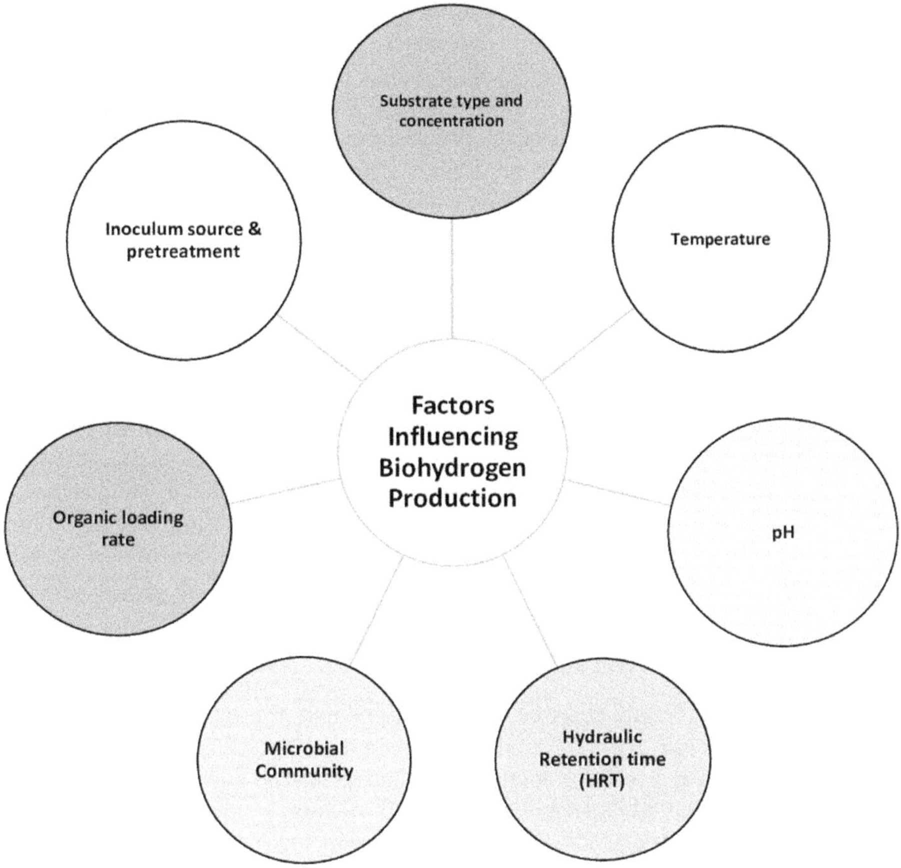

FIGURE 7.4 Factors influencing hydrogen production via dark fermentation.

temperature are also very important as they influence enzyme activity and the balance of metabolic pathways. Acidic conditions of a low pH usually favor volatile acid formation pathways leading to hydrogen generation, whereas unsuitable environmental conditions can cause a change in metabolism toward the hydrogen-retaking or solvent-producing path (Alavi-Borazjani et al., 2021; Ziara et al., 2019). A combination of all these parameters needs to be well coordinated in order to ensure that the hydrogen output is maximized. Figure 7.4 presents a conceptual design of the main aspects that influence the performance of fermentation in darkness.

7.4.1 Inoculum: Source, Pretreatment, and Ratio

The microbial inoculum is the primary factor that determines the success of the dark fermentation process, as it basically defines the diversity and the activity of the organisms that produce hydrogen. Usually, the sources of inoculum are the sludge from an anaerobic digester, the manure of farm animals, compost, and the effluent of a food waste processing plant. All of these are the places that naturally contain mixed microbial consortia that are dominated by several species, among which *Clostridium* is the one that is very likely to be the most efficient hydrogen producer (Singh et al., 2023).

It is a common practice to carry out pretreatment of an inoculum to obtain more hydrogen. One of such things to be done is thermal shock, which, among other things, kills methanogens and different non-hydrogen-producing microbes, thus leaving hydrogen producers that are able to form spores. Similar to the case in acid or alkaline conditions, pretreatment can also change community composition to the ones that are most favorable organisms (Machhirake et al., 2024). It has also been proven that simultaneously pretreating of inoculum and substrate leads to a substantial increase in hydrogen production in some situations (Alavi-Borazjani et al., 2024).

The ratio of inoculum to substrate (ISR) is just as important as other considerations. A ratio, which is very low, might lead to insufficient microbial activity. On the other hand, a very high ratio may result in the excess formation of biomass rather than hydrogen. Most of the time, the maximum level of hydrogen production takes place at 0.2 and 0.4 (Ahmad et al., 2024), but the optimum ratio depends on the substrate complexity and the design of the reactor.

7.4.2 pH: Directing Microbial Pathways

pH is the main factor that changes the dark fermentation process. In fact, it is the parameter that decides the microbial metabolism way. Typically, the effort to hold the initial pH around 5.5–6.0 is vital for graphene paths, in particular, those resulting in acetate and butyrates formation. The acid medium is good for the competing pathways, such as lactate production, which uses reducing power and thus decreases hydrogen yield (Ahmed et al., 2021).

Various experiments have demonstrated that pH control within this physiological range not only results in higher hydrogen production but also lessens the accumulation of inhibitory VFAs.

7.4.3 Temperature: Microbial Activity and Process Stability

Temperature is one of the major factors that changes the metabolic rate of microbes drastically. It also influences the activity of hydrogenase as well as other enzymes that are released during fermentation. The majority of hydrogenase enzymes in bacteria producing hydrogen show their highest activity at normal temperature ranges (35–40°C), and it is a compromise between microbial stability and hydrogen productivity (Zhou et al., 2023).

If the thermophilic temperatures are not too high (50–60°C or more), they can cause the reaction to be faster and the substrate to be broken down with ease. Nevertheless, the production of VFAs may become so high that these acids can cause the inhibition of hydrogen-producing microbes. Gao et al. (2024) demonstrated that not only are thermophilic strains like *Clostridium thermocellum* tolerant, but they can also efficiently digest lignocellulosic substrates such as Avicel;

therefore, the development of thermophilic systems with a complex substrate becomes feasible.

Therefore, the temperature regulation depends on the feedstock, the hydrogen yield target, and the system arrangement.

7.4.4 Substrate Composition and Organic Loading Rate

Chemical composition and concentration of the substrate are the substrate components that have the most significant influence on hydrogen production. Glucose and sucrose are simple sugars that can be very easily fermented, and as a rule, they lead to high hydrogen yields. However, complex substrates, particularly lignocellulosic biomasses, have to go through pretreatment processes, thus allowing cellulose, hemicellulose, or lignin to be broken down into fermentable sugars (Zhang et al., 2024). Besides that, the organic loading rate (OLR) should be kept within reasonable limits. Extremely high OLRs can lead the microbial community to become heavily stressed, thus acidification will be very rapid, and hydrogen production will decline. On the other hand, very low OLRs can cause productivity to be limited. As a result, maximum hydrogen production was recorded for OLRs ranging from 22 to 120 g COD/L·day, and this depends on the reactor configuration and the hydraulic retention time (Patel et al., 2024). Under ideal conditions, including a pH value between 5.5 and 6.0 and a mesophilic temperature of 37°C, hydrogen yields of up to 40 L H_2/L·day have been achieved.

7.4.5 Integration with Advanced Systems

In the recent studies, researchers looked for various means to improve dark fermentation by linking it to other biological or electrochemical systems. One such method is the dark fermentation merged with photofermentation, in which organic acids can be further transformed into hydrogen, thus the total fuel recovery is increased (Al-Mamun et al., 2020). Another approach is the integration of dark fermentation with MECs that is capable of providing additional hydrogen release if a slight voltage is applied to the system in order to induce electrochemical reactions.

Further, the improvements in machine learning (ML) and artificial intelligence (AI) have the potential to affect dark fermentation optimally. These instruments can forecast the changes in the buildup of VFAs, rates of hydrogen production, and the stability of the system, thus they allow the immediate intervention to get the process efficient (Wang et al., 2024).

7.4.6 Nanotechnology and Membrane Bioreactors

With the help of nanotechnology, there is an immense scope to improve the membrane bioreactor (MBR) system performance in dark fermentation. It has been found by the researchers that the introduction of nanomaterials like metal oxide nanoparticles or carbon-based nanostructures into the membrane makes the membrane antifouling, increases the permeability, and makes the attachment of the microorganism more favorable (García-Depraect et al., 2025).

MBRs in biohydrogen generation are capable of holding the hydrogen-producing microorganisms in the reactor for a longer time; thus, substrate utilization and hydrogen yield are improved. Nanotechnology combined with these systems can accomplish even more work by enabling microbial activity to a higher extent and lessening membrane clogging (Biswal et al., 2023). Such a technique is in harmony with the grand objectives of the circular bioeconomy and the integration of waste-to-energy systems.

7.5 ENVIRONMENTAL AND ECONOMIC ASPECTS OF DARK FERMENTATION

Before we can think of dark fermentation as a sustainable hydrogen production method, one of the major things to be grasped is how the dark fermentation process impacts the environment and the economy. The environmental side of the dark fermentation is certainly green as it employs organic waste (a common source of environmental pollution) to generate clean energy. On the other hand, the economic feasibility of such a method depends mostly on the energy balances, hydrogen yield, reactor performance, feedstock availability, and the cost of the downstream processing, among other factors.

Their life cycle assessment (LCA) methods are implemented to find the sources of greenhouse gases, the consumption of resources, and the overall environmental footprint of systems based on dark fermentation (Yin et al., 2024). These three together are conditions for changes in system design, process improvement, and policy development.

7.5.1 Energy Conversion Characteristic Indices

An energy conversion analysis is the primary focus of the work that seeks to determine how dark fermentation can be utilized in an energy-efficient way to produce hydrogen from waste biomass. Such assessments take into account a range of energies: the chemical energy of the substrate, the physical energy of, for example, heating or stirring, and the energy that is lost due to various inefficiencies like friction or heat dissipation, etc.

A system of various performance indices is used to present the overall performance of the system. These include:

- Total and net energy input/output
- Energy contributions from the recycling process
- Hydrogen production efficiency
- Energy recovery and conversion ratios
- The net energy ratio that relates total output to total energy consumed

As the generation of hydrogen is the principal objective of the process, the efficiency and purity of the hydrogen output are considered the most important performance indicators. More definitions, formulas, and related background information for these indices can be found in the supplementary materials accompanying this chapter.

7.5.2 Sensitivity Analysis

Sensitivity analysis serves as a main methodological instrument in determining the parameters that have the major influence on system performance. For instance, in the case of dark fermentation, the life cycle inventory (LCI) precision depends on factors like fermentation efficiency, reactor operation, and organic waste composition. Changes in these variables will have a direct impact on energy consumption, carbon emissions, and total hydrogen productivity.

Through the alteration of single input parameters and the maintenance of other parameters at a constant level, sensitivity analysis makes it possible for scientists to locate the variables that have the greatest effect on environmental or energy conversion results. The use of this method is necessary for the creation of strong and flexible designs, particularly when there is a need for system scaling or the integration of several processes such as photofermentation, microbial electrolysis, or anaerobic digestion.

7.5.3 Economic Viability and Scale-Up Barriers of Dark Fermentation–Based Biohydrogen Production

Dark fermentation is generally seen as an excellent method to generate hydrogen from organic waste materials. Nevertheless, its implementation on a large scale is still limited due to numerous barriers of both economic and technical nature.

7.5.3.1 Capital and Operational Costs

The profitability of dark fermentation is basically a function of CAPEX (capital expenditures) and OPEX (operational expenditures). The investment usually covers bioreactors, the units for pretreatment, the devices for gas purification, and the systems for storage. As per Wang and Wan (2009), capital expenses for the pilot-scale dark fermentation installation are typically between USD 1,200 and USD 2,000 per cubic meter of the bioreactor volume. The operating costs are combined with activities such as substrate handling, heating, and maintenance, and services provided by personnel and control systems.

The feedstock cost is extremely important. Running the process on low-cost or negative-cost feedstocks such as municipal solid waste, sewage sludge, or food waste would make the project economically viable. The placement of fermentation units near facilities for waste treatment is beneficial in that both can save the operating costs, but the variability of feedstock may cause a lack of stability in the process.

7.5.3.2 Hydrogen Yield and Biological Limitations

Usually, hydrogen yields are limited by biological pathways and are within the range of 1.0–2.5 mol H_2 per mol hexose (Matsakas et al., 2017). Pathways that generate alcohols or VFAs instead of hydrogen also exist, thus cutting the yield of hydrogen (Kapdan & Kargi, 2006). Though the use of co-cultures, genetic manipulation, and pH control may bring about improvements in performance, the drawbacks of these methods are that they increase the complexity and cost of the system.

7.5.3.3 Gas Separation and Purification Requirements

Hydrogen obtained from dark fermentation is a mix of carbon dioxide, hydrogen sulfide, and small quantities. To provide the fuel cells with pure hydrogen, separation procedures such as pressure swing adsorption (PSA), membrane separation, or cryogenic units are used (Dufour et al., 2012). The purification operations that must be followed in a process raise the energy consumption and the cost of installing the plant.

7.5.3.4 Scale-Up Challenges

The commercial scale of the work done in a laboratory is not directly hand-overable, but rather it comes with challenges that must be resolved first:

- **Reactor stability:** large-scale operations require keeping the same levels of microbial activity that is stable over time.
- **Uniformity:** proper regulating of parameters like pH, temperature, and mixing in a uniform manner is difficult in large volumes.
- **Feedstock variability:** high-solids feedstocks can clog some parts of the system or interfere with gas release, causing problems.
- **Process integration:** as a result of added infrastructure and the need for more skilled labor, combined systems like DF + photo fermentation or DF + MEC come with disadvantages.

7.5.3.5 Policy and Market Limitations

While the concept of the hydrogen economy is spreading worldwide, the biohydrogen sector is still very small in the global scene. The few changes in regulation that lack enough incentives, together with a market that is still at the very early stages, are the main reasons for this slow implementation of biohydrogen technologies. Although the first green hydrogen certification, carbon pricing, and waste-to-energy credits could serve as a big lever to dark fermentation, these instruments are only at their nascent stages of development or adoption.

7.5.4 Techno-Economic Outlook and Future Strategies

The techno-economic analyses of the dark fermentation process suggest that the minimum selling price of hydrogen from the process is around USD 4–8 per kg. The exact figures will be influenced by various factors, such as the scale, feedstock, and system design. Currently, the price is three to four times that of steam methane reforming hydrogen (USD 1–2 per kg), but the clean energy factor and the possibility of carbon credit incentives make biohydrogen a promising product for the far future.

- **Increasing hydrogen yield:** there are options such as metabolic engineering, trained inocula, and enhanced process control.
- **Utilization of cheap feedstocks:** substrate costs will be low or even zero if DF is integrated into the current waste management systems.
- **By-product valorization:** the capturing of VFAs, biofertilizers, or biopolymers can open up new revenue streams.

- **The advancement of modular and scalable reactor architectures:** this will reduce the capital investment risk and make it easier to install decentralized systems.
- **Connection with renewable energy systems:** the use of solar or waste heat for heating and mixing, which are the most energy-consuming process operations, will be beneficial for the energy balance.

All of these steps, if taken simultaneously, may elevate the technology of dark fermentation to a new level where it could be used as a source of clean and affordable hydrogen.

7.6 WASTE AND BY-PRODUCT MANAGEMENT OF DARK FERMENTATION

Well-regulated biohydrogen and its co-products are essential elements that help to preserve both the environmental safety and the operational sustainability of dark fermentation systems. In essence, biohydrogen is an excellent clean fuel; however, its physical and chemical characteristics still render it a potential risk source, which is why these risks should be alleviated with utmost caution, particularly when biohydrogen is being produced, stored, or transported. Alongside that, the fermentation process raises the issue of sound management of the solid residues, biomass, and liquid wastes. Should these by-products be properly treated and recycled, they will be sources of materials for the CE and will contribute to reducing the total environmental impact.

The current section first lists the principal safety problems that result from the use of biohydrogen and then explains how its production aligns with the CE goals.

7.6.1 Safe Handling and Storage Practices for Biohydrogen

Biohydrogen via one of the following processes: dark fermentation, photolysis, or microbial photosynthesis, is going to be a clean fuel alternative. At the same time, since hydrogen is a highly flammable gas and is prone to leak, there is a need for strict safety measures to be observed during its production, handling, and storage. Such safety measures are necessary in research labs as well as in industries.

7.6.1.1 Fundamental Safety Aspects

Hydrogen has a few properties that set it apart from the rest of the gases. One of such properties is its flammability range, which is very broad, about 4–75% in the air, and it also needs almost no ignition energy to burn. As its molecular weight is very low, the dispersion is very fast, but at the same time, it can penetrate the materials and escape through very small openings. Even more, the flame of hydrogen is almost invisible, thus leaks can hardly be detected unless you have the suitable instrument. That is why it is very important to always keep it under surveillance and have a well-designed system for containment.

7.6.1.2 Handling Procedures

7.6.1.2.1 Personnel Training and Safety Protocols

Plant workers at hydrogen production sites should be trained in detailed standard operating procedures, including the correct wearing of personal protective equipment (PPE), the use of the reactors and storage systems, and the implementation of the rescue operations. To support safety regulations, the organization of drills as well as refresher courses is necessary.

7.6.1.2.2 Leak Prevention and Specialized Equipment

Every part of a system must be hydrogen-compatible in order to avoid the degradation of materials. Generally, the use of stainless steel, particularly composite materials, and certain plastics resistant to hydrogen, which are certified, is recommended. The checking of the valves, the seals, the pipelines, and the joints should be done regularly. In order to detect a leak at an early stage, the location where the leak occurred should be fitted with detectors that are based on electrochemical or thermal conductivity sensors.

7.6.1.2.3 Controlled Operating Environments

The hydrogen-handling areas have to be equipped with sufficient air circulation in order to prevent the accumulation of gases. The electrical appliances should be of the explosion-proof type, and very tight rules have to be followed in order to give off no sparks or static discharge. In all places where there is hydrogen, the use of open flame is strictly prohibited.

7.6.1.2.4 Safe Storage Methods

Hydrogen is able to be stored in only three different ways, but each, unlike the other, has its own specific safety precautions:

7.6.1.3 High-Pressure Gas Storage

The storage of compressed hydrogen gas is carried out at pressures of up to 700 bar. The tanks should be properly fastened in an upright position, ventilated areas where there are no ignition sources should be used to store them, and the tanks should be easily noticeable with the help of signs. It is necessary to perform a regular control of the valves and the pressure equipment, and the performance of the valves has to be checked as well.

7.6.1.4 Cryogenic Storage

After intensive cooling, hydrogen can be stored in liquid form at temperatures near –253°C. The cryogenic vessels, which are thermally isolated, have to be equipped with devices that allow the gas to be discharged as well as the pressure relief devices.

7.6.1.5 Solid-State Storage Options

Once fully developed, systems utilizing metal hydrides and porous absorbents, among other technologies, will be not only safer and low-pressure alternatives but also more compact and with less chance of leakage.

7.6.1.5.1 Transportation and Distribution

Transport systems should comply with regulatory guidelines for hydrogen containers; thus, they must have safety valves of the right kind, cylinders that have been tested and certified for the pressure, and signs that indicate the nature of the hazard clearly. Besides the drivers/operators getting a thorough training program, they also need to make sure their equipment is regularly inspected.

7.6.1.5.2 Emergency Preparedness

In consequence of the fact that hydrogen fires produce almost no visible light, thermal imaging devices or flame detectors have to be used in locating them. In case of fire, dry chemical agents are preferred, while carbon dioxide should never be applied to a hydrogen fire. In case of leaks, evacuating the area and increasing the air exchange rate are the primary measures to be taken. The use of automated shutoff valves is yet another step in ensuring the safest handling of biohydrogen. All these actions together form a safety net that makes sure biohydrogen is a green technology that can be incorporated safely into the energy systems, while operational risk is being kept at a minimum.

7.6.2 The Role of Biohydrogen in the Circular Economy

The CE model is primarily about being more resource efficient, producing less waste, and most importantly, continually reusing materials within interconnected systems. Since biohydrogen is the technology that first of all produces the clean energy needed from organic residues, and second, it makes it possible for the by-products to be regenerated into agricultural or industrial cycles, theoretically, biohydrogen is a perfect fit within this paradigm.

7.6.2.1 Transforming Waste into Energy and Resources

The most important factor relating biohydrogen to CE principles is the ability of the biological system to produce hydrogen from wastes, which include agricultural residues, food scraps, and industrial effluents. This is achieved by microbial pathways. This practice not only lessens the environmental load that characterizes the usual methods of waste disposal but also releases the energy that has been locked for years. Basically, dark fermentation is a capacity that can handle a variety of waste sources such as municipal solid waste, dairy by-products, and sewage sludge, and the like (Kumar et al., 2020).

Actually, along with hydrogen, dark fermentation is also replete with nutrient-rich leftovers such as digestate and VFAs, which, if utilized as soil amendments or feedstock for the next bioprocess, can turn into fertilizer. In this way, the CE objective of prolonging the lifecycle of biological materials is achieved.

7.6.2.2 Supporting Low-Carbon and Renewable Energy Systems

When made in a sustainable way, biohydrogen is a carbon-neutral fuel and produces only water when burnt. As opposed to hydrogen derived from fossil sources, biohydrogen is quite a clean energy carrier, as it leads to significant reductions in greenhouse gas emissions, which makes it a perfect candidate for low-carbon transitions in industries and transport sectors.

Moreover, the combination of waste management, agriculture, and renewable energy systems not only makes the cycle of resource flows more symbiotic but also creates a host of new activities bringing the economy into the circular system, which places this interconnectedness at its core.

7.6.2.2.1 Strengthening the Circular Bioeconomy

First of all, the circular bioeconomy is about the use of biological resources in a sustainable way, and in this context, biohydrogen helps by allowing the production of high-value energy of high value without the need for natural reserves to be exhausted. During the fermentation process, the residual digestate is one of the substances that not only can improve soil fertility but, at the same time, can support regenerative agriculture and the closure of nutrient loops.

7.6.2.2.2 Opportunities and Existing Barriers

Even though it is compatible with CE objectives, to occupy a larger space in biohydrogen's role, a few problems have to be resolved:

- Limited hydrogen yields lead to an economically less attractive situation.
- Compared to fossil or electrolysis-based hydrogen, the technological maturity level is still very low.
- Operational costs and system complexity are still at the highest level.

However, progress in microbial engineering, bioreactor design, and integrated energy systems is helping to overcome these challenges and decrease expenses. Local and national rules like waste-to-energy requirements and CE strategies could be the main factors that speed up the deployment of the full operating cycle. Biohydrogen powered with decentralized energy systems or fuel cells can be a major source of local circular energy networks.

REFERENCES

Ahmad, A., Krishnamoorthy, R., Hasan, S. W., Show, P. L., & Banat, F. (2024). Biohydrogen production through dark fermentation: Recent trends and advances in transition to a circular bioeconomy. *International Journal of Hydrogen Energy, 52*, 335–357. https://doi.org/10.1016/J.IJHYDENE.2023.05.161

Akhlaghi, N., & Najafpour-Darzi, G. (2020). A comprehensive review on biological hydrogen production. *International Journal of Hydrogen Energy, 45*(43), 22492–22512. https://doi.org/10.1016/J.IJHYDENE.2020.06.182

Alavi-Borazjani, S. A., Tarelho, L. A., da,C., & Capela, M. I. (2021). Parametric optimization of the dark fermentation process for enhanced biohydrogen production from the organic fraction of municipal solid waste using Taguchi method. *International Journal of Hydrogen Energy, 46*(41), 21372–21382. https://doi.org/10.1016/J.IJHYDENE.2021.04.017

Al-Mamun, A., Jafary, T., Baawain, M. S., Rahman, S., Choudhury, M. R., Tabatabaei, M., & Lam, S. S. (2020). Energy recovery and carbon/nitrogen removal from sewage and contaminated groundwater in a coupled hydrolytic-acidogenic sequencing batch reactor and denitrifying biocathode microbial fuel cell. *Environmental Research, 183*, 109273. https://doi.org/10.1016/J.ENVRES.2020.109273

Ayodele, D. T., Ogunbiyi, O. D., Akamo, D. O., Otun, K. O., Akinpelu, D. A., Adegoke, J. A., Fapojuwo, D. P., & Oladoye, P. O. (2023). Factors affecting biohydrogen production: Overview and perspectives. *International Journal of Hydrogen Energy, 48*(71), 27513–27539. https://doi.org/10.1016/J.IJHYDENE.2023.04.001

Azwar, M. Y., Hussain, M. A., & Abdul-Wahab, A. K. (2014). Development of biohydrogen production by photobiological, fermentation and electrochemical processes: A review. *Renewable and Sustainable Energy Reviews, 31*, 158–173. https://doi.org/10.1016/J.RSER.2013.11.022

Balachandar, G., Varanasi, J. L., Singh, V., Singh, H., & Das, D. (2020). Biological hydrogen production via dark fermentation: A holistic approach from lab-scale to pilot-scale. *International Journal of Hydrogen Energy, 45*(8), 5202–5215. https://doi.org/10.1016/J.IJHYDENE.2019.09.006

Biswal, T., Shadangi, K. P., & Sarangi, P. K. (2023). Application of nanotechnology in the production of biohydrogen: A review. *Chemical Engineering & Technology, 46*(2), 218–233. https://doi.org/10.1002/ceat.202000565

Das, D., & Veziroğlu, T. (2001). Hydrogen production by biological processes: A survey of literature. *International Journal of Hydrogen Energy, 26*, 13–28. https://doi.org/10.1016/S0360-3199(00)00058-6

Dincer, I. (2000). Renewable energy and sustainable development: A crucial review. *Renewable and Sustainable Energy Reviews, 4*(2), 157–175. https://doi.org/10.1016/S1364-0321(99)00011-8

Dufour, J., Serrano, D. P., Gálvez, J. L., González, A., Soria, E., & Fierro, J. L. G. (2012). Life cycle assessment of alternatives for hydrogen production from renewable and fossil sources. *International Journal of Hydrogen Energy, 37*(2), 1173–1183. https://doi.org/10.1016/J.IJHYDENE.2011.09.135

Gao, Y., Yun, S., Wang, K., Liu, L., Abbas, Y., Saqlain Jamil, M., Ke, T., Liu, J., & Zhang, X. (2024). Understanding of the potential role of carbon dots in promoting interspecies electron transfer process in anaerobic co-digestion under magnetic field: Focusing on methane and hydrogen production. *Chemical Engineering Journal, 489*, 151381. https://doi.org/10.1016/J.CEJ.2024.151381

García-Depraect, O., Vargas-Estrada, L., Muñoz, R., & Castro-Muñoz, R. (2025). Membrane-assisted dark fermentation for integrated biohydrogen production and purification: A comprehensive review. *Fermentation, 11*(1). https://doi.org/10.3390/fermentation11010019

Ghimire, A., Frunzo, L., Pirozzi, F., Trably, E., Escudie, R., Lens, P. N. L., & Esposito, G. (2015). A review on dark fermentative biohydrogen production from organic biomass: Process parameters and use of by-products. *Applied Energy, 144*, 73–95. https://doi.org/10.1016/J.APENERGY.2015.01.045

Hallenbeck, P. C., & Benemann, J. R. (2002). Biological hydrogen production; Fundamentals and limiting processes. *International Journal of Hydrogen Energy, 27*(11–12), 1185–1193. https://doi.org/10.1016/S0360-3199(02)00131-3

Hawkes, F. R., Hussy, I., Kyazze, G., Dinsdale, R., & Hawkes, D. L. (2007). Continuous dark fermentative hydrogen production by mesophilic microflora: Principles and progress. *International Journal of Hydrogen Energy, 32*(2), 172–184. https://doi.org/10.1016/J.IJHYDENE.2006.08.014

Javed, M. A., Zafar, A. M., Aly Hassan, A., Zaidi, A. A., Farooq, M., El Badawy, A., Lundquist, T., Mohamed, M. M. A., & Al-Zuhair, S. (2022). The role of oxygen regulation and algal growth parameters in hydrogen production via biophotolysis. *Journal of Environmental Chemical Engineering, 10*(1), 107003. https://doi.org/10.1016/J.JECE.2021.107003

Kapdan, I. K., & Kargi, F. (2006). Bio-hydrogen production from waste materials. *Enzyme and Microbial Technology, 38*(5), 569–582. https://doi.org/10.1016/J.ENZMICTEC.2005.09.015

Kotay, S. M., & Das, D. (2009). Novel dark fermentation involving bioaugmentation with constructed bacterial consortium for enhanced biohydrogen production from pretreated sewage sludge. *International Journal of Hydrogen Energy*, *34*(17), 7489–7496. https://doi.org/10.1016/J.IJHYDENE.2009.05.109

Kumar, V., Kothari, R., & Singh, S. (2015). Dark fermentation: A green way to produce hydrogen and methane. *International Journal of Science, Technology & Society*, *1*, 39–42.

Lee, D. J., Show, K. Y., & Su, A. (2011). Dark fermentation on biohydrogen production: Pure culture. *Bioresource Technology*, *102*(18), 8393–8402. https://doi.org/10.1016/J.BIORTECH.2011.03.041

Liu, Y., Cruz-Morales, P., Zargar, A., Belcher, M. S., Pang, B., Englund, E., Dan, Q., Yin, K., & Keasling, J. D. (2021). Biofuels for a sustainable future. *Cell*, *184*(6), 1636–1647. https://doi.org/10.1016/j.cell.2021.01.052

Łukajtis, R., Hołowacz, I., Kucharska, K., Glinka, M., Rybarczyk, P., Przyjazny, A., & Kamiński, M. (2018). Hydrogen production from biomass using dark fermentation. *Renewable and Sustainable Energy Reviews*, *91*, 665–694. https://doi.org/10.1016/J.RSER.2018.04.043

Machhirake, N. P., Vanapalli, K. R., Kumar, S., & Mohanty, B. (2024). Biohydrogen from waste feedstocks: An energy opportunity for decarbonization in developing countries. *Environmental Research*, *252*, 119028. https://doi.org/10.1016/J.ENVRES.2024.119028

Matsakas, L., Gao, Q., Jansson, S., Rova, U., & Christakopoulos, P. (2017). Green conversion of municipal solid wastes into fuels and chemicals. *Electronic Journal of Biotechnology*, *26*, 69–83. https://doi.org/10.1016/J.EJBT.2017.01.004

Mohanakrishna, G., & Pengadeth, D. (2024). Mixed culture biotechnology and its versatility in dark fermentative hydrogen production. *Bioresource Technology*, *394*, 130286. https://doi.org/10.1016/J.BIORTECH.2023.130286

Mokhtarani, B., Zanganeh, J., & Moghtaderi, B. (2025). A review on biohydrogen production through dark fermentation, process parameters and simulation. *Energies*, *18*(5). https://doi.org/10.3390/en18051092

Ndayisenga, F., Yu, Z., Zheng, J., Wang, B., Liang, H., Phulpoto, I. A., Habiyakare, T., & Zhou, D. (2021). Microbial electrohydrogenesis cell and dark fermentation integrated system enhances biohydrogen production from lignocellulosic agricultural wastes: Substrate pretreatment towards optimization. *Renewable and Sustainable Energy Reviews*, *145*, 111078. https://doi.org/10.1016/J.RSER.2021.111078

Palanivel, P. D., Jovita S, A., Babu, K., & Naina Mohamed, S. (2025). Harnessing dark fermentation: Mechanisms, metabolic pathways, and nanoparticle innovations for biohydrogen enhancement. *International Journal of Hydrogen Energy*, *109*, 383–396. https://doi.org/10.1016/J.IJHYDENE.2025.02.033

Panwar, N. L., Kaushik, S. C., & Kothari, S. (2011). Role of renewable energy sources in environmental protection: A review. *Renewable and Sustainable Energy Reviews*, *15*(3), 1513–1524. https://doi.org/10.1016/J.RSER.2010.11.037

Patel, G. H., Havukainen, J., Horttanainen, M., Soukka, R., & Tuomaala, M. (2024). Climate change performance of hydrogen production based on life cycle assessment. *Green Chem*, *26*(2), 992–1006. https://doi.org/10.1039/D3GC02410E

Rangel, C. J., Hernández, M. A., Mosquera, J. D., Castro, Y., Cabeza, I. O., & Acevedo, P. A. (2021). Hydrogen production by dark fermentation process from pig manure, cocoa mucilage, and coffee mucilage. *Biomass Conversion and Biorefinery*, *11*(2), 241–250. https://doi.org/10.1007/s13399-020-00618-z

Singh, T., Sehgal, A., Singh, R., Sharma, S., Pal, D. B., Tashkandi, H. M., Raddadi, R., Harakeh, S., Haque, S., Srivastava, M., Aly Hassan, A., Srivastava, N., & Gupta, V. K. (2023). Algal biohydrogen production: Impact of biodiversity and nanomaterials induction. *Renewable and Sustainable Energy Reviews*, *183*, 113389. https://doi.org/10.1016/J.RSER.2023.113389

Sinha, P., & Pandey, A. (2011). An evaluative report and challenges for fermentative biohydrogen production. *International Journal of Hydrogen Energy, 36*(13), 7460–7478. https://doi.org/10.1016/J.IJHYDENE.2011.03.077

Solomon, B. D. (2010). Biofuels and sustainability. *Annals of the New York Academy of Sciences, 1185*(1), 119–134. https://doi.org/10.1111/j.1749-6632.2009.05279.x

Srivastava, P., García-Quismondo, E., Palma, J., & González-Fernández, C. (2024). Coupling dark fermentation and microbial electrolysis cells for higher hydrogen yield: Technological competitiveness and challenges. *International Journal of Hydrogen Energy, 52*, 223–239. https://doi.org/10.1016/J.IJHYDENE.2023.04.293

Stabel, M., Haack, K., Lübbert, H., Greif, M., Gorenflo, P., Aliyu, H., & Ochsenreither, K. (2022). Metabolic shift towards increased biohydrogen production during dark fermentation in the anaerobic fungus *Neocallimastixcameroonii* G341. *Biotechnology for Biofuels and Bioproducts, 15*(1), 96. https://doi.org/10.1186/s13068-022-02193-z

Tiegam Tagne, R. F., Costa, P., Casella, S., & Favaro, L. (2024). Optimization of biohydrogen production by dark fermentation of African food-processing waste streams. *International Journal of Hydrogen Energy, 49*, 266–276. https://doi.org/10.1016/J.IJHYDENE.2023.07.348

Vidal, A., Mohiuddin, O., Chance, E., Serrano-Blanco, S., Howard, T. P., Muñoz-Muñoz, J., Velasquez-Orta, S., & Rios-Solis, L. (2025). Biohydrogen production through dark fermentation of agricultural waste: Novel strain and feedstock characterisation. *Bioresource Technology, 434*, 132839. https://doi.org/10.1016/J.BIORTECH.2025.132839

Wang, J., & Wan, W. (2009). Factors influencing fermentative hydrogen production: A review. *International Journal of Hydrogen Energy, 34*(2), 799–811. https://doi.org/10.1016/J.IJHYDENE.2008.11.015

Wang, R., Zhang, H., Zhang, J., Zhou, C., Zhang, X., Yan, X., Yu, F., & Zhang, J. (2024). Comparison of calcium magnesium ferrite nanoparticles for boosting biohydrogen production. *Bioresource Technology, 395*, 130410. https://doi.org/10.1016/J.BIORTECH.2024.130410

Yin, T., Wang, W., Guo, W., Zhuo, S., Cao, G., Ren, H., Li, J., Xie, G., Ding, J., & Liu, B. (2024). Enhanced thermophilic hydrogen production by an enriched novel acetic-acid-type fermentative bacterium from inoculum sludge with nonheat pretreatment. *Energy & Fuels, 38*(10), 8749–8761. https://doi.org/10.1021/acs.energyfuels.3c04696

Zhang, Q., Jiao, Y., He, C., Ruan, R., Hu, J., Ren, J., Toniolo, S., Jiang, D., Lu, C., Li, Y., Man, Y., Zhang, H., Zhang, Z., Xia, C., Wang, Y., Jing, Y., Zhang, X., Lin, R., Li, G., & Tahir, N. (2024). Biological fermentation pilot-scale systems and evaluation for commercial viability towards sustainable biohydrogen production. *Nature Communications, 15*(1), 4539. https://doi.org/10.1038/s41467-024-48790-4

Zhou, C., Zhang, J., Pei, Y., Tian, K., Zhang, X., Yan, X., & Yang, J. (2023). Molten salt strategy to activate biochar for enhancing biohydrogen production. *Bioresource Technology, 385*, 129466. https://doi.org/10.1016/J.BIORTECH.2023.129466

Ziara, R. M. M., Miller, D. N., Subbiah, J., & Dvorak, B. I. (2019). Lactate wastewater dark fermentation: The effect of temperature and initial pH on biohydrogen production and microbial community. *International Journal of Hydrogen Energy, 44*(2), 661–673. https://doi.org/10.1016/J.IJHYDENE.2018.11.045

8 Bioethanol and Biobutanol

Production, Optimization, and Disposal Guidelines

Panchamoorthy Saravanan, R. Rajeshkannan, and M. Rajasimman

8.1 INTRODUCTION

The worldwide demand for energy has surged significantly due to rapid population growth and the heightened influence of industrialization, transportation, and living standards, among other factors. Consequently, the continually rising energy needs can be addressed by leveraging biofuels as an alternative source, which boasts zero environmental impact (Bhaskar et al., 2011). A multitude of research projects are currently being conducted across the globe, and the International Energy Agency projects that biofuel production levels are expected to rise by 25% by the year 2024 (Suresh et al., 2024). Research indicates that biofuels perform as effectively as traditional fossil fuels, and their prolonged usage could alleviate the strain caused by fossil fuel depletion. Biofuels are classified as liquid or gaseous fuels, primarily derived from biomass. They offer numerous benefits compared to conventional fuels, including the readily available biomass for energy generation, minimal carbon dioxide (CO_2) emissions during combustion, environmental sustainability, economic viability, and biodegradability (Cherubini & Strømman, 2011).Cellulose is the most abundant renewable natural biological resource, and the production of bio-based products and bioenergy from less costly renewable lignocellulosic materials in the form of agro-industrial biological waste. The cost of enzyme production can be significantly reduced if low value biological substrates like fruit processing waste were used. The cost of lignocellulosic raw material influences the economy of many fermentation processes, hence the cost play a decisive role in future and scope of industries employing fermentation processes. Discarded plant part like a peel of fruits and agro harvest materials, huge waste in the civil area and generates an intolerable odor, so it can be effectively utilized for the production of cellulase enzymes. The recognition that environmental pollution is a worldwide threat to public health has given rise to a new massive industry for environmental restoration. Submerged and solid state fermentation systems have been used to produce compounds of industrial interest from lignocellulose, as an alternative for valorization of these wastes and also to solve environmental problems caused by their improper disposal. When submerged

 DOI: 10.1201/9781003642589-8

fermentation systems are used, a previous stage of hydrolysis for separation of the lignocellulose constituents is required. The major constituents of lignocellulose are cellulose, hemicellulose and lignin, the polymers that are closely associated with each other constituting the cellular complex of the vegetal biomass (Lange, 2007). It can be degraded by cellulase enzyme complex and is deceptively simple chemically, a polymer consisting only of glucose linked only by β-1, 4 bonds. Cellulase refers to a class of enzymes produced chiefly by fungi, bacteria, and protozoa that catalyze cellulolysis (i.e., hydrolysis of cellulose). So with help of valuable microbe agro-industrial waste which is rich in lignocelluloses product could be used as media for enzyme production (Khanal, 2008).

8.2 PRE-TREATMENT

Cellulose and hemicelluloses act as the main source of C6 and C5 fermentable sugars in lignocellulosic biomass; nature has developed woody structures to efficiently withstand microbial attack. This emphasizes the importance of utilizing suitable pre-treatment methods to make these sugars available to microorganisms. Investigation into the pre-treatment of lignocellulosic biomass has been conducted for many years, with a variety of thermal, mechanical, chemical, and biological pre-treatment techniques analyzed and recorded in scientific studies (Yüksel & Yüksel, 2004). It is essential that the selected pre-treatment technology fulfills various requirements: (1) improve the enzymatic accessibility of the lignocellulosic substance; (2) reduce the loss of possible sugars; (3) prevent the formation of substances that may hinder microbial breakdown or enzymatic functions; (4) make sure the pre-treatment approach is economical, allowing the transformation of biomass into simple sugars to be a feasible technology. Pre-treatment aims to remove lignin from the biomass to enhance the accessibility of cellulose in the hydrolysis process. This procedure leads to the isolation of lignin and hemicellulose from cellulose, while also enhancing the internal surface area of the fibers, which promotes more efficient enzymatic hydrolysis (Tsai et al., 2020). Typical pre-treatment methods consist of steam explosion, alkali treatment, and acid treatment. Steam explosion is often used, whereas alkali pre-treatment has demonstrated better lignin removal. The amount of solids is crucial for the economic feasibility of a dilute acid pre-treatment and enzymatic hydrolysis process (Yiin et al., 2021). Studies have been performed on solid loadings of 30% for dilute acid pre-treatment. More investigation is necessary to comprehend the connection between enzymatic digestion and the structural properties of pre-treated materials to gain a clearer understanding of the elements affecting cellulose hydrolysis. Renewable and sustainable energy sources like bioethanol, biodiesel, biohydrogen, and methane are derived from different feedstocks to address energy and environmental concerns. From a scientific perspective, biofuels can be classified into four unique categories according to the feed stocks utilized (Yeong et al., 2018).

In 2023, global bioethanol production is projected to reach 150 billion liters, an increase from approximately 120 billion liters in 2019. The vast majority of global bioethanol production is illustrated in Figure 8.1. Due to the biofuel policies of both the United States and the European Union, which promote the development of

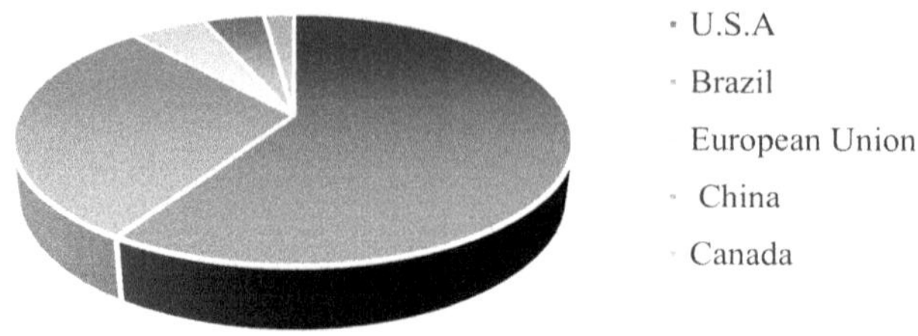

FIGURE 8.1 Worldwide bioethanol production.

cellulose-based biofuels on a global scale, the commercialization of cellulosic bioethanol from agricultural wastes is expected to enhance the involvement of second-generation biofuels in the global bioethanol and biobutanol market (Yeong et al., 2018).

8.3 FIRST-GENERATION BIOETHANOL

In Figure 8.2, the evolution of bioethanol production, often referred to as generation, is presented. The procedure for bioethanol synthesis and its fundamental components are also illustrated in Figure 8.2. First-generation bioethanol is a liquid biofuel produced from agricultural crops rich in starch and sugar, aimed for automotive use. Since *Saccharomyces cerevisiae* is unable to decompose complex carbohydrates, both starchy and lignocellulosic materials need to undergo hydrolysis (Cherubini, 2010). Figure 8.3 showcases the process of first-generation bioethanol synthesis along with its raw materials. As a result, major concerns arise about the social and environmental impacts of large-scale production. The first-generation bioethanol biofuel requires water and arable land for food cultivation, leading to over-fertilization, which may result in resource depletion, including reduced water availability and degradation of soil and water resources (Koller et al., 2005). Additionally, food costs for grains, crops, vegetable oils, and livestock feed could escalate as a consequence. Moreover, the emission of greenhouse gases from biofuels is lower compared to that from fossil fuel combustion (Khuong et al., 2017). Biodiesel, produced by transesterifying vegetable oils, is another type of first-generation biofuel alongside food-based bioethanol.

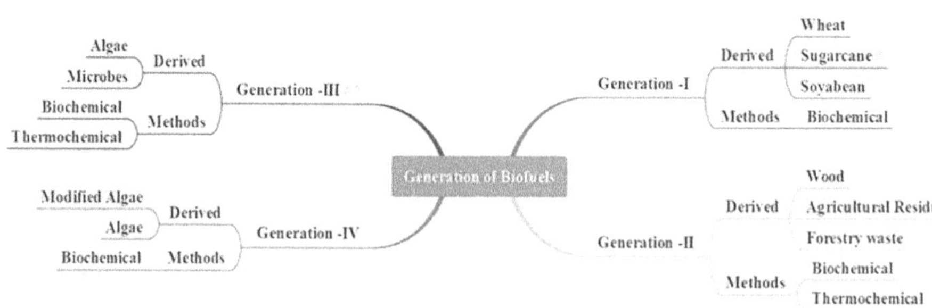

FIGURE 8.2 Generation of bioethanol production with raw materials.

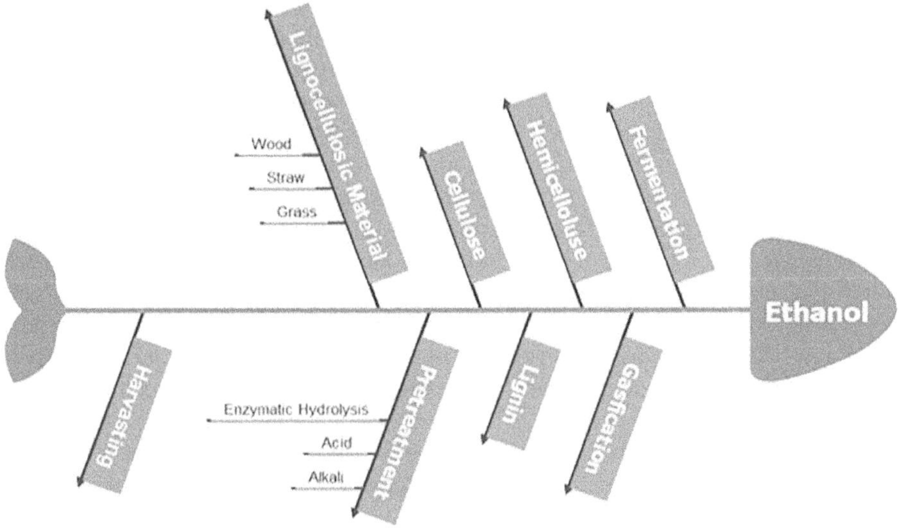

FIGURE 8.3 Second-generation bioethanol production methods.

8.4 SECOND-GENERATION BIOETHANOL

Biofuels from the second generation and beyond, including bioethanol do not compete with animal feed supplies because they are derived from non-food feedstocks. The main feedstock for second-generation bioethanol production is lignocellulosic biomass, although industrial by-products such as whey or crude glycerol can also be utilized. Generally, this type of biomass is both inexpensive and abundant in specific regions (Barakat et al., 2015). The availability of lignocellulosic biomass is influenced by climatic conditions. Converting lignocellulose into reducing sugars is notably more challenging than starch conversion (Bansal et al., 2021). Researchers have explored the utilization of various types of plant biomass for biofuel production. Biofuels from the second and subsequent generations, including bioethanol, do not contend with feed supplies because they are generated from non-feedstock materials. The process of second-generation ethanol production using thermophilic bacteria is represented in Figure 8.3.

8.5 THIRD-GENERATION BIOETHANOL

The cultivation of microalgae or unicellular microorganisms (like Cyanidium, Caldarium, or Synechococcus) from eukaryotes and prokaryotes (cyanobacteria) represents the third generation of biofuels (Van Thuijl & Deurwaarder, 2006). Active microalgal biomass, like live biocatalysts, can utilize nutrients such as carbon, nitrogen, phosphate, or sulfur from industrial effluents as substrates to generate high biomass levels. Effluent gases, wastewater, hydrolysis products of organic waste, and digestate from industrial power plants represent some of the waste streams associated with biogas production. Consequently, producing third-generation biofuels can help decrease waste outputs from various sectors. Algae, for instance, serve as a superb alternative fuel due to their rich lipid and carbohydrate levels. Through alkaline transesterification with methanol, oils from microalgae serve as a potential feedstock for vegetable oils routinely utilized in biodiesel production (Prieur-Vernat & His, 2007). Advancements in technology, like improved oil extraction from cellular biomass, are necessary for producing biodiesel from microalgae. Residual biomass from microalgae can be transformed into sustainable energy carriers, such as bioethanol and biogas for long-term use (biofuels), and is also applicable for methane fermentation and biohydrogen production. The anaerobic fermentation of starch and starch-like polysaccharides generates bioethanol from algal biomass. Although the process is complex, the yields for bioethanol production from biomass occur at a lower rate. The by-product biohydrogen results from methane production derived from synthetic organic acids during the acidogenic phase of anaerobic digestion of organic waste conversion (Li et al., 2014). However, since the amount of biohydrogen produced by algae is currently limited, the biohydrogen production process needs improvement. Cultivating algae is simple and can occur in different aquatic settings (Balat & Balat, 2009).

8.6 RAW MATERIALS UTILIZED IN THE PRODUCTION OF BIOETHANOL AND BIOBUTANOL

8.6.1 Raw Materials Based on Sugar

Figure 8.4 shows raw materials utilized in the manufacturing of bioethanol. Cane sugar and sugar beet are the two most suitable plants for manufacturing sugar on the planet. Cane sugar accounts for global sugar production, whereas sugar beet contributes to the remaining one-third (Festel et al., 2013). The majority of *Saccharomyces species* produce the enzyme invertase, capable of hydrolyzing them effectively. Consequently, producing bioethanol from sugar (sucrose) sources does not need pretreatment, rendering it more feasible than starch sources. To extract sugar, sugar crops must be processed through milling and subsequently moved to a fermentation medium, where ethanol is directly obtained from the juice or molasses (Zhao et al., 2015). Figure 8.5 shows the benefits of sugar cane. Bioethanol or bio-based products can be produced from raw, thin and thick juices, while high purity sugar crystals are also generated during sugar beet processing. The production of bioethanol also

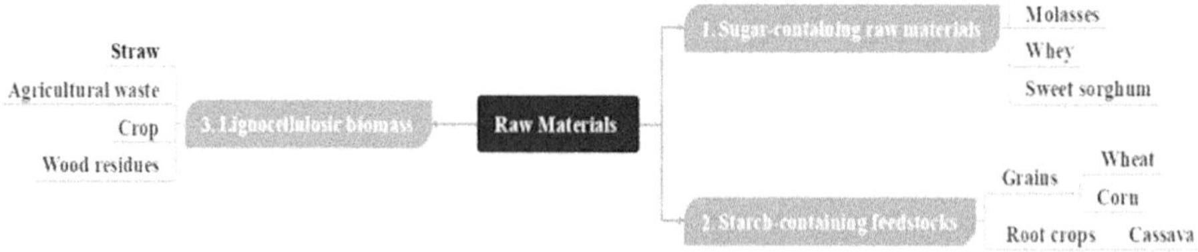

FIGURE 8.4 Raw materials utilized in the manufacturing of bioethanol.

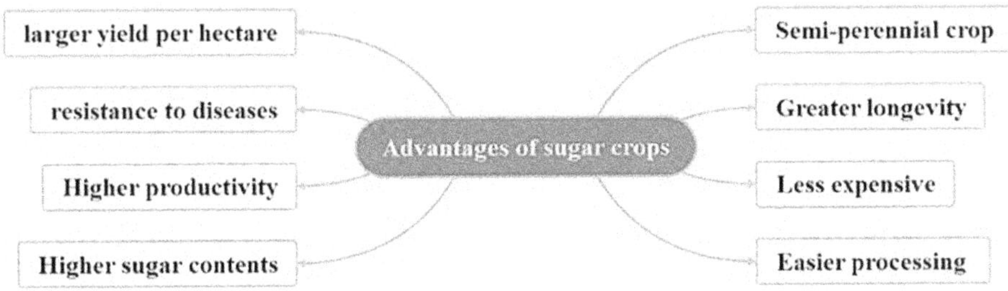

FIGURE 8.5 Benefits of sugar crops for bioethanol production.

utilizes sugar beet cossettes in their unprocessed form as a substrate (Baz et al., 2021). The use of sugar processing intermediates has influenced the bioprocess setup, microbial stability, and transport properties. Both sugar syrup and granulated sugar can serve as bioethanol substrates throughout the year and can also act as precursors for various chemical intermediates and final products. Molasses, a common by-product of sugar production, is utilized as a raw material for the production of biochemicals and as animal feed. Molasses features a primary residual sugar concentration of 50–60% (m/V), with sucrose making up about 60% of that, making it a suitable substrate for extensive bioethanol production (Mussatto et al. 2010). Figure 8.5 shows benefits of sugar crops for bioethanol production.

8.6.2 Raw Materials Based on Starch

Figure 8.6 depicts bioethanol production using starchy raw materials. Starch consists of a combination of linear (amylose) and branched (amylopectin) polyglucans. The enzyme α-amylase, a starch hydrolase, acts on the –1,4 linkages but not the –1,6 connections in amylopectin (Tomás-Pejó et al., 2011). Starch is hydrolyzed (primarily by α-amylase and glucoamylase) into glucose syrup, which is subsequently transformed into ethanol by the yeast *Saccharomyces cerevisiae*, essential for bioethanol production from feedstocks containing starch. This step is costlier when compared to bioethanol production from feedstocks that contain sugar (Linoj et al., 2006). Microorganisms like Bacillus licheniformis and modified organisms generate α-amylase and glucoamylase.

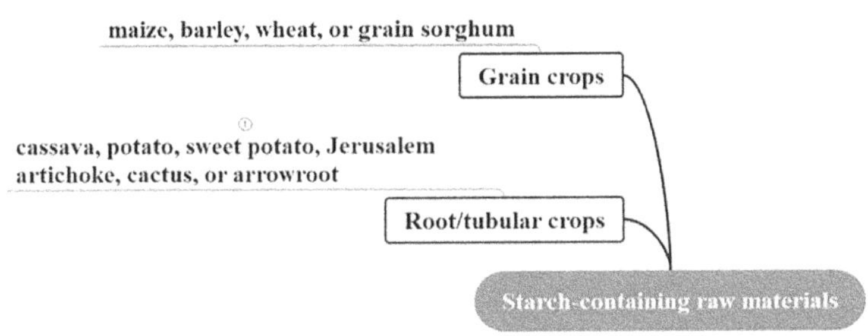

FIGURE 8.6 Bioethanol production using starchy raw materials.

The yeast *S. cerevisiae* converts glucose into ethanol in the absence of oxygen. The efficiency for mass conversion is approximately 51% in producing the maximum yield of glucose into ethanol. Yeast requires glucose for cell development and the creation of various metabolic products, though its effectiveness in optimal conversion is diminished. In practice, glucose is transformed into ethanol at a mass conversion efficiency of 40–48%. Ethanol produced from starch improves enzyme functionality and yeast strains that exhibit high ethanol tolerance relative to sugar-derived ethanol.

Microalgae convert CO_2 into lipids and polysaccharides, rendering them a potential renewable biomass source for biofuel generation. As a result, CO_2 utilized in industries is required for the growth of microalgae as part of a strategy to decrease atmospheric CO_2 emissions. Microalgae also achieve the accumulation of starch as a storage polysaccharide following pre-treatment, which is then utilized to produce bioethanol (third generation). Additionally, residual biomass (which includes organic material and minerals) can serve as a biofertilizer after the production of bioethanol. Consequently, it is somewhat clear that adopting the biorefinery approach could significantly enhance microalgae bioethanol production (İçöz et al., 2008).

8.6.3 Raw Materials Based on Lignocellulose

On average, lignocellulosic biomass contains 43% cellulose, 27% lignin, 20% hemicellulose, and 10% additional components. Figure 8.7 depicts the processes in which lignocellulosic biomass plays a crucial role in bioethanol production. Lignocellulosic biomass demands more complex chemical processes due to its diverse structure, unlike the uniform and consistent raw materials usually employed in the chemical

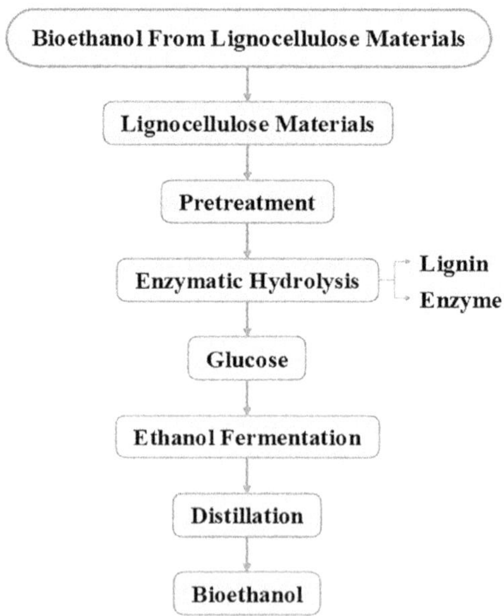

FIGURE 8.7 Shows the steps involved in producing bioethanol from lignocellulosic biomass.

industry (Bharathiraja et al., 2017). Moreover, the collection of lignocellulosic crops is often not practical throughout the year, making the tasks of biomass providers more complex. Thus, this matter must be tackled by effectively stabilizing biomass, especially to ensure long-term storage and to uphold continuous biorefinery operations moving forward. Figure 8.7 demonstrates that the four types of pre-treatment techniques include physical, chemical, physicochemical, and biological categories. Nevertheless, not all the techniques are completely feasible for industrial-scale application (Pavlečić et al., 2010).

8.7 PRE-TREATMENT METHODS

8.7.1 Physical Preparation of Lignocellulose-Containing Raw Materials

Methods such as irradiation (gamma ray, electron beam, microwave), milling (e.g., two-roll, ball or hammer, colloid or vibroenergy), and other techniques (e.g., hydrothermal, expansion, extrusion, pyrolysis) represent forms of physical (mechanical) preparation. Lignocellulosic biomass requires physical pre-treatment to decrease particle size and crystallinity, ultimately increasing specific surface area and reducing the degree of polymerization (Pavlečić et al., 2017). The particle size must, however, during milling, have energy consumption optimized considerably, as the negative effects on pre-treatment (very small particles are undesirable). The majority of physical technologies that achieve size reduction are not economically feasible because they require significant energy consumption. A novel physical technique known as extrusion is employed to convert biomass into fermentable sugars, showing significant potential. Figure 8.8 illustrates the physical preparation of lignocellulose-containing raw materials for bioethanol and biobutanol production. Some of the key benefits of an extruder include the capacity to maintain a high shear rate, rapid heat transfer, and effective mixing. It is compared to extrusion as a pre-treatment for transforming soybean husks into fermentable sugars with two typical pre-treatments. They are weak acid (1% by mass H_2SO_4) and base (1% by mass NaOH). The authors

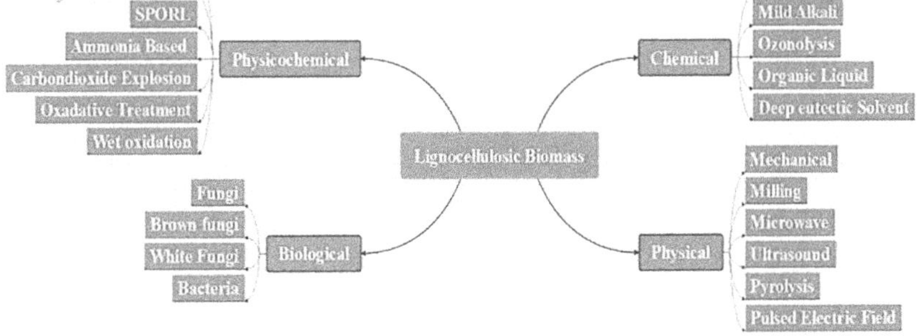

FIGURE 8.8 Physical preparation of lignocellulose-containing raw materials for bioethanol and biobutanol production.

found that under optimal conditions, the mass conversion rates of cellulose to glucose can reach a maximum of 95% (Senthilkumar & Gunasekaran, 2009).

8.7.2 Chemical Processing of Lignocellulose–Comprising Raw Materials

The given figure demonstrates the Lignocellulose-containing raw materials processed by chemical pre-treatment for the production of bioethanol and biobutanol. Figure 8.9 shows chemical treatment methods for bioethanol and biobutanol production

The illustrated figure shows the raw materials with lignocellulose that have undergone chemical pre-treatment for bioethanol and biobutanol production. It also presents detailed chemical treatment techniques for producing bioethanol and biobutanol. Acid pre-treatment has a primary objective: to break down the hemicellulose, facilitating enzyme access to the cellulose (Manikandan et al., 2008). NH_4OH, $Ca(OH)_2$, NaOH, and KOH are frequently utilized alkaline pre-treatments for lignocellulosic feedstocks. In these preparations, lower temperatures and pressures are utilized compared to the traditional pre-treatment techniques. The alkaline pre-treatment enhances lignin solubilization and improves cellulose digestibility. A major drawback is the steep cost of alkalis. The decrease in lignin levels results from ozonolysis (with hemicellulose slightly reduced and cellulose remaining the same), which produces no harmful residues, though it is expensive due to higher ozone requirements (Jobling, 2004).

The mixture of large organic cations and small inorganic anions comprises salt ionic liquids (ILs), which have been thoroughly researched as solvents for pre-treating lignocellulosic feedstock. The large hydrogen bonds in polysaccharides can be broken by ILs, leading to their dissolution. They provide excellent solvation properties for different materials that are thermally and chemically stable, non-flammable, and possess a broad liquid temperature range (Solomon et al., 2007). ILs are called "green" solvents since they do not produce harmful or explosive vapors when used. The report indicates that the pre-treatment of switchgrass with ILs significantly enhances the enzymatic breakdown of bothxylan and cellulose.

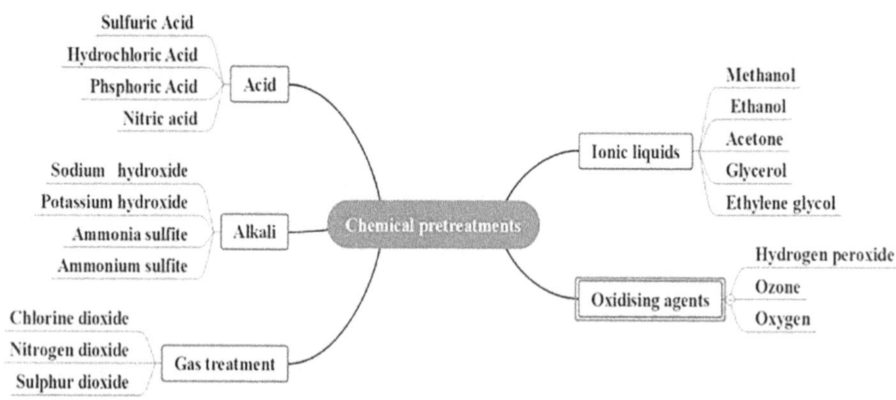

FIGURE 8.9 Chemical treatment methods for bioethanol and biobutanol production.

The Organosolv technique uses organic or water-based solvents to extract lignin and enhance the availability of cellulose. During this pre-treatment, different proportions of solvents are mixed with water, which is then added to the biomass and heated to a temperature of 100–250°C. Conversely, the reactor needs to have solvents removed, which are subsequently evaporated, condensed, and reused, leading to certain pre-treatment costs. The yield obtained from Organosolv extraction of sugarcane bagasse is 29.1% by weight of sugars that can be converted under optimal conditions (Mousdale, 2008)

8.7.3 Physicochemical Treatments of Raw Materials Containing Lignocellulose

In the production of bioethanol, a synchronized process of saccharification and fermentation was performed, incorporating steam explosion in sugar cane bagasse at temperatures of 200, 215, and 230°C for duration of 5 minutes (Soccol et al., 2011). The resulting product yielded a glucose mass of 86.8% at the suggested pre-treatment temperature (215°C for 5 minutes). The method utilizing both chemical (ammonia) and physical (high temperature and pressure) effectively generates feedstock hydrolysis, commonly referred to as ammonia fiber explosion (AFEX). AFEX improves the surface accessibility of the processed feedstock for hydrolysis, increases cellulose de-crystallization, facilitates the depolymerization of some hemicellulose, and reduces lignin recalcitrance. Nonetheless, this approach is ineffective for biomass with high lignin concentration (Ding et al., 2019).

Substances in a gaseous state that have been compressed to a liquid-like density above their critical temperature are referred to as supercritical fluids. The supercritical fluids commonly utilized include water, CO_2, and ammonia. Supercritical CO_2 is widely utilized as an extraction solvent because it is safe, non-flammable, cost-effective, and readily available. It also does not produce any inhibitory substances. The substrate, which comprises cellulose and hemicellulose, is fragmented by the sudden release of CO_2 pressure in a supercritical CO_2 explosion, enhancing the accessible surface area of the substrate for enzymes. Using supercritical CO_2 and ultrasound enhances the efficiency of enzymatic hydrolysis in the pre-treatment of sugar cane bagasse (Farmanbordar et al., 2018).

An effective method is required for the sulfite pre-treatment of woody biomass (both hardwoods and softwoods) to address lignocellulose recalcitrance (SPORL; 82–85). SPORL works efficiently at temperatures between 160°C and 190°C for durations ranging from 10 to 30 minutes. The addition of sulfite raises the medium's pH, leading to a reduction in fermentation inhibitors as a consequence (Shigechi et al., 2004). Partial sulfonation of lignin with sulfite ensures the softening of wood, significantly reducing the energy required for pre-treatment. A comparison of SPORL with explosion via acid-catalyzed steam reveals that SPORL is 30 times more efficient in terms of energy. Hydrolysis of gymnosperm substrates, pre-treated using SPORL with 15 FPU/g of cellulose enzyme loading, achieved a mass conversion rate of 95% after 48 hours. Microwaves can be used for the pre-treatment of raw materials that contain lignocellulose. Numerous studies indicate that microwaves create localized heating in feedstock, disrupting lignocellulosic structure and enhancing the accessibility of hemicellulose and cellulose for enzymatic hydrolysis (Lee et al. 2007).

8.7.4 Biological Processing of Raw Materials That Contain Lignocellulose

Utilizing enzymes to hydrolyze unprocessed lignocellulosic substances is also a component of biological pre-treatment. The speed at which lignocellulosic feedstock breaks down into fermentable sugars influences the overall efficiency of the bioprocess. Cellulose enzymatic hydrolysis through cellulase enables the fermentation of reducing sugars into ethanol by utilizing yeasts or bacteria. Endoglucanases, exoglucanases, and glucosidases represent the three primary categories of cellulases that play a role in the hydrolysis of cellulose (Schubert, 2006). The two current stages of enzymatic hydrolysis are primary and secondary. In the initial hydrolysis phase, the release of oligosaccharides into the liquid phase is facilitated by the activity of both endoglucanases and exoglucanases on the solid substrate surface. In the secondary hydrolysis phase, oligosaccharides are broken down further into cellobiose and glucose. Cellulases cannot reach lignin because it is strongly attached to cellulose. Peroxidases are crucial for lignin degradation, with manganese peroxidase (known as Mn-dependent peroxidase) and lignin peroxidase (referred to as ligninase) being the primary enzymes involved. These enzymes, identified in *Phanerochaete chrysosporium*, are recognized as true ligninases because of their high redox potential. Lignin peroxidase generates cation radicals when nonphenolic lignin substructures undergo oxidation (by losing one electron) and are chemically broken down (Tanadul et al., 2014).

8.8 PRODUCTION OF BIOETHANOL AND BIOBUTANOL FROM SUGAR-BASED RAW MATERIALS

The microorganism *S. cerevisiae* is frequently utilized for bioethanol production from sugar-rich materials due to its capability to convert sucrose into hexoses (glucose and fructose). Since *S. cerevisiae* cells require minimal oxygen for fatty acid and sterol synthesis during bioethanol production, a key bioprocess parameter is aeration. *Saccharomyces cerevisiae* is unable to tolerate higher levels of sugar and salts, along with increased temperatures in the environment. The media with cane molasses has a high osmolarity, restricting ethanol production because of the concentrations of sugar and salt present in the medium. Numerous tests were conducted to create *S. cerevisiae* strains that can endure elevated salt and temperature levels (Laureano-Perez et al., 2005). The feed-batch method requires low substrate levels initially and the extraction of yeast cells from the broth prior to distillation. Once the bioreactor and broth are prepared, the isolated yeast cells are often utilized in a new fed-batch bioethanol production method. This is a widely utilized method in Brazil's industrial technology for bioethanol production because of its superior volumetric productivity in bioprocesses. In this bioprocess, the enhancement of ethanol yield and productivity is specified by modifying the feeding process. In this batch

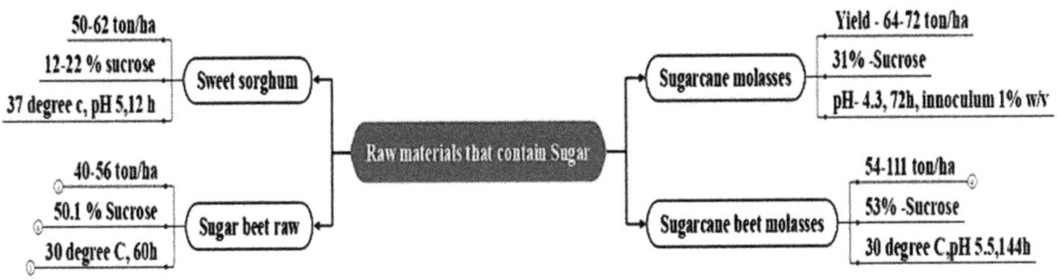

FIGURE 8.10 Depicts the bioethanol production process using sugar as a starting material.

fermentation method, the use of a flocculating microbial strain is essential. After the initial batch process is finished and the clarified broth is removed, yeast cell settling takes place at the bottom of the bioreactor. In the upcoming batch, incorporating an equal volume of new media into the bioreactor leads to low ethanol inhibition and elevated cell densities. This batch is reused until the yeast cells lose their activity and viability, after which a new inoculum must be prepared for reinoculating the system (Mosier et al., 2005). Figure 8.10 depicts the bioethanol production process using sugar as a starting material.

In ongoing bioethanol production processes, a series of continuous bioreactors is typically utilized to mitigate ethanol inhibition. This concept presupposes that the ethanol generated in the initial bioreactor is easily moved to the following bioreactors, leading to decreased ethanol inhibition. One method to enhance bioprocess productivity is by continuously extracting ethanol from the broth using vacuum or membrane systems throughout the active bioprocess, despite the rise in capital expenses. An enhanced supply of air in a continuous bioethanol production system can boost the viability, yield, and concentration of yeast cells. The continuous bioprocesses offer specific benefits compared to batch bioprocesses for bioethanol production, including lower bioreactor construction expenses, reduced maintenance and operational costs, enhanced process control, and improved efficiencies (Pan et al., 2005).

The fed-batch operational mode continues to be utilized by most bioethanol production facilities in Brazil because of its benefits at an industrial scale. Owing to the benefits of increased yeast cell concentrations, continuous bioprocess systems are utilized in 30% of industrial plants in Brazil for bioethanol production. Immobilize, recuperate and reuse yeast cells or control yeast proliferation to enhance yeast cell density.

In continuous bioprocesses for bioethanol production, the concentration of immobilized cells is notably high, allowing for easier control of bioprocesses at elevated dilution rates, which enhances bioprocess productivities (Chandra et al., 2007).

8.9 RAW MATERIALS THAT CONTAIN STARCH FOR THE PRODUCTION OF BIOETHANOL AND BIOBUTANOL

Yeasts (*Saccharomyces cerevisiae*, *Saccharomyces pastorianus*, *Schizosaccharomyces pombe*, and *Kluyveromyces* sp., etc.) capable of metabolizing starch hydrolysates are utilized in the dry-grind (67%) and wet-mill (33%) methods for producing bioethanol from maize starch, respectively. In the United States, dry milling is commonly employed for bioethanol production because of its lower capital and operational expenses (Huang et al., 2008). The entire corn is crushed and turned into a mash through mixing with water. Additionally, in a jet cooker maintained at 80–90°C, the mash is cooked for 15–20 minutes. The enzyme α-amylase is added in minimal amounts to improve liquefaction during the jet cooking process. In secondary liquefaction, additional α-amylase is introduced, which requires 90 minutes at 95°C.

Some modifications to the dry-grind method were implemented prior to fermentation to recover the maize germ and fiber. Wet milling generates a range of value-added co-products (including fiber, germ, starch, and gluten) prior to fermentation, resulting in an energy-efficient and cost-effective process. The demands of wet milling processes involve clean, soaked, and degermed maize to effectively obtain oil from corn germ. The defibration of maize occurs to obtain the fibers, gluten, and starch. The process of bioethanol production encompasses the same stages as the dry-grind technique (Kucera et al., 2017). The wet-mill yield is reported as 0.2919, whereas the dry-grind yield stands at 0.3235 per kilogram of corn. For bioethanol production, the process that involves cultivating yeast (from active dry yeasts) in the bioreactor during initial saccharification is utilized as well. Various terms used for this method include simultaneous saccharification, yeast growth, and fermentation.

8.10 RAW MATERIALS CONTAINING LIGNOCELLULOSE FOR BIOETHANOL AND BIOBUTANOL PRODUCTION

Biotechnology is establishing its presence in numerous industrial sectors such as healthcare, agricultural production, non-food applications of crops, and other products (e.g., biodegradable plastics, vegetable oils, biofuels, etc.) as well as environmental applications. Given the swiftly increasing economic demand, it is essential to utilize all local resources to their fullest potential to achieve the objective. Generating value-added products from agro-industrial waste can partially reduce production costs.

A vast quantity of agro-based waste and by-products is generated annually worldwide, and their inadequate management leads to environmental pollution. As a co-partner, the food industry plays a major role in generating a substantial portion of agro-based biological waste. The removal of various pollutants and waste is essential to promote a sustainable and eco-friendly environment. Industrialization greatly impacted the environment in terms of the introduction of pollutants and environmental toxins. Biotechnology significantly contributes to the removal of pollutants and utilizes the enhanced flexibility of microorganisms to break down or transform these substances. Agro-horticulture and industrial by-products abundant in lignocelluloses hold significant potential for enzyme production. Their conversion into beneficial products could improve environmental issues. Certain agro-wastes such as cereal straw, leaves, corncobs, etc. (New et al., 2022), as well as horticultural food processing waste, and other cellulose-dense waste materials like paper, coir, and wood extract waste. In many regions of the country, these materials are primarily utilized as livestock feed or contribute to environmental pollution (New et al., 2022).

From an economic perspective, cellulose and wood products, including the wood itself, are the most significant industrial materials, aside from foodstuffs, that are influenced by microorganisms. The creation of wood products like pulp, paper, and textiles from natural fibers such as cotton, flax, and jute is improved by microorganisms, particularly fungi. Cellulose, constituting roughly 40–50% of plant structure, is the most prevalent organic material on the planet. Effective use of these wastes in the environment will reduce pollution and transform them into valuable by-products is a significant issue.

8.11 INSIGHTS OF BIOBUTANOL PRODUCTION

Biobutanol production is increasingly recognized as a renewable alternative to fossil fuels, primarily through the acetone-butanol-ethanol (ABE) fermentation utilizing Clostridium species. However, the large-scale commercialization of biobutanol encounters several challenges, including low yields, elevated production costs, inefficient substrate utilization, and energy-intensive recovery processes. Although lignocellulosic biomass presents a promising feedstock, its efficient conversion continues to pose significant challenges (Lorenci Woiciechowski et al., 2020). At present, industrial adoption remains limited, with only a handful of commercial plants operating globally. Future advancements in metabolic engineering and synthetic biology are aimed at enhancing butanol yields, microbial tolerance, and substrate flexibility. Improved fermentation techniques, such as continuous processing and in situ product recovery, are anticipated to increase production efficiency. The utilization of lignocellulosic and waste biomass is expected to reduce costs, while the integration of biobutanol production into biorefineries may enhance economic viability (Preethi et al., 2021). Furthermore, the growing demand for biofuels in the aviation and transportation sectors, coupled with supportive policies such as carbon credits and renewable fuel mandates, could propel commercialization efforts. With ongoing research and industrial investments, biobutanol is positioned to become a significant biofuel and chemical feedstock, contributing to a more sustainable and low-carbon future. Artificial intelligence (AI) and optimization models are playing an essential role in enhancing fermentation performance and predicting critical parameters in biobutanol production. AI-driven methodologies, including machine learning and deep learning, have the capacity to analyze extensive datasets to optimize microbial growth conditions, substrate utilization, and fermentation kinetics. By discerning patterns within complex biological processes, AI models may facilitate real-time monitoring and predictive control, thereby reducing trial-and-error experimentation and enhancing process efficiency (Borah et al., 2019). Optimization models, such as response surface methodology, genetic algorithms, and artificial neural networks (ANNs), are instrumental in fine-tuning key variables including pH, temperature, nutrient availability, and bioreactor conditions. These models contribute to maximizing butanol yields, minimizing by-product formation, and bolstering microbial tolerance to solvents. Moreover, AI can enhance in situ product recovery strategies by predicting solvent toxicity thresholds and optimizing extraction techniques such as pervaporation and gas stripping. The integration of AI with process automation and sensor technologies enables adaptive control of fermentation parameters, thereby reducing production costs and increasing scalability. By harnessing the capabilities of AI and optimization models, biobutanol production can evolve into a more predictable, resource-efficient, and economically viable process, thereby accelerating its significance in the biofuel industry and the circular bioeconomy (MohdSabri et al., 2022).

8.12 IMPACT OF OPERATIONAL PARAMETERS ON BIOETHANOL AND BIOBUTANOL PRODUCTION

The operational factors, such as agitation, medium pH, incubation duration, temperature, and the influence of toxins, significantly affect the success of each fermentation process.

8.12.1 Agitation

The rotation of the incubator shaker is essential for achieving complete mixing of the medium ingredients, ensuring uniform oxygen transfer rates. The agitation speed is crucial for attaining the best biobutanol concentration and yield. The effect of rotation is crucial for efficient fermentation, as it ensures proper mixing and enhances mass and heat transfer. It assists in the transfer of mass between two separate phases and guarantees an even distribution of microbial cells in a uniformly mixed medium. An increased agitation speed lowers temperature variations, enhances the uniformity of the broth, and promotes production. Nonetheless, too much agitation can damage the cells, so it's crucial to identify the optimal speed for improved productivity (Sahoo et al., 2022).

8.12.2 pH

The pH level is an important element that significantly affects the generation of biobutanol and bioethanol. The acids generated during fermentation establish a buffering effect, resulting in a consistent pH level in media that begins with a high initial pH. This scenario may obstruct fermentation microorganisms from adjusting to swift environmental shifts. On the other hand, if the starting pH is excessively low, the conditions might turn unfavorable for butanol producers. However, with certain modifications and minor self-regulation to environmental conditions like pH, these microorganisms can slowly start to produce butanol at a moderate pace. Studies indicate diverse outcomes across a wide pH spectrum, with the best production found in the pH range of 4.5–5.5. Moreover, various microbial cultures can influence both the effect and the output. High acid production in the acidogenic phase can also inhibit acid crashes. Numerous documented experiments indicate that a reduced pH increases butanol production (Shabbirahmed et al., 2023).

8.12.3 Incubation Period

A key element to take into account is the incubation duration, which distinguishes the stages of the fermentation process into solventogenic and acidogenic phases. The duration of fermentation affects the concentrations of biobutanol and bioethanol, since microorganisms will continue to multiply in a fermenting solution medium that can convert substrate into biobutanol over time. An elevated level of microorganisms leads to enhanced biobutanol production during fermentation. Nonetheless, fermentation is restricted to a maximum period of 7 days. As microorganisms enter a death phase, their ability to transform substrate into biobutanol decreases, even though biobutanol levels rise. Ninety hours after inoculation signals the start of the solventogenesis phase, lasting for an additional 90 hours. After inoculation, the acidogenesis stage begins and continues for about 30 hours, resulting in the production of different acids. These stages depend on the initial glucose availability at particular pH levels and temperatures (Chandra et al., 2021)

8.12.4 Temperature

A crucial factor affecting the economics of a procedure is the upkeep of an ideal process temperature. Temperature influences microbial cell proliferation, spore development, germination, and general microbial function, which in turn affect biobutanol generation. In 1983, a study by Lin et al. exposed the natural strain *Clostridium acetobutylicum* ATCC 824 to different butanol concentrations (0.0%, 1.0%, 1.5%, v/v) and temperatures (22°C, 37°C, 42°C). The original strain displayed more fluid membranes, whereas the mutant strain exhibited no notable reaction to the butanol levels ranging from 22°C to 37°C. Both strains showed an increased ratio of saturated fatty acids to unsaturated fatty acids at lower temperatures. Nonetheless, the adverse impacts of the increased temperature (42°C) were seen exclusively in the mutant strain. This study emphasizes the capacity to create a stable system in the presence of butanol (Debnath et al., 2021).

8.12.5 Effect of Toxins

Redox potential influences the balance of electron transfer and, therefore, the cellular metabolism of fermentation by-products. This potential is affected by temperature and the proportion of oxidizing to reducing agents. They found increased butanol production at a redox potential of 460 mV, equating to 13.2 g/l. The metabolic transition toward solvent production can be associated with a rise in ATP availability and a decrease in the use of nicotinamide adenine dinucleotide (NADH). A rise in NADH generation or a reduction in its consumption could elevate redox potential, since the fermentation process of butanol needs 4 moles of NADH for every mole of butanol created. The internal metabolic flow needs to adjust according to variations in the NADH/NAD+ ratio. The redox equilibrium is preserved by the activation of secondary alcohol dehydrogenase and butanol dehydrogenase (Bao et al., 2020)

8.13 DISPOSAL GUIDELINES

Through extensive research on pre-treatment methodologies, fermentation techniques, and microbial strains, significant advancements have been made in the yield and production efficiency of lignocellulosic bioethanol and biobutanol. Nevertheless, achieving bioethanol and biobutanol production from lignocellulosic biomass at an industrial scale remains challenging due to the high economic and environmental costs involved. Given the current technological landscape, it is nearly impossible to reduce these costs while maintaining high efficiency in the production of lignocellulosic bioethanol and biobutanol (Arregoitia-Sarabia et al., 2020). Consequently, considerable efforts have been directed toward the development of recycling techniques. These include the recycling of waste materials to mitigate secondary pollution, the recycling of cellulases and yeasts to reduce costs, and the recycling of residues for the production of by-products, all aimed at enhancing both economic and environmental efficiency (Adesanya et al., 2022). This study discusses nearly all recycling techniques associated with conversion processes, including pre-treatment, enzymatic hydrolysis, fermentation, and distillation, while identifying the primary challenges faced by these recycling methods: (1) The generation and accumulation of toxic substances during the recycling process, such as furfural, 5-HMF, and secondary degradation products of lignocellulose, are unavoidable (Abel et al., 2022). This accumulation can lead to reduced efficiency or even the cessation of recycling efforts, particularly in the case of waste liquid recycling. Therefore, the rapid testing and specific removal of these toxins are critical for ensuring continuous recycling. (2) Reduced efficiency often arises from diminished concentrations of recycling targets (for pre-treatment chemicals) or their activity levels (for yeast and cellulase). Consequently, the effective recycling times for nearly all the techniques summarized remain low, highlighting the urgent need to improve recycling efficiency and sustainability in the future. (3) Economic feasibility is often constrained by the additional investments required, especially concerning by-product production. Therefore, conducting mass balance and economic analyses of recycling techniques at either pilot or industrial scales is essential moving forward. (4) Most recycling techniques or strategies tend to concentrate on one or two stages of the lignocellulosic bioethanol conversion process, which hinders the comprehensive utilization of feedstock and the complementarity of various recycling processes. Thus, the integration of diverse recycling techniques may represent a promising avenue for enhancing lignocellulosic bioethanol conversion (Zhao et al. 2022). In light of the examination of recycling methods and their associated challenges, a comprehensive integration concept for recycling techniques is introduced. This concept primarily encompasses three key elements: (1) Comprehensive utilization of all components: in addition to converting cellulose into bioethanol, it is essential to fully utilize other components of lignocellulosic biomass, such as hemicellulose and lignin. (2) Recycling of high-value additives: while lignocellulosic biomass is relatively inexpensive, the additives required for conversion, such as yeast and cellulase, are costly. Therefore, it is crucial to reuse high-value additives multiple times, and the proposed recycling technique offers several strategies to facilitate this. (3) Prevention of waste generation: despite the complete utilization of lignocellulosic biomass and high-value additives, the generation of final waste is unavoidable, which significantly impacts both economic and environmental performance. Consequently, the proposed recycling technique takes into account all waste and feedstock materials. To the best of our understanding, no strategy currently exists that can fulfill all three requirements at the same time. Nevertheless, it is anticipated that the suggested recycling methods will be implemented with advancements in recycling efficiency and the emergence of new technologies.

8.14 CONCLUSION AND OUTLOOK

Energy security, climate change, and the escalating prices of oil, among other factors, significantly impact the global population. Furthermore, the demands posed by the continuously growing population place considerable pressure on energy resources. It is evident that the valorization of agricultural and related wastes has the potential to generate valuable products, such as bioenergy. These approaches not only mitigate waste accumulation on the planet but also propose a sustainable energy source accessible to all. The principles of the circular economy further facilitate advancements in biotechnology. Compared to conventional fuels, biofuels derived from agricultural wastes offer lower pollution rates and may prove beneficial in the long term. The reduction in biowaste accumulation, coupled with the transition from conventional fuels to biofuels, represents a progressive step toward environmental conservation. Additionally, the efficiency and cost-effectiveness of various biofuels, including bioethanol, and biobutanol, can be enhanced through rigorous research, advancing the pursuit of a more sustainable environment.

REFERENCES

Abel, S., Tesfaye, J. L., Gudata, L., Nagaprasad, N., Subramanian, K., Mani, M., Shanmugam, R., Dwarampudi, L. P., Roy, A., Stalin, B., & Krishnaraj, R. (2022). Biobutanol preparation through sugar-rich biomass by *Clostridium saccharoperbutylacetonicum* conversion using ZnO nanoparticle catalyst. *Biomass Conversion and Biorefinery*. https://doi.org/10.1007/S13399-022-02424-1

Adesanya, Y., Atiyeh, H. K., Olorunsogbon, T., Khanal, A., Okonkwo, C. C., Ujor, V. C., Shah, A., & Ezeji, T. C. (2022). Viable strategies for enhancing acetone-butanol-ethanol production from non-detoxifed switchgrass hydrolysates. *Bioresource Technology, 344*(Pt A),126167. https://doi.org/10.1016/j.biortech.2021.126167

Arregoitia-Sarabia, C., González-Revuelta, D., Fallanza, M., Gorri, D., & Ortiz, I. (2020). Polymer inclusion membranes containing ionic liquids for the recovery of n-butanol from ABE solutions by pervaporation. *Separation and Purification Technology, 248*, 117101. https://doi.org/10.1016/J.SEPPUR.2020.117101

Balat, M., & Balat, H. (2009). Recent trends in global production and utilization of bioethanol fuel. *Applied Energy, 86*(11), 2273–2282. https://doi.org/10.1016/j.apenergy.2009.03.015

Bansal, S., Jadaun, J. S., & Singh, S. P. (2021) Resource recovery from the abundant agri-biomass. *Bioprospecting of Plant Biodiversity for Industrial Molecules*131–152. https://doi.org/10.1002/9781119718017.CH7

Bao, T., Feng, J., Jiang, W., Fu, H., Wang, J., & Yang, S. T. (2020). Recent advances in n-butanol and butyrate production using engineered *Clostridium tyrobutyricum*. *World Journal of Microbiology and Biotechnology, 36*, 1–14. https://doi.org/10.1007/S11274-020-02914-2/METRICS

Barakat, A., Monlau, F., Solhy, A., & Carrere, H. (2015). Mechanical dissociation and fragmentation of lignocellulosic biomass: Effect of initial moisture, biochemical and structural proprieties on energy requirement. *Applied Energy, 142*, 240–246. https://doi.org/10.1016/j.apenergy.2014.12.076

Baz, K., Cheng, J., Xu, D., Abbas, K., Ali, I., Ali, H., & Fang, C. (2021). Asymmetric impact of fossil fuel and renewable energy consumption on economic growth: A nonlinear technique. *Energy, 226*, 120357. https://doi.org/10.1016/J.ENERGY.2021.120357

Bharathiraja, B., Jayamuthunagai, J., Sudharsanaa, T., Bharghavi, A., Praveenkumar, R., Chakravarthy, M., & Yuvaraj, D. (2017). Biobutanol – An impending biofuel for future: A review on upstream and downstream processing techniques. *Renewable and Sustainable Energy Reviews*, *68*, 788–807. https://doi.org/10.1016/J.RSER.2016.10.017.

Bhaskar, T., Bhavya, B., Singh, R., Naik, D. V., Kumar, A., & Goyal, H. B. (2011). Thermochemical conversion of biomass to biofuels. In A. Pandey, C. Larroche, S. C. Ricke, C. G. Dussap, & E. Gnansounou (Eds.), *Biofuels—Alternative feedstocks and conversion processes* (pp. 51–77). Academic Press. https://doi.org/10.1016/B978-0-12-385099-7.00003-6

Borah, A. J., Roy, K., Goyal, A., & Moholkar, V. S. (2019). Mechanistic investigations in biobutanol synthesis via ultrasound-assisted ABE fermentation using mixed feedstock of invasive weeds. *Bioresource Technology*, *272*, 389–397. https://doi.org/10.1016/J.BIORTECH.2018.10.063

Chandra, M. S., Srinivasulu, M., Yadav, P. S., Ramesh, B., Narasimha, G., & Chandrasekhar, T. (2021) Value added products from agriculture, paper and food waste: A source of bioenergy production. *Bioenergy Research: Commercial Opportunities & Challenges* 91–126. https://doi.org/10.1007/978-981-16-1190-2_3

Chandra, R. P., Bura, R., Mabee, W. E., Berlin, A., Pan, X., & Saddler, J. N. (2007). Substrate pretreatment: The key to effective enzymatic hydrolysis of lignocellulosics? In L. Olsson (Ed.),*Biofuels: Advances in biochemical engineering/biotechnology* 108, 67–93). Springer. https://doi.org/10.1007/10_2007_064

Cherubini, F. (2010). The biorefinery concept: Using biomass instead of oil for producing energy and chemicals. *Energy Conversion and Management*, *51*(7), 1412–1421. https://doi.org/10.1016/j.enconman.2010.01.015

Cherubini, F., & Strømman, A. H. (2011). Principles of biorefining. In A. Pandey, C. Larroche, S. C. Ricke, C. G. Dussap, & E. Gnansounou (Eds.), *Biofuels—Alternative feedstocks and conversion processes* (pp. 3–24). Academic Press. https://doi.org/10.1016/B978-0-12-385099-7.00001-2

Debnath, C., Bandyopadhyay, T. K., Bhunia, B., Mishra, U., Narayanasamy, S., & Muthuraj, M. (2021). Microalgae: Sustainable resource of carbohydrates in third-generation biofuel production. *Renewable and Sustainable Energy Reviews*, *150*, 111464. https://doi.org/10.1016/J.RSER.2021.111464

Ding, J., Xu, M., Xie, F., Chen, C., & Shi, Z. (2019). Efficient butanol production using cornstarch and waste *Pichia pastoris* semi-solid mixture as the substrate. *Biochemical Engineering Journal*, *143*, 41–47. https://doi.org/10.1016/j.bej.2018.12.017

Farmanbordar, S., Amiri, H., & Karimi, K. (2018). Simultaneous organosolvpretreatment and detoxification of municipal solid waste for efficient biobutanol production. *Bioresource Technology*, *270*, 236–244. https://doi.org/10.1016/J.BIORTECH.2018.09.017

Festel, G., Würmseher, M., Rammer, C., Boles, E., & Bellof, M. (2013). *Modelling production cost scenarios for biofuels and fossil fuels in Europe*. Discussion paper No. 13–075. ZEW—Centre for European Economic Research. https://ftp.zew.de/pub/zew-docs/dp/dp13075.pdf

Huang, H. J., Ramaswamy, S., Tschirner, U. W., & Ramarao, B. V. (2008). A review of separation technologies in current and future biorefineries. *Separation and Purification Technology*, *62*(1), 1–21. https://doi.org/10.1016/j.seppur.2007.12.011

İçöz, E., Tuğrul, K. M., Saral, A., & İçöz, E. (2008). Research on ethanol production and use from sugar beet in Turkey. *Biomass and Bioenergy*, *33*(1), 1–7. https://doi.org/10.1016/j.biombioe.2008.05.005

Jobling, S. (2004). Improving starch for food and industrial applications. *Current Opinion in Plant Biology*, *7*(2), 210–218. https://doi.org/10.1016/j.pbi.2003.12.001

Khanal, S. K. (2008). *Anaerobic biotechnology for bioenergy production: Principles and applications*. Wiley-Blackwell. https://doi.org/10.1002/9780813804545

Khuong, L. S., Masjuki, H. H., Zulkifli, N. W. M., NizaMohamad, E., Kalam, M. A., & Alabdulkarem, A. (2017). Effect of gasoline-bioethanol blends on the properties and lubrication characteristics of commercial engine oil. *RSC Advances, 7(25),*15005–15019. https://doi.org/10.1039/C7RA00357A

Koller, M., Bona, R., Hermann, C., Horvat, P., Martinz, J., & Neto, J. (2005). Biotechnological production of poly (3-hydroxybutyrate) with Wautersiaeutropha by application of green grass juice and silage juice as additional complex substrates. *Biocatalysis Biotransformation*, *23*(5), 329–337. https://doi.org/10.1080/10242420500292252

Kucera, D., Benesova, P., Ladicky, P., Pekar, M., Sedlacek, P., & Obruca, S. (2017). Production of polyhydroxyalkanoates using hydrolyzates of spruce sawdust: Comparison of hydrolyzates detoxification by application of overliming, active carbon, and lignite. *Bioengineering (Basel)*, *4*(2).https://doi.org/10.3390/bioengineering4020053

Lange, J. P. (2007). Lignocellulose conversion: An introduction to chemistry, process and economics. *Biofuels, Bioproducts and Biorefining*, *1*(1), 39–48. https://doi.org/10.1002/bbb.7

Laureano-Perez, L., Teymouri, F., Alizadeh, H., & Dale, B. E. (2005). Understanding factors that limit enzymatic hydrolysis of biomass. *Applied Biochemistry Biotechnology*, *121–124*(1–3), 1081–1099. https://doi.org/10.1385/ABAB:124:1-3:1081

Lee, S., Speight, J. G., & Loyalka, S. K. (Eds.). (2007). *Handbook of alternative fuel technologies*. CRC Press. https://doi.org/10.1201/9781420014518

Li, K., Liu, S., & Liu, X. (2014). An overview of algae bioethanol production. *International Journal of Energy Research*, *38*(8), 965–977. https://doi.org/10.1002/er.3164

Linoj, K. N. V., Dhavala, P., Goswami, A., & Maithel, S. (2006). Liquid biofuels in South Asia: Resources and technologies. *Asian Biotechnology and Development Review*, *8*(2), 31–49.

Lorenci Woiciechowski, A., Dalmas Neto, C. J., de Souza, P., Vandenberghe, L., de Carvalho Neto, D. P., Novak Sydney, A. C., Letti, L. A. J., Karp, S. G., Zevallos Torres, L. A., & Soccol, C. R. (2020). Lignocellulosic biomass: Acid and alkaline pretreatments and their effects on biomass recalcitrance – Conventional processing and recent advances. *Bioresource Technology*, *304*, 122848. https://doi.org/10.1016/J.BIORTECH.2020.122848

Manikandan, K., Saravanan, V., & Viruthagiri, T. (2008). Kinetics studies on ethanol production from banana peel waste using mutant strain of *Saccharomyces cerevisiae. Indian Journal of Biotechnology*, *1*(7), 83–88.

MohdSabri, M. N. I., Ahmad Mokhtar, A. M., Shafq, M. D., Alkarimiah, R., & MohdZainiMakhtar, M. (2022) Different types of pre-treatments for renewable bioresources.*Green Energy and Technology*,45–78. DOI: 10.1007/978-981-16-9314-4_3

Mosier, N., Wyman, C., Dale, B. D., Elander, R., Lee, Y. Y., & Holtzapple, M. (2005). Features of promising technologies for pretreatment of lignocellulosic biomass. *Bioresource Technology*, *96*(6), 673–686. https://doi.org/10.1016/j.biortech.2004.06.025

Mousdale, D. M. (2008). *Biofuels. Biotechnology, chemistry and sustainable development*. CRC Press.

Mussatto, S. I., Dragone, G., Guimarães, P. M. R., Silva, J. P. A., Carneiro, L. M., & Roberto, I. C. (2010). Technological trends, global market, and challenges of bio-ethanol production. *Biotechnology Advances*, *28*(6), 817–830. https://doi.org/10.1016/j.biotechadv.2010.07.001

New, E. K., Tnah, S. K., Voon, K. S., Yong, K. J., Procentese, A., Shak, K. P. Y., Subramonian, W., Cheng, C. K., & Wu, T. Y. (2022). The application of green solvent in a biorefinery using lignocellulosic biomass as a feedstock. *The Journal of Environmental Management*, *307*, 114385.

Pan, X., Xie, D., Gilkes, N., Gregg, D. J., & Saddler, J. N. (2005). Strategies to enhance the enzymatic hydrolysis of pretreated softwood with high residual lignin content. *Applied Biochemistry Biotechnology*, *121–124*, 1069–1079. https://doi.org/10.1385/ABAB:124:1-3:1069

Pavlečić, M., Rezić,T., IvančićŠantek, M., Horvat, P., & Šantek, B. (2017). Bioethanol production from raw sugar beet cossettes in horizontal rotating tubular bioreactor. *Bioprocess and Biosystems Engineering*, *40*(11), 1679–1688. https://doi.org/10.1007/s00449-017-1823-x

Pavlečić, M., Vrana, I., Vibovec, K., IvančićŠantek, M., Horvat, P., & Šantek, B. (2010). Ethanol production from different intermediates of sugar beet processing. *Food Technology and Biotechnology*, *48*(3), 362–367.

Preethi, G. M., Kumar, G., Karthikeyan, O. P., Varjani, S., & Rajesh Banu, J. (2021). Lignocellulosic biomass as an optimistic feedstock for the production of biofuels as valuable energy source: Technoeconomic analysis, environmental impact analysis, breakthrough and perspectives. *Environmental Technology & Innovation*, *24*, 102080. https://doi.org/10.1016/J.ETI.2021.102080

Prieur-Vernat, A., & His, S. (2007). *Biofuels worldwide*. Panorama 2007. IFP-Innovation Energy Environment. https://www.firp.ula.ve/archivos/material_web_4xx/07_IFP_Biofuels.pdf

Sahoo, A., Kumar, S., & Mohanty, K. (2022). A comprehensive characterization of non-edible lignocellulosic biomass to elucidate their biofuel production potential. *Biomass Conversion and Biorefinery*, *12*,5087–5103. https://doi.org/10.1007/s13399-020-00924-6

Schubert, C. (2006). Can biofuels finally take center stage?*Nature Biotechnology*, *24*, 777–784. https://doi.org/10.1038/nbt0706-777

Senthilkumar, V., & Gunasekaran, P. (2009). Bioethanol from biomass production of ethanol from molasses. In A.Pandey (Ed.), *Handbook of plant-based biofuels*(pp. 73–86). CRC Press.

Shabbirahmed, A. M., Joel, J., Gomez, A., Patel, A. K., Singhania, R. R., & Haldar, D. (2023). Environment friendly emerging techniques for the treatment of waste biomass: A focus on microwave and ultrasonication processes. *Environmental Science and Pollution Research*, *30*(33), 79706–79723.

Shigechi, H., Koh, J., Fujita, Y., Matsumoto, T., Bito, Y., & Ueda, M. (2004). Direct production of ethanol from raw corn starch via fermentation by use of a novel surface-engineered yeast strain codisplaying glucoamylase and α-amylase. *Applied and Environment Microbiology*, *70*(8), 5037–5040. https://doi.org/10.1128/AEM.70.8.5037-5040.2004

Soccol, C. R., Faraco, V., Karp, S., Vandenberghe, L. P. S., Thomaz-Soccol, V., & Woiciechowski, A. (2011). Lignocellulosic bioethanol: Current status and future perspectives. In A. Pandey, C. Larroche, S. C. Ricke, C. G. Dussap, & E. Gnansounou (Eds.), *Biofuels—Alternative feedstocks and conversion processes*(pp. 101–122). Academic Press. https://doi.org/10.1016/B978-0-12-385099-7.00005-X

Solomon, B. D., Barnes, J. R., & Halvorsen, K. E. (2007). Grain and cellulosic ethanol: History, economics, and energy policy. *Biomass and Bioenergy*, *31*(6), 416–425. https://doi.org/10.1016/j.biombioe.2007.01.023

Suresh, A. R., Alphonse Mani, A. S., & Muthuvelu, K. S. (2024). Investigation of pre-treatment techniques on spent substrate of *Pleurotus ostreatus* for enhanced biobutanol production using *Clostridiumacetobutylicum* MTCC 11274. *Bioresource Technology*, *394*, 130228. https://doi.org/10.1016/j.biortech.2023.130228.

Tanadul, O. U., VanderGheynst, J. S., Beckles, D. M., Powell, A. L. T., & Labavitch, J. M. (2014). The impact of elevated CO2 concentration on the quality of algal starch as a potential biofuel feedstock. *Biotechnology and Bioengineering*, *111*(7), 1323–1331. https://doi.org/10.1002/bit.25203

Tomás-Pejó, E., Alvira, P., Ballesteros, M., & Negro, M. J. (2011). Pretreatment technologies for lignocellulose-to-bioethanol conversion. In A. Pandey, C. Larroche, S. C. Ricke, C. G. Dussap, & E. Gnansounou (Eds.), *Biofuels—Alternative feedstocks and conversion processes*(pp. 149–176). Academic Press. https://doi.org/10.1016/B978-0-12-385099-7.00007-3

Tsai, T. Y., Lo, Y. C., Dong, C. D., Nagarajan, D., Chang, J. S., & Lee, D. J. (2020). Biobutanol production from lignocellulosic biomass using immobilized *Clostridium acetobutylicum*. *Applied Energy*, *277*, 115531. https://doi.org/10.1016/J.APENERGY.2020.115531

Van Thuijl, E., & Deurwaarder, E. P. (2006). *European biofuels policies in retrospect. Petten, the Netherlands: Energy Research Centre of the Netherlands (ECN).* https://www.ecn.nl/publications/PdfTetch.aspx?nr=ECN-C-06-016

Yeong, T. K., Jiao, K., Zeng, X., Lin, L., Pan, S., & Danquah, M. K. (2018). Microalgae for biobutanol production – Technology evaluation and value proposition. *Algal Research*, *31*, 367–376. https://doi.org/10.1016/J.ALGAL.2018.02.029

Yiin, C. L., Yap, K. L., Ku, A. Z. E., Chin, B. L. F., Lock, S. S. M., Cheah, K. W., Loy, A. C. M., & Chan, Y. H. (2021). Recent advances in green solvents for lignocellulosic biomass pretreatment: Potential of choline chloride (ChCl) based solvents. *Bioresource Technology*, *333*, 125195. https://doi.org/10.1016/J.BIORTECH.2021.125195

Yüksel, F., & Yüksel, B. (2004). The use of ethanol–gasoline blend as a fuel in an SI engine. *Renewable Energy*, *29*(7), 1181–1191. https://doi.org/10.1016/j.renene.2003.11.012

Zhao, L., Sun, Z. F., Zhang, C. C., Nan, J., Ren, N. Q., Lee, D. J., & Chen, C. (2022) Advances in pretreatment of lignocellulosic biomass for bioenergy production: Challenges and perspectives. *Bioresource Technology*,*343*, 126123. https://doi.org/10.1016/J.BIORTECH.2021.126123

Zhao, L., Zhang, X., Xu, J., Ou, X., Chang, S., & Wu, M. (2015). Techno-economic analysis of bioethanol production from lignocellulosic biomass in China: Dilute-acid pretreatment and enzymatic hydrolysis of corn stover. *Energies*, *8*(5), 4096–4117. https://doi.org/10.3390/en8054096

9 Biogas Process Optimization for Green Methane and Biohydrogen Generation and Sustainable Disposal Strategies

Mumna Munir, Mushtaq Ahmad, Mamoona Munir, Shazia Sultana, and Nigar Mursal

9.1 INTRODUCTION

The world's population is growing at an accelerated rate, which is making environmental problems worse and putting more strain on energy infrastructure. Global energy is still mostly derived from fossil fuels, and demand is being greatly increased by growing industrialization and urbanization. Even with the rise of renewable energy, fossil fuels still account for a significant amount of the world's energy production, which the International Energy Agency (IEA) estimates was 14,485. Approximately 30.9%, 26.8%, and 23.2% of the world's energy consumption in that year came from oil, coal, and natural gas, respectively (IEA, 2019). China was the world's top producer of coal by 2021, and the United States, European Union, India, and China were its main users. China and the United States were the biggest users of oil, while Saudi Arabia, Russia, and the United States produced the most. Similar to this, the United States and Russia produced the majority of natural gas, which satisfied the high demand in the United States, the Middle East, the European Union, Russia, and China (Hossain, 2023).

According to the 2021 IEA Key World Energy Statistics (IEA, 2021), renewable choices including bioenergy (8.5%), nuclear (4.9%), hydro (2.6%), and others like solar and geothermal (2.2%) were still underdeveloped despite the growing availability of alternative energy sources. The economic and environmental effects of relying too heavily on fossil fuels, as well as the greenhouse gas (GHG) emissions that contribute to climate change, continue to worry the world community. As a result, nations have been urged by global frameworks such as the Paris Agreement to adopt climate action plans every five years in order to cut emissions and prepare for climate change (Bodansky, 2016).

 DOI: 10.1201/9781003642589-9

Renewable energy sources provide a more environmentally friendly and sustainable option than fossil fuels. For example, geothermal, solar, and biomass energy all lead to far reduced emissions. Among the cleanest energy sources include hydroelectric, wind, and solar. According to the International Renewable Energy Agency, solar and biomass energy are important to the global transition to sustainable energy. The IEA predicts that renewable electricity capacity will increase by 50% between 2019 and 2024. Estimates estimate that biofuel output would increase to 180 billion liters by 2025, despite a drop from 165 to 142 billion liters between 2019 and 2020 (IEA, 2021; IEA, 2024).

Biodiesel, bioethanol, biogas, hydrogen, and biobutanol are examples of organic-based biofuels. Their value is increased by their broad availability and adaptability, especially in the transportation and industrial sectors. Converting waste biomass into biofuels is an efficient waste management approach that also meets energy needs, particularly in nations switching from coal and oil to domestic biomass. Biofuels have a number of benefits over fossil fuels, such as cheaper production costs for heating and power, off-grid energy options, and less of an impact on the environment. They see the transition to a low-carbon economy to be advantageous from an economic standpoint. One critical component of a modern, sustainable circular economy is the effective generation of biofuels from renewable biomass sources (IEA, 2021). Biogas and biohydrogen are developing as important waste valorization processes.

Additionally, the residual from the first biohydrogen reactor can be fed into a second biomethane reactor to achieve zero-waste methods. The feedstocks used to produce biofuel fall into a variety of categories. Edible plant oils and crops high in carbohydrates, such as corn and sugarcane, are examples of first-generation feedstocks. In contrast, the second generation concentrates on lignocellulosic biomass and non-food waste (FW). The third generation investigates algae and aquatic biomass that do not threaten food sources. Fourth-class feedstocks are crucial and equally vital for the production of high-value electrofuels, photobiological, and electrofuels using new protocols (Varling et al., 2023).

The stepwise synthesis of biohydrogen, biogas, and biohythane increases biomass energy recovery as well as waste reduction (Amato et al., 2023). Determining functional units and examining every phase, from raw material collection to conversion, distribution, usage, and waste management, are all part of the life cycle assessment of such systems. Inputs include energy, raw materials, and auxiliary materials. Gaseous energy products and byproducts like digestate, sludge, and pollutants are examples of outputs. Lastly, a significant step toward sustainable energy and waste reduction is the enhanced capacity to produce biohydrogen and biomethane through integrated, effective biogas systems. In order to support the circular bioeconomy principles, this chapter emphasizes the sustainable use and disposal of digestate as well as creative methods for boosting the production of green methane (CH_4) and biohydrogen.

9.2 FEEDSTOCK SELECTION AND PRE-TREATMENT STRATEGIES

The three primary structural components of lignocellulosic biomass are cellulose, hemicellulose, and lignin. Unlike cellulose, which can be both crystalline and amorphous, hemicellulose is totally amorphous and easier to hydrolyze. Lignin-carbohydrate complexes are chemically linked inside plant cell walls. Monlau et al. (2012) developed a model to predict the biochemical methane potential of lignocellulosic materials based on structural and compositional characteristics.

Bartacek et al. (2007) Numerous pretreatment procedures have also been investigated and classed as physical, chemical, biological, or a mix of these in order to boost production yield and digestibility of complicated substrates. These pretreatments are classified into numerous categories, including thermal, mechanical, microwave, ultrasonic, and pulsed electric field. Ozonation, ionic liquids (ILs), acidic or alkaline solutions, and the Fenton process are all chemical pretreatment methods that effectively break down large macromolecules into smaller ones. Microorganisms (aerobic or anaerobic, enzymes, or fungi) are used in biological pretreatments. The literature reveals that each of these techniques has benefits and drawbacks when used alone; nevertheless, their simultaneous or sequential application can boost CH_4 yields.

Pure carbohydrates such as glucose, energy crops such as Miscanthus, solid organic wastes such as FW, and industrial effluents such as those from the pulp and paper industry are the four substrates applied in fermentative hydrogen production. A very large selection of substrates has been tested for the biohydrogen production. For instance, domestic organic waste has been investigated at varying temperatures, and thermophilic conditions have been found to produce more hydrogen yields (Kim et al., 2009).

Wastewaters with high carbohydrate content and residual biomass have also proved suitable for dark fermentation processes. Molasses and cheese whey are two substrates that have been studied in immobilized configurations and continuous stirred-tank reactors. Sediment layers from breweries, sugar beet, cellulose, xylose, starch (soluble and particulate), and effluent from sugar beet processing have been studied as other substrates (Fan et al., 2006; Singh et al., 2015).

Agricultural crops, crop residues, lignocellulosic biomass (wood and waste products thereof), food process by-products, aquatic plants, algae, and domestic and industrial effluents are a few examples of the sources of biomass for fermentative biohydrogen production. Microbial fermentation is also employed to generate hydrogen from biomass containing high moisture content materials like animal manures and sewage sludge. In particular for agriculturally rich nations, these biomass feedstocks are increasingly being recognized as valuable feedstocks for renewable hydrogen, where their integration into bioenergy systems can enhance sustainability and economic development.

Feedstocks with high simple or complex sugars are particularly suited for hydrogen evolution because carbohydrates are the main catalyzers of fermentative hydrogen processes (Ntaikou et al., 2010). When selecting the best ones, consider pricing, availability, carbohydrate content, and biodegradability. Sugars like glucose, lactose, and sucrose are chosen because they are highly biodegradable, but their commercial use is limited due to cost constraints. As a result, inexpensive, renewable feedstocks are essential for economically feasible hydrogen production (Das & Veziroğlu, 2001). Pretreatment is a key step in the biohydrogen generation process because it degrades lignin, increases sugar release, and reduces cellulose crystallinity. Effective pretreatment also facilitates feedstock handling, digester loading, and the avoidance of difficulties such as floating material layers during digestion in industrial applications.

Aerobic and anaerobic processes are used in the biological processing of lignocellulosic wastes, but not municipal wastes. Although fungal treatment can lead to longer pretreatment periods and cellulose degradation, white rot fungi have also been investigated as a biological pretreatment agent. Another method that enhances hydrolysis and increases VS removal and CH_4 production is temperature-phased anaerobic digestion(AD) (Carrère et al., 2010). Industrial fermentation can also be employed as a pretreatment accelerant, but their long retention times and efficiency make them less desirable than other treatment methods.

Acidic pretreatment decomposes lignocellulosic substrates to monosaccharides but at a cost that makes it less cost-effective compared to alkaline pretreatments (Ariunbaatar et al., 2014). Alkaline pretreatment, often using ammonia or hydroxide compounds, swells the fibers, disrupting the structure of lignin and making the substrate available to enzymatic breakdown. Mechanical pretreatment intends to decrease particle size of wastes, augmenting surface area. Some mechanical pretreatments for decreasing substrate size include ultrasound pretreatment, collision plate pretreatment, milling, and thermal pretreatment. Thermal pretreatment temperature influences the effectiveness of the pretreatment process since high temperatures can inhibit VS destruction and promote toxic melanoidins formation due to the Maillard reaction (Kim et al., 2016; Li et al., 2012).

9.3 FUNDAMENTALS OF ANAEROBIC DIGESTION BIOGAS COMPOSITION

Methane and carbon dioxide constitute the bulk of biogas, with trace levels of ammonia, hydrogen sulfide (H_2S), and water vapor. It is slightly less dense than air, by about 20%, compared to air. Its calorific value is the sole contribution of the CH_4 content, which burns in a clean blue flame and produces temperatures distinctly higher than those of conventional biomass-based cooking fires. Biogas is not liquefiable at room temperature and pressure, similar to LPG. The energy output is around 21.5 MJ (5.97 kWh) when 1 m^3 raw biogas containing 60% CH_4 is combusted under standard temperature and pressure, whereas pure CH_4 is capable of producing up to 35.8 MJ (9.94 kWh) under the same conditions (Muylaert et al., 2000).

There are various applications of biogas. It is commonly utilized within homes for illumination, heating, and cooking. It may be treated to meet natural gas quality specifications for broader utility, upgraded to biomethane for fuel for transportation, or applied in combined heat and power (CHP) systems. Biogas can be upgraded and supplied to end-use locations better due to biomethanation technologies such as (Holm-Nielsen et al., 2009). Massive plants have been constructed in several developing countries to utilize CHP engines or fuel cells to produce electricity. Biogas has been utilized in India to operate electric generators with altered diesel engines. Biogas can be compressed in cylinders for application in cars by removing CO_2. The improved biomethane is suitable for any compressed natural gas (CNG) use because its CH_4 concentration is similar to that of CNG (Kapdi et al., 2005). The most widespread use of biogas globally is cooking, followed by lighting. Biogas lights employing special mantles typically employ 0.07–0.14 m^3 of gas per hour and perform optimally at gas pressures of 70–84 mm water column (Dhakal, 2008).

AD has emerged as a feasible answer to issues around rural development in India. The first anaerobic digester was put in place for lighting in 1859. In 1910, AD on a cellulose basis was employed to process animal manure and retain nitrogen in farming (Fowler & Joshi, 1920). The large-scale use of fertilizers derived from fossil fuels was preceded by the setup of biogas facilities utilizing animal dung to retain nitrogen by 1935. To accelerate adoption, a national program of biogas development was initiated in 1982. Simultaneous efforts in countries such as China, where by 1990 over 5.5 million biogas plants existed, proved the impetus behind the adoption of biogas throughout the world (Vijay, 2012). Biogas replaced traditional biomass fuels in the kitchen, especially in rural areas, due to the fuelwood shortage in the 1970s. AD is better for the environment and economy compared to composting, incineration, or combined composting-digester systems.

It boosts energy efficiency and minimizes emissions of volatile organic compounds like CH_4, ammonia, aldehydes, and ketones, according to Bharathiraja et al. (2016). Small-scale anaerobic digesters were initially developed to meet the energy requirements of rural communities. They employed farming waste and animal manure to yield 2–4 m^3 of gas per day, enough for simple cooking and lighting. An estimated 500 million tons of non-woody farming residues and 44 million tons of urban trash are generated each year in India alone. Whereas rural residues can be excellent organic manure, urban refuse is rich in methanogenic, biodegradable components. Small biogas plants are used by tens of thousands of families in countries such as China, India, Nepal, and Vietnam. Farm-scale biogas plants are also increasingly found in North America and Europe. More than 3,700 farm-scale biogas plants were operating in Germany by 2007 (Al Seadi, 2008).

9.4 OPTIMIZATION PROCESS PARAMETERS IN ANAEROBIC DIGESTION

Biochemical oxygen demand (BOD) is an established measure of the amount of biodegradable organic material present in sludge. BOD is employed as a baseline measurement to assess the operational performance of anaerobic digesters. BOD is determined by the measurement of microbial oxygen consumption over a specified incubation period of five days. BOD measures the metabolic activity required for the oxidation of organic materials under aerobic conditions. One special one, carbonaceous BOD, measures the amount of oxygen required to oxidize carbon alone while using nitrification inhibitors to stop the oxidation of ammonia, nitrogen, and nitrite.

Chemical oxygen demand, or COD, is the total quantity of oxygen that materials, whether biodegradable or not, can use during chemical oxidation. COD is an important metric for calculating the total organic content and assessing the efficiency of AD. Even though the test takes five days, COD levels are usually greater than BOD due to their larger oxidative effect. Accordingly, COD is particularly useful when assessing a substrate for anaerobic treatment (Van Lier et al., 2008). The use of AD to treat the organic component of municipal solid waste would minimize CO_2 emissions. As a result, given the Kyoto Summit's commitment to reducing emissions, environmental concerns may outweigh economic ones (Callaghan et al., 2002). Given the large range of waste solids/slurries suitable for anaerobic biodegradation, a network of centralized digestion centers accepting a diverse range of wastes could be proposed. Co-digestion has been studied as a method for a variety of waste combinations, as well as the idea of a centralized facility that co-digested a base material (Wang et al., 2014).

AD is a multistep biotechnological process that produces no accumulating intermediates such as H_2. Anaereobic digestion has recently acquired popularity as a method of producing hydrogen (Mei et al., 2016). This is due to the fast rate of H_2 production and the ability to use a variety of feedstocks as a substrate. Traditional AD consumes H_2 instantly, for example, by hydrogenotrophic methanogens, producing CH_4 and CO_2 (Esposito et al., 2012; American Public Health Association (APHA) et al., 2005).

The most crucial elements in creating biogas from wastewater through AD are the hydraulic retention time (HRT) and organic loading rate (OLR), which are inversely proportional and closely associated. A proper HRT increases AD efficiency by limiting the formation of gas pockets, disengaging the biomass from the gas, and providing sufficient contact between the inoculum and the substrate (Shi et al., 2017). Controlling the carbon to nitrogen (C/N) ratio increases the rate of microbial growth and substrate use, which enhances the aerobic digestion process. On the contrary, substrates that are slower degrading like lignocellulosic biomass are more tolerant of extended HRTs (Mao et al., 2015).

The amount of organic matter fed into the digester on a given day is termed the OLR. VFA buildup and risk of digester failure can be induced by overloading, which can interfere with the digestion process. However, in certain instances, the capacity of the microbial community to adapt to greases high-load substrates enhanced digester performance (Rincón et al., 2007). Total solids measurement provides data regarding the dry matter content in sludge. Its benefits, such as lower volume of digester and less energy required for heat, have been of interest (Duan et al., 2012). The two temperature ranges on which anaerobic digesters typically operate are mesophilic (around 35°C) and thermophilic (around 55°C). Mesophilic digestion produces lesser biogas and lower reaction rates, but is stable and economical (Moset et al., 2015). In contrast, thermophilic systems are more efficiently converted and better at pathogen removal but use more energy and possibly can be more subject to fluctuations in operation. Fluctuations in the ambient temperature will result in seasonal fluctuations in CH_4 production in some digesters that lack temperature control (Mao et al., 2015).

9.5 STRATEGIES FOR ENHANCED METHANOGENESIS

Methane output can be greatly increased by optimizing the conversion of acetic acid, a crucial substrate in the methanogenic phase of AD. Maintaining ideal control over AD systems' operational parameters is crucial to optimizing their CH_4 output efficiency. By reducing the oxidation-reduction potential (ORP), microelectrolysis-based technologies, including microbial electrolysis cells (MECs), especially those that use iron-graphite electrodes, create an environment that is favorable to microbial growth. Both facultative and anaerobic bacteria can flourish as a result of this alteration. A sufficient dosage of magnetite, which raises methanogen populations and lowers effluent pH, can also result in further lowering circumstances. In expanded granular sludge bed reactors, mediators such as granular activated carbon and nanoscale zero-valent iron greatly increase CH_4 outputs while maintaining low ORP levels. To function optimally, anaerobic bacteria require a drop in ORP and oxygen levels (Zhang et al., 2018).

Methane production and the stability of the sludge digestion process have both been demonstrated to be enhanced by biochar. Because of its porosity, ammonia adsorption is encouraged, which lessens the inhibitory effect and increases methanogen immobilization. Although the particular mechanism remains controversial, research shows that ammonia represses CH_4-producing enzymes, disrupts proton gradients, and lowers potassium levels. By means of adsorption of total ammonia nitrogen, augmentation of precipitation reactions, and raising buffering capacity in continuous digesting systems, biochar prevents such effects (Fagbohungbe et al., 2017). Alkalinity of the digester can be enhanced and necessary ingredients for methanogen development can be provided by biochar materials that contain high levels of calcium, magnesium, and iron, including those produced from pine or corn stover. Biochar can also elevate levels of cobalt and nickel, which are essential cofactors for the enzymes present in the process of methanogenesis, in environments that are rich in iron. Additionally, biochar porosity makes it suitable for biofilm production, which increases adhesion, activity, and stress tolerance of microorganisms. Nutrient availability to microbial metabolism is critical during AD. Trace metals, specifically iron (Chen et al., 2008), favor hydrogen-consuming methanogens involved in the removal of hydrogen—a thermodynamically less favorable intermediate in methanogenesis.

Alkalinity or buffer capacity of substrates is another parameter influencing effective methanogenesis. By keeping pH levels constant, waste activated sludge pretreatment using alkaline chemicals can help stimulate CH_4 production. Active methanogenic biomass may be inhibited or even destroyed by very low or very high pH levels, however. Acetoclastic methanogens of the Methanosarcinaceae and Methanosaetaceae families are particularly efficient in the methanogenesis of acetic and formic acids, studies suggest. High substrate alkalinity supports such microorganisms and enhances reactor stability (Dai et al., 2016). Microbial system and community stability are greatly affected by operational conditions like temperature and OLR. Abrupt microbial shifts at high OLRs would lower CH_4 productivity even when rising temperatures increase the metabolic reaction rate as well as pathogen inactivation. Therefore, constant production of gases and organic degradation rates are dependent on careful regulation of OLR (Gonzalez et al., 1998). In order to support CH_4 production, feedstocks must have a balanced carbon-to-nitrogen (C/N) ratio. Very high C/N ratios suppress methanogenesis due to acute nitrogen loss, and low ratios lead to ammonia buildup.

Co-digestion strategies, like mixing swine manure with distillers' grains, can maximize CH_4 production at approximately a 30:1 C/N (Wang et al., 2013). Furthermore, key nutrients like carbohydrates, which are a vital source of carbon, promote syntrophic cooperation between bacteria and archaea. Purple photosynthetic non-sulfur bacteria may be able to compensate for carbon deficits in low carbon/nitrogen feedstock systems (Hülsen et al., 2018). Methanogens, particularly acetoclastic methanogens, are poisoned by AD-produced H_2S. Sulfate-reducing bacteria (SRB) exacerbate this problem by consuming the substrate before methanogens. Techniques like alkaline pretreatment, in which the sludge is taken down to pH 10 for a couple of days, can increase CH_4 production by preventing SRB activity and lowering H_2S content (Deng et al., 2012). Such conditions also favor microbial communities like Firmicutes and Bacteroidetes. In addition, the production of CH_4 can be hindered by antibiotics in wastewater, especially those that are tetracycline sources. Antibacterial activities of such chemicals decline and microbial activity returns once they degrade effectively (Li et al., 2019).

9.6 PROCESS INTENSIFICATION USING BIOAUGMENTATION AND ENZYMES

The primary objective of bioaugmentation is to boost the capability of an indigenous microbial community to produce biomethane by specifically adding special bacteria, either extremely tolerant stress or extremely metabolic efficiency bacteria. Nzila et al. (2016) stated that this approach has been promising in aerobic biodegradation of soil and wastewater recalcitrant pollutants, particularly those generally resistant to biodegradation. One of the most significant examples is bioaugmentation of a consortium of cellulolytic and hemicellulolytic bacteria, such as *Pseudobutyrivibrio xylanivorans*, *Fibrobacter succinogenes*, *Clostridium cellulovorans*, and *Ruminococcus flavefaciens*, into anaerobic digesters to treat brewery residual grain. Bioaugmentation with the first three bacteria individually or in combinations increased CH_4 production by 4–18% in 1-liter batch reactors inoculated with sludge from a brewery wastewater treatment system. In a surprise finding, *R. flavefaciens*, used individually or in combination, could not stimulate biomethane production, allegedly due to the inhibition of its growth in the reactor condition (Čater et al., 2015; Nzila, 2017).

Another similar research utilized a two-phase sequencing batch system, phase one of which utilized a commercial blend of Clostridium species for improving the hydrolysis of corn-sweet residue. Phase two made use of anaerobic sludge from a municipal wastewater treatment plant. Single hydrolysis phase only was utilized for bioaugmentation, and one-time and repeated bioaugmentation processes were experimented with by the researchers. Regular augmentation was 15% better than one-time inoculation, and the CH_4 yield was 56% greater compared to the control, announces the findings. These results demonstrate the advantage of repeated microbial addition as opposed to a single treatment and confirming the operation of bioaugmentation in its ability to increase CH_4 production (Martin-Ryals et al., 2015). Another attractive technique for maximizing AD efficiency, especially where complex feedstock is involved, involves enzymatic hydrolysis.

Provided ecoenzymatic conversion types, dosage levels, and environmental factors such as temperature and pH are optimized to the optimum, with the process being fast, safe for the environment, and efficient under controlled conditions. Cell-free enzymes have some benefits over whole-cell microbial therapy, including the ability to withstand the presence of inhibiting or toxic compounds and the ability to work in a broad spectrum of conditions (Gianfreda & Rao, 2004). Enzymes have a better chance to interact with substrates than do microbiological cells since they are small and very mobile. With the use of organic wastes, such as crop residues, animal wastes, and municipal solid waste, AD is a green way of producing biogas, a clean fuel to replace fossil fuels. Nevertheless, because of the recalcitrant nature of lignocellulosic biomass, the hydrolysis step tends to be the limiting factor for its efficiency.

It has been established that the use of hydrolytic enzymes on biological pretreatments greatly enhances the conversion of recalcitrant biomass into fermentable sugars. Increased production of biogas has been attained through the application of lipases on lipid-rich wastage and cellulases and hemicellulases on agricultural wastage. In the biodegradation of high-strength organic wastages, low-cost enzymatic products such as lipases, cellulases, and proteases are of particular interest due to the fact that the utilization of costly commercial enzymes would be uneconomical. Enzymatic pretreatment of municipal sludge has also gained interest due to its capability to promote solubilization, biodegradability, dewatering, and hygiene quality. Hydrolysis of complex biomass requires the combination of a variety of enzymes, dosing schedules well-managed, and optimized conditions for optimal efficiency (Ahuja et al., 2004; Parawira, 2012).

9.7 INTEGRATED APPROACHES TO OPTIMIZE BIOHYDROGEN PRODUCTION FROM ORGANIC WASTE

High energy density, minimal environmental footprint, and its capability to make a transition towards a low-carbon energy system have brought biohydrogen (Bio-H_2) into the limelight as a green, renewable energy carrier (Rizvi et al., 2023). AD is currently employed on a large scale as a biological process because it is capable of producing valuable byproducts like biogas and biofertilizers and transforming organic waste, such as FWs, into biohydrogen in a sustainable way (Sahrin et al., 2022).

Direct power and heat generation can be achieved with biogas as it is mostly made up of CO_2 and CH_4, but in its natural state, it tends to need upgrading in order to increase quality. FW carries a high energy content and is readily available, making it ideal as feedstock for the production of AD-based biogas (Harun et al., 2019). As the use of biogas results in a considerable reduction of the release of GHG, the hydrogen obtained from it comes into the category of clean energy source. Methane is a far more aggressive GHG compared to CO_2, and therefore CH_4 reforming into H_2 through the right channels can significantly reduce the global warming potential of the gas stream and lead to the eradication of climate change (Chen et al., 2023).

Upgrading methods of high technology must be combined with AD to make biogas a greener, more energy-efficient transport. Those include processes for cleaning impurities like CO_2, CH_4 concentration for use as grid injection or transport fuel. ILs have exhibited remarkable potential among the purification processes because of their high CO_2 selectivity, chemically designable nature, and stability under different operational conditions (Pellegrini et al., 2018). Low volatility is responsible for long-term sustainability and recyclability. Regeneration of IL usually includes selective CO_2 uptake and desorption.

For carbon capture and utilization, purified separated CO_2 must be liquefied. To achieve this, the CO_2 must be compressed and cooled to a supercritical state before it can be converted into a liquid form. A closed-loop approach in carbon management, liquefied CO_2 can be stored and transported more conveniently and used in most industries, such as geological sequestration, refrigeration, and enhanced oil recovery. Recovering hydrogen from enhanced biogas has advantages for the environment and the economy.

The basic technologies for on-site hydrogen supply are fossil fuel reformation to produce syngas, followed by downstream units for CO_2 extraction and hydrogen purification. Natural gas and refinery off-gases account for approximately 80% of the world's hydrogen needs. Steam methane reforming (SMR) is a popular commercially viable reforming technology for producing syngas ($H_2 + CO$) from natural gas and other hydrocarbons. The SMR process is utilized in most industrial-scale hydrogen facilities to convert hydrocarbons into H_2, CO, and CO_2 (Salkuyeh et al., 2017).

Water electrolysis is a method that uses electricity to divide water into hydrogen and oxygen. The energy source heavily influences the technology's efficiency and environmental impact; renewable energy sources such as solar and wind make it significantly more sustainable (Rumayor et al., 2022). Electrolyzers use a variety of technologies, including proton exchange membranes and alkaline electrolysis, all of which have advantages and disadvantages. Hydrogen can also be produced through the partial oxidation of hydrocarbons, gasification of coal or biomass, or thermochemical water splitting. The main process for producing hydrogen is steam reforming of CH_4. Consequently, nonrenewable and environmentally undesirable technologies are used to manufacture 96% of hydrogen. Sustainable hydrogen can be produced by thermochemical conversion of biomass or water electrolysis (Gholkar et al., 2021).

9.8 BIOGAS UPGRADING TECHNIQUES FOR GREEN METHANE

Biogas upgrading is a multi-step gas separation process that includes compressing, removing minor components, drying the gas, and extracting the primary component CO_2. The end product is called biomethane, and it usually has a CH_4 level of more than 95% (v/v) and very few dangerous impurities. Currently, burning in an on-site CHP gas engine, which produces both heat and electricity, is the primary method of valuing the energy content of biogas. The electrical efficiency of modern gas engines ranges from 33% for small units to 40% for bigger ones. The overall efficiency can reach 83–88% if thermal heat is also utilized (Petersson & Wellinger, 2009). Biogas electricity is rarely competitive in the free electricity market; instead, it must be sold as positive and negative balancing energy, or green power. Thermal energy can be marketed as industrial or district heat. Upgrading biogas to a specific quality and using it as a local fuel (compressed biomethane, or CBM) or a sustainable natural gas substitute (injection into the natural gas grid) is an alternate strategy that has emerged during the last 20 years (Awe et al., 2017; Miltner et al., 2017).

Moreover AD works best when gases like oxygen and nitrogen are not present. However, depending on the feedstock and process variables, the quantity of extra pollutants can change. Adsorption, membrane separation, gas cooling, compression, or chemical absorption can be used to remove these contaminants, which include nitrogenous compounds and water vapor. In order to handle siloxanes and volatile methylsiloxanes, which are commonly found in biogas from wastewater treatment plants, commercially available technologies like adsorption, absorption, and cryogenic separation are widely used. The development of biological treatments for these substances is still in its infancy. Prior to the upgrading stage and at the digester outflow, solid particles are typically physically screened (Miltner et al., 2017). Because raw biogas contains a lot of moisture, wet dedusting methods, such as electrostatic precipitators and wet scrubbers, are frequently used to remove particles. Usually, halogenated chemicals are eliminated along with CO_2 (Singh & Shukla, 2014).

In order to improve the CH_4 concentration and create biomethane that may be injected into natural gas pipelines or used as fuel for automobiles, biogas upgrading is the process of refining raw biogas, especially by removing CO_2. The elimination of other trace gases may also be the goal of this upgrading procedure. There are two types of biogas upgrading technologies: fully established commercial systems and new methods that are still being tested in labs or on a pilot scale. These technologies can be divided into five primary groups according to the gas separation mechanism: phase addition, solid sorbents, membrane barriers, phase production (like condensation), and those that depend on external gradients or forces (like pressure or temperature changes) (Seader et al., 2016).

9.9 INTEGRATION WITH CIRCULAR BIOECONOMY AND WASTE MANAGEMENT

Greater levels of waste and nonrenewable resource use are placing greater stresses on environmental systems. Consequently, greater emphasis on the establishment of a bioeconomy and the utilization of substitute biofuels has taken place within recent decades. Within the most fundamental principles of the movement is that of the circular economy, which rebalances trash as an economic asset (Bose et al., 2022). The viability of such circular bioeconomy models, however, relies not only on technology but also on local conditions, economic viability, and cooperation among value chain stakeholders. The involvement of local stakeholders is especially critical to ensuring coordination and long-term success. Secure supply and demand relationships in a multi-stakeholder circular system, where stakeholders cooperate to ensure the entire chain—raw material sourcing to end-use energy applications—enable long-term profitability (Angouria-Tsorochidou et al., 2021). While biological material use in a circular economy is of major potential, efficient performance depends on well-managed processes and good waste treatment (Awasthi et al., 2020). Biomethane has been cited as a good case of a circular economy (Le Pera et al., 2022).

Production of biohydrogen from organic waste streams is a twin solution for renewable energy production and valorization of wastes. In order to produce more hydrogen, new technologies such as biomass gasification, MECs, and dark fermentation are being explored. Cost problems, scalability limitations, and system efficiency hurdles still remain, however. Some of those limitations are being softened by integrated solutions that have been newly developed. Different paths of hydrogen production have different efficiencies; MECs generate 15–98 g/kg, biological pathways (dark fermentation) generate 4–44 g/kg, and thermochemical processes (gasification) yield 40–190 g/kg. Both routes involve issues of low yields, CO_2 release, and expense. While most hydrogen currently produced still depends on natural gas, hybrid technologies and increasing synergy with circular economy technologies will be needed in order to deliver additional contributions of the energy through renewables (Kundu et al., 2025). AD and co-digestion are effective ways of treating FW to produce biogas, a clean, renewable alternative fuel to fossil fuel-based ones.

These comparatively straightforward processes have a massive positive impact on public health and the environment through production of high-purity CH_4 for uses including heating, transport fuel, and power. The operations in the food-to-biomethane stream involve waste collection, pre-treatment, biogas generation, gas cleaning, and upgrading to biomethane. The entire cycle reflects the idea of the circular bioeconomy. Higher biomethane productions are due to new technologies for AD and biogas upgrading. Novel upgrading methods in use are five primary upgrading methods, which are physical scrubbing, chemical scrubbing, pressure swing adsorption, membrane separation, and cryogenic separation. When combined, these factors can have a substantial impact on the transition to a circular, sustainable energy economy (Devi et al., 2023). Biomethane plays an important role in the circular economy concept. To fully fulfill its potential, cross-sector cooperation is required in a variety of sectors, including waste management, infrastructure development, energy generation, regulation, and public awareness. Feed-in tariffs and stimulus plans are examples of government initiatives that could encourage investment in biomethane infrastructure and technology, allowing it to be integrated into a green energy strategy (Catalano et al., 2024).

However, there are still some challenges with technology. These consist of higher production costs compared to conventional fossil fuels, a low-carbon green gas grid supply system, and resistance from more pro-electricity industries like transport. Some sustainability objectives are countered by existing energy policies, which have a tendency to favor electrification at the expense of the potential provided by waste-to-energy systems (Morone et al., 2021). Promotion of biomethane as an abundant fuel can significantly accelerate the transition in energy. The RepowerEU plan plans to meet this goal by 2030 through a huge ramp-up of biomethane production capacity (Sulewski et al., 2023).

9.10 SUSTAINABLE DISPOSAL AND ENVIRONMENTAL RISK MITIGATION

The environmental advantage of AD is the primary reason for increased demand as a green FW treatment technology globally (Zhou et al., 2018). While effective, AD does not fully stabilize FW and often leaves behind a nutrient-rich residue in liquid or solid digestate form. As Peng et al. (2020) indicate, AD increases the quantity of solid digestate that needs to be handled and disposed of with due care.

Solid digestate mainly consists of the insoluble organic wastes, intermediate metabolites, and residual microbial mass from AD. Solid digestate is normally utilized as an organic fertilizer because it contains high levels of major elements: nitrogen, potassium, and phosphorus. However, it is likely to possess certain contaminants like pathogenic microbes and heavy metals such as copper and zinc, which would pose risks to environmental safety and human health if not contained suitably. Therefore, proper treatments of solid digestate represent one of the most significant parameters for the longevity and feasibility of AD plants in the long term.

In China, the solid digestate from FW was generally disposed of by landfilling, incineration, or composting. Composting is widely recognized as one of the most feasible strategies to regenerate nitrogen cycles and improve soil health. Composting has been widely used in North America, Europe, and around the world. However, composting results in a number of problems due to its slow-degradable components and high water content that requires the frequent use of bulking or drying agents. Wojnowska-Baryła et al. (2018) showed that composting can lead to contamination with heavy metals, unpleasant odors, and increased emissions of GHGs. Another method for value addition to increase the organic matter and energy value of digestate is by incineration. Landfilling is a low-cost and straightforward option when combined with already available MSW facilities (Ma et al., 2018). Both of these options—while in most scenarios not in accordance with material recycling principles—are preferable in China because they are relatively simple to be superimposed on existing MSW facilities. But when one considers the entire FW management lifecycle, their economic and environmental sustainability remain to be assessed.

An interest in integrated biogas systems has been felt since the adoption of circular economy principles. Research has shown that solid digestate treatment plays a critical role in determining the overall environment and wider impact of such systems. For example, Nordahl et al. (2020) have found that when solid digestate of dry AD is only sent to landfills, the GWP100 (100-year carbon dioxide global warming potential) could be up to 40 kg CO_2 per ton of organic waste. Tiwary et al. (2015) indicated that the eutrophication and acidity can be controlled through digestate composting. The financial and environmental performance of a number of digestate treatment systems has also been investigated in a number of comparative studies (Herbes et al., 2020). Efficient digestate management is found by the study to decrease FW-based biogas production systems' energy demands and impacts on the environment by a great extent.

9.11 FUTURE TRENDS AND RESEARCH DIRECTIONS

The quest to maximize biogas output has been fast-tracked by the current global energy crisis, a strong emphasis being placed on enhancing pretreatment methods and updating technologies. Approximately half of global biogas output by 2022 was generated in Europe, with Germany producing 20% on its own. Agricultural waste is the primary fuel in several European countries, including Denmark, Italy, France, and Germany. As much as 49% more CH_4 has been generated due to pretreatment technologies, for example, membrane-based separation of swine dung. In addition, due to their scalability and low energy needs, innovative and environmentally friendly upgrading methods like MECs and microalgae biofixation are promising. Low-cost and environmentally friendly production methods are urgently required in the future to enhance the broader application of biogas technology.

It is possible to realize substantial environmental advantages from biogas technology, particularly in the area of organic waste management and added value, including FW, municipal solid waste, and animal and industrial residues. In particular,

AD achieves a maximum reduction of GHG emissions while using low-cost feedstocks (Daniel-Gromke et al., 2015). Yet additional research is required to optimize electrical efficiency and achieve the lowest CH_4 concentrations required for piston engine operation. To compete with traditional internal combustion systems, biogas-based power must become less expensive and fuel cell life expectancies need to be increased (Bora et al., 2021). AD is beneficial in the recovery of nutrients, but especially nitrogen and phosphorus, from organic waste and municipal waste, making it beneficial in agriculture and reducing the utilization of artificial fertilizers. While domestic digesters are more common in industrial regions, medium- and large-scale digesters are typically used for municipal and agricultural waste treatment (Scarlat et al., 2018).

AD systems are effective in rural settings, especially for farmers who have constant access to crop residues and animal manure, ensuring a consistent input for the production of biogas. The digestate by-product can be converted into a nutrient-rich fertilizer (Lin et al., 2021). Biogas upgrading produces CO_2-rich off-gases with untapped potential as well as CH_4. Bio-CO_2 can be employed, say, for modified environment packaging or grain fumigation. Although cost-effective energy supply remains an obstacle, the power-to-gas idea wherein biogas is combined with renewable energy resources such as wind and solar power is becoming popular. The production of higher alcohols from biogas and methanol synthesis has been the interest of recent research. Just like this, biogas is gaining interest in being converted to syngas, a gas rich in hydrogen, which can be used in fuel cells or to synthesize urea, alcohol, and dimethyl ether (dos Santos et al., 2018).

There is also ongoing research on the catalytic conversion of biogas to syngas and carbon nanofibers. These carbon materials are employed in numerous devices, including transistors, capacitors, and solar cells (Al-Timimi & Yaakob, 2022). Another fascinating area is applying acetogens and methanotrophs in biological processes to transform CH_4 into liquid fuels.

However, no large-scale industrial studies have been carried out so far, and further development is required for its application. Biogas and biomethane investments could support the nation's technology and economy. This technology can provide jobs in rural areas when co-operated with environmental and social activities. Other worth recovering by-products using biomass degrading technology include H_2S, volatile organic compounds containing sulfur, ketones, amines, and aldehydes. Ammonia is produced when proteins break down into amino acids; the quantity may vary from 18 to 150 g per ton of sludge. The main challenges faced by this industry are improving biodegradability, knowledge of composition of FW, increasing microbial activity, optimizing CH_4 production, and assuring nutritional availability and balance (Perin et al., 2020).

9.12 CONCLUSION

One promising answer to the urgent energy and environmental problems of today is the biogas produced from trash. Biogas is a renewable energy source, much more ecological compared with fossil fuels, as it decreases carbon emissions and enhances energy self-sufficiency. AD is a waste-to-energy approach that has several benefits for the environment, economically creating clean electricity. Thanks to technological advances in pretreatment technologies, judicious use of additives, and maintenance of optimum operating conditions, the biogas systems of today achieve the highest efficiency possible, especially in green CH_4 and biohydrogen production. All these technology advancements promote higher yields of gas and overall cost efficiency. Further evidence of the flexibility of biogas systems in the attainment of overall sustainability objectives is shown in its integration with other systems like hydroponic agriculture, hydrogen production technologies, and electrothermal power generation.

It is likely that in the path to fully exploiting the potential of biogas, a wide range of research on operational and technological challenges will be needed. First lies the low-cost, environmentally benign pretreatment technology; second, constant and reliable feedstock supply; and third, increased automation for process optimization. The waste heat will be used to build the biogas system and the new digestor synthesis components. Scientific research and technical innovation will thus play a crucial role in the development of a greener economy and hasten the movement toward more resilient and sustainable energy futures by optimizing biogas potential. High operation costs, along with technological limitations in chemical absorption and membrane separation, result in inadequate regulatory support, all acting as hurdles to commercial growth despite increasing demand. Another positive trend is the adoption of environment-friendly legislative patterns, as evidenced by the increasing emphasis that is being placed on circular economy systems-a proof that the concept of biogas systems is changing from a waste management system to an integral part of the future energy sector. Careful legislation and innovative industry developments have resulted in biogas making slow contributions to the global energy revolution.

REFERENCES

Ahuja, S. K., Ferreira, G. M., & Moreira, A. R. (2004). Utilization of enzymes for environmental applications. *Critical Reviews in Biotechnology*, *24*(2–3), 125–154.

Al Seadi, T. (2008). *Biogas handbook.* ISBN 978-87-992962-0-0 Published by University of Southern Denmark Esbjerg, Niels Bohrs Vej 9-10, DK-6700 Esbjerg, Denmark. https://www.lemvigbiogas.com/BiogasHandbook.pdf

Al-Timimi, B. A., & Yaakob, Z. (2022). Catalysts for the simultaneous production of syngas and carbon nanofilaments via catalytic decomposition of biogas. In M. T. Ravanchi (Ed.), *Natural gas: New perspectives and future developments.* IntechOpen. https://doi.org/10.5772/intechopen.101320).

Amato, A., Tsigkou, K., Becci, A., Beolchini, F., Ippolito, N. M., & Ferella, F. (2023). Life cycle assessment of biomethane vs. fossil methane production and supply. *Energies*, *16*(12), 4555. https://doi.org/10.3390/en16124555

American Public Health Association (APHA), American Water Works Association (AWWA), & Water Environment Federation (WEF). (2005). *Standard methods for the examination of water and wastewater* (21st ed.). American Public Health Association..

Angouria-Tsorochidou, E., Teigiserova, D. A., & Thomsen, M. (2021). Limits to circular bioeconomy in the transition towards decentralized biowaste management systems. *Resources, Conservation and Recycling*, *164*, 105207.

Ariunbaatar, J., Panico, A., Esposito, G., Pirozzi, F., & Lens, P. N. (2014). Pretreatment methods to enhance anaerobic digestion of organic solid waste. *Applied Energy*, *123*, 143–156.

Awasthi, M. K., Sarsaiya, S., Patel, A., Juneja, A., Singh, R. P., Yan, B., & Taherzadeh, M. J. (2020). Refining biomass residues for sustainable energy and bio-products: An assessment of technology, its importance, and strategic applications in circular bio-economy. *Renewable and Sustainable Energy Reviews, 127*, 109876.

Awe, O. W., Zhao, Y., Nzihou, A., Minh, D. P., & Lyczko, N. (2017). A review of biogas utilisation, purification and upgrading technologies. *Waste and Biomass Valorization, 8*, 267–283.

Bartacek, J., Zabranska, J., & Lens, P. N. (2007). Developments and constraints in fermentative hydrogen production. *Biofuels, Bioproducts and Biorefining, 1*(3), 201–214.

Bharathiraja, B., Sudharsanaa, T., Bharghavi, A., Jayamuthunagai, J., & Praveenkumar, R. (2016). Biohydrogen and biogas – An overview on feedstocks and enhancement process. *Fuel, 185*, 810–828.

Bodansky, D. (2016). The Paris climate change agreement: A new Hope?*American Journal of International Law, 110*(2), 288–319.

Bora, D., Barbora, L., Borah, A. J., & Mahanta, P. (2021). A comparative assessment of biogas upgradation techniques and its utilization as an alternative fuel in internal combustion engines. In *Alternative fuels and advanced combustion techniques as sustainable solutions for internal combustion engines* (pp. 95–115). Springer Singapore.

Bose, A., O'Shea, R., Lin, R., Long, A., Rajendran, K., Wall, D., & Murphy, J. D. (2022). Evaluation of a biomethane, food and biofertiliser polygeneration system in a circular economy system. *Renewable and Sustainable Energy Reviews, 170*, 112960.

Callaghan, F. J., Wase, D. A. J., Thayanithy, K., &Forster, C. F. (2002). Continuous co-digestion of cattle slurry with fruit and vegetable wastes and chicken manure. *Biomass and Bioenergy, 22*(1), 71–77.

Carrère, H., Dumas, C., Battimelli, A., Batstone, D. J., Delgenes, J. P., Steyer, J. P., & Ferrer, I. (2010). Pretreatment methods to improve sludge anaerobic degradability: A review. *Journal of Hazardous Materials, 183*(1–3), 1–15.

Catalano, G., D'Adamo, I., Gastaldi, M., Nizami, A. S., & Ribichini, M. (2024). Incentive policies in biomethane production toward circular economy. *Renewable and Sustainable Energy Reviews, 202*, 114710.

Čater, M., Fanedl, L., Malovrh, Š., & Logar, R. M. (2015). Biogas production from brewery spent grain enhanced by bioaugmentation with hydrolytic anaerobic bacteria. *Bioresource Technology, 186*, 261–269.

Chen, W. H., Chen, W. H., Chein, R. Y., Hoang, A. T., Manatura, K., & Naqvi, S. R. (2023). Optimization of hydrogen purification via vacuum pressure swing adsorption. *Energy Conversion and Management: X, 20*, 100459.

Chen, Y., Cheng, J. J., & Creamer, K. S. (2008). Inhibition of anaerobic digestion process: A review. *Bioresource Technology, 99*(10), 4044–4064.

Dai, X., Li, X., Zhang, D., Chen, Y., & Dai, L. (2016). Simultaneous enhancement of methane production and methane content in biogas from waste activated sludge and perennial ryegrass anaerobic co-digestion: The effects of pH and C/N ratio. *Bioresource Technology, 216*, 323–330.

Daniel-Gromke, J., Liebetrau, J., Denysenko, V., & Krebs, C. (2015). Digestion of bio-waste-GHG emissions and mitigation potential. *Energy, Sustainability and Society, 5*(1), 3.

Das, D., & Veziroğlu, T. N. (2001). Hydrogen production by biological processes: A survey of literature. *International Journal of Hydrogen Energy, 26*(1), 13–28.

Deng, Y., Zhang, Y., Gao, Y., Li, D., Liu, R., Liu, M., & Yang, M. (2012). Microbial community compositional analysis for series reactors treating high level antibiotic wastewater. *Environmental Science & Technology, 46*(2), 795–801.

Devi, M. K., Manikandan, S., Kumar, P. S., Yaashikaa, P. R., Oviyapriya, M., & Rangasamy, G. (2023). A comprehensive review on current trends and development of biomethane production from food waste: Circular economy and techno-economic analysis. *Fuel*, 351, 128963.

Dhakal, N. H. (2008). *Financing domestic biogas plants in Nepal.* Stichting Nederlandse Vrijwilligers.
dos Santos, R. O., de Sousa Santos, L., & Prata, D. M. (2018). Simulation and optimization of a methanol synthesis process from different biogas sources. *Journal of Cleaner Production*, *186*, 821–830.
Duan, N., Dong, B., Wu, B., & Dai, X. (2012). High-solid anaerobic digestion of sewage sludge under mesophilic conditions: Feasibility study. *Bioresource Technology*, *104*, 150–156.
Esposito, G., Frunzo, L., Liotta, F., Panico, A., & Pirozzi, F. (2012). Bio-methane potential tests to measure the biogas production from the digestion and co-digestion of complex organic substrates. *The Open Environmental Engineering Journal*, *5*(1), 1–8.
Fagbohungbe, M. O., Herbert, B. M., Hurst, L., Ibeto, C. N., Li, H., Usmani, S. Q., & Semple, K. T. (2017). The challenges of anaerobic digestion and the role of biochar in optimizing anaerobic digestion. *Waste Management*, *61*, 236–249.
Fan, K. S., Kan, N. R., & Lay, J. J. (2006). Effect of hydraulic retention time on anaerobic hydrogenesis in CSTR. *Bioresource Technology*, *97*(1), 84–89.
Fowler, G. J., & Joshi, G. V. (1920). Studies in the fermentation of cellulose. *Journal of the Indian Institute of Science*, *3*, 39–60.
Gholkar, P., Shastri, Y., & Tanksale, A. (2021). Renewable hydrogen and methane production from microalgae: A techno-economic and life cycle assessment study. *Journal of Cleaner Production*, *279*, 123726.
Gianfreda, L., & Rao, M. A. (2004). Potential of extracellular enzymes in remediation of polluted soils: A review. *Enzyme and Microbial Technology*, *35*(4), 339–354.
Gonzalez, J. S., Rivera, A., Borja, R., & Sanchez, E. (1998). Influence of organic volumetric loading rate, nutrient balance and alkalinity: COD ratio on the anaerobic sludge granulation of an UASB reactor treating sugar cane molasses. *International Biodeterioration & Biodegradation*, *41*(2), 127–131.
Harun, N., Hassan, Z., Zainol, N., Ibrahim, W. H. W., &Hashim, H. (2019). Anaerobic digestion process of food waste for biogas production: A simulation approach. *Chemical Engineering & Technology*, *42*(9), 1834–1839.
Herbes, C., Roth, U., Wulf, S., & Dahlin, J. (2020). Economic assessment of different biogas digestate processing technologies: A scenario-based analysis. *Journal of Cleaner Production*, *255*, 120282.
Holm-Nielsen, J. B., Al Seadi, T., & Oleskowicz-Popiel, P. (2009). The future of anaerobic digestion and biogas utilization. *Bioresource Technology*, *100*(22), 5478–5484.
Hossain, E. (2023). Energy mixaround the world. In *The sun, energy, and climate change* (pp. 349–399). Springer Nature Switzerland.
Hülsen, T., Hsieh, K., Lu, Y., Tait, S., & Batstone, D. J. (2018). Simultaneous treatment and single cell protein production from agri-industrial wastewaters using purple phototrophic bacteria or microalgae – A comparison. *Bioresource Technology*, 254, 214–223.
IEA. (2019). *World energy balances – 2019.* International Energy Agency.
IEA. (2021). *Key world energy statistics 2021.* International Energy Agency.
IEA. (2024). *World energy outlook.* International Energy Agency.
Kapdi, S. S., Vijay, V. K., Rajesh, S. K., & Prasad, R. (2005). Biogas scrubbing, compression and storage: Perspective and prospectus in Indian context. *Renewable Energy*, *30*(8), 1195–1202.
Kim, D. H., Kim, S. H., & Shin, H. S. (2009). Hydrogen fermentation of food waste without inoculum addition. *Enzyme and Microbial Technology*, *45*(3), 181–187.
Kim, J. S., Lee, Y. Y., & Kim, T. H. (2016). A review on alkaline pretreatment technology for bioconversion of lignocellulosic biomass. *Bioresource Technology*, *199*, 42–48.
Kundu, P., Vineetha, S. V., Mohan, A., & Ravikumar, A. (2025). Bio-hydrogen production from various waste resources through circular economy: Current technologies and future perspective. *Journal of Material Cycles and Waste Management*, *27*, 1263–1282. https://doi.org/10.1007/s10163-025-02183-x

Le Pera, A., Sellaro, M., Bencivenni, E., & D'Amico, F. (2022). Environmental sustainability of an integrated anaerobic digestion-composting treatment of food waste: Analysis of an Italian plant in the circular bioeconomy strategy. *Waste Management, 139*, 341–351.

Li, H., Li, C., Liu, W., & Zou, S. (2012). Optimized alkaline pretreatment of sludge before anaerobic digestion. *Bioresource Technology, 123*, 189–194.

Li, Y., Chen, Y., & Wu, J. (2019). Enhancement of methane production in anaerobic digestion process: A review. *Applied Energy, 240*, 120–137.

Lin, R., O'Shea, R., Deng, C., Wu, B., & Murphy, J. D. (2021). A perspective on the efficacy of green gas production via integration of technologies in novel cascading circular bio-systems. *Renewable and Sustainable Energy Reviews, 150*, 111427.

Ma, H., Guo, Y., Qin, Y., & Li, Y. Y. (2018). Nutrient recovery technologies integrated with energy recovery by waste biomass anaerobic digestion. *Bioresource Technology, 269*, 520–531.

Mao, C., Feng, Y., Wang, X., & Ren, G. (2015). Review on research achievements of biogas from anaerobic digestion. *Renewable and Sustainable Energy Reviews, 45*, 540–555.

Martin-Ryals, A., Schideman, L., Li, P., Wilkinson, H., & Wagner, R. (2015). Improving anaerobic digestion of a cellulosic waste via routine bioaugmentation with cellulolytic microorganisms. *Bioresource Technology, 189*, 62–70.

Mei, R., Narihiro, T., Nobu, M. K., Kuroda, K., & Liu, W. T. (2016). Evaluating digestion efficiency in full-scale anaerobic digesters by identifying active microbial populations through the lens of microbial activity. *Scientific Reports, 6*(1), 34090.

Miltner, M., Makaruk, A., & Harasek, M. (2017). Review on available biogas upgrading technologies and innovations towards advanced solutions. *Journal of Cleaner Production, 161*, 1329–1337.

Monlau, F., Sambusiti, C., Barakat, A., Guo, X. M., Latrille, E., Trably, E., & Carrere, H. (2012). Predictive models of biohydrogen and biomethane production based on the compositional and structural features of lignocellulosic materials. *Environmental Science & Technology, 46*(21), 12217–12225.

Morone, P., Yilan, G., & Imbert, E. (2021). Using fuzzy cognitive maps to identify better policy strategies to valorize organic waste flows: An Italian case study. *Journal of Cleaner Production, 319*, 128722.

Moset, V., Poulsen, M., Wahid, R., Højberg, O., & Møller, H. B. (2015). Mesophilic versus thermophilic anaerobic digestion of cattle manure: Methane productivity and microbial ecology. *Microbial Biotechnology, 8*(5), 787–800.

Muylaert, M. S., Ambram, R., Campos, C., Montez, E., Oliveira, L., Pereira, A., & Reis, M. (2000). Energy consumption and warming of the planet—Clean development mechanism analysis (CDM).In *Kyoto Protocol—Case studies*. COPPE Press.

Nordahl, S. L., Devkota, J. P., Amirebrahimi, J., Smith, S. J., Breunig, H. M., Preble, C. V., & Scown, C. D. (2020). Life-cycle greenhouse gas emissions and human health trade-offs of organic waste management strategies. *Environmental Science & Technology, 54*(15), 9200–9209.

Ntaikou, I., Antonopoulou, G., & Lyberatos, G. (2010). Biohydrogen production from biomass and wastes via dark fermentation: A review. *Waste and Biomass Valorization, 1*, 21–39.

Nzila, A. (2017). Mini review: Update on bioaugmentation in anaerobic processes for biogas production. *Anaerobe, 46*, 3–12.

Nzila, A., Razzak, S. A., & Zhu, J. (2016). Bioaugmentation: An emerging strategy of industrial wastewater treatment for reuse and discharge. *International Journal of Environmental Research and Public Health, 13*(9), 846.

Parawira, W. (2012). Enzyme research and applications in biotechnological intensification of biogas production. *Critical Reviews in Biotechnology, 32*(2), 172–186.

Pellegrini, L. A., De Guido, G., & Langé, S. (2018). Biogas to liquefied biomethane via cryogenic upgrading technologies. *Renewable Energy, 124*, 75–83.

Peng, W., Lü, F., Hao, L., Zhang, H., Shao, L., & He, P. (2020). Digestate management for high-solid anaerobic digestion of organic wastes: A review. *Bioresource Technology*, *297*, 122485.
Perin, J. K. H., Borth, P. L. B., Torrecilhas, A. R., da Cunha, L. S., Kuroda, E. K., & Fernandes, F. (2020). Optimization of methane production parameters during anaerobic co-digestion of food waste and garden waste. *Journal of Cleaner Production*, *272*, 123130.
Petersson, A., & Wellinger, A. (2009). Biogas upgrading technologies – Developments and innovations. *IEA Bioenergy*, *20*, 1–19.
Rincón, B., Travieso, L., Sánchez, E., de los Ángeles Martín, M., Martín, A., Raposo, F., & Borja, R. (2007). The effect of organic loading rate on the anaerobic digestion of two-phase olive mill solid residue derived from fruits with low ripening index. *Journal of Chemical Technology & Biotechnology*, *82*(3), 259–266.
Rizvi, S. M. M., Taqvi, S. A. A., Khan, M. M., Zabiri, H., Qadir, D., & Metwally, A. S. M. (2023). Techno-economic sustainability assessment for bio-hydrogen production based on hybrid blend of biomass: A simulation study. *Fuel*, *347*, 128458.
Rumayor, M. A., Corredor, J. U., Rivero, M. J., & Ortiz, I. (2022). Prospective life cycle assessment of hydrogen production by waste photoreforming. *Journal of Cleaner Production*, *336*, 130430.
Sahrin, N. T., Khoo, K. S., Lim, J. W., Shamsuddin, R., Ardo, F. M., Rawindran, H., & Cheng, C. K. (2022). Current perspectives, future challenges and key technologies of biohydrogen production for building a carbon–neutral future: A review. *Bioresource Technology*, *364*, 128088.
Salkuyeh, Y. K., Saville, B. A., & MacLean, H. L. (2017). Techno-economic analysis and life cycle assessment of hydrogen production from natural gas using current and emerging technologies. *International Journal of Hydrogen Energy*, *42*(30), 18894–18909.
Scarlat, N., Dallemand, J. F., & Fahl, F. (2018). Biogas: Developments and perspectives in Europe. *Renewable Energy*, *129*, 457–474.
Seader, J. D., Henley, E. J., & Roper, D. K. (2016). *Separation process principles: With applications using process simulators*. John Wiley & Sons.
Shi, X. S., Dong, J. J., Yu, J. H., Yin, H., Hu, S. M., Huang, S. X., & Yuan, X. Z. (2017). Effect of hydraulic retention time on anaerobic digestion of wheat straw in the semicontinuous continuous stirred-tank reactors. *BioMed Research International*, *2017*(1), 2457805.
Singh, A., Sevda, S., Abu Reesh, I. M., Vanbroekhoven, K., Rathore, D., & Pant, D. (2015). Biohydrogen production from lignocellulosic biomass: Technology and sustainability. *Energies*, *8*(11), 13062–13080.
Singh, R., & Shukla, A. (2014). A review on methods of flue gas cleaning from combustion of biomass. *Renewable and Sustainable Energy Reviews*, *29*, 854–864.
Sulewski, P., Ignaciuk, W., Szymańska, M., & Wąs, A. (2023). Development of the biomethane market in Europe. *Energies*, *16*(4), 2001.
Tiwary, A., Williams, I. D., Pant, D. C., & Kishore, V. V. N. (2015). Emerging perspectives on environmental burden minimisation initiatives from anaerobic digestion technologies for community scale biomass valorisation. *Renewable and Sustainable Energy Reviews*, *42*, 883–901.
Van Lier, J. B., Mahmoud, N., & Zeeman, G. (2008). Anaerobic wastewater treatment. In M. Henze, M. C. M. van Loosdrecht, G. A. Ekama, & D. Brdjanovic (Eds.), *Biological wastewater treatment: Principles, modelling and design* (pp. 415–456). IWA Publishing.
Varling, A. S., Christensen, T. H., & Bisinella, V. (2023). Life cycle assessment of alternative biogas utilisations, including carbon capture and storage or utilisation. *Waste Management*, *157*, 168–179. https://doi.org/10.1016/j.wasman.2022.12.005
Vijay, V. K. (Ed.). (2012). *Biogas production, upgradation and slurry management*. Narosa Publishing.
Wang, Q., Peng, L., & Su, H. (2013). The effect of a buffer function on the semi-continuous anaerobic digestion. *Bioresource Technology*, *139*, 43–49.

Wang, X., Lu, X., Li, F., & Yang, G. (2014). Effects of temperature and carbon-nitrogen (C/N) ratio on the performance of anaerobic co-digestion of dairy manure, chicken manure and rice straw: Focusing on ammonia inhibition. *PLOS One*, *9*(5), e97265.

Wojnowska-Baryła, I., Bernat, K., & Sartowska, S. (2018). Biological stability of multi-component agri-food digestates and post-digestates. *Waste Management*, *77*, 140–146.

Zhang, Z., Gao, P., Cheng, J., Liu, G., Zhang, X., & Feng, Y. (2018). Enhancing anaerobic digestion and methane production of tetracycline wastewater in EGSB reactor with GAC/NZVI mediator. *Water Research*, *136*, 54–63.

Zhou, Z., Tang, Y., Dong, J., Chi, Y., Ni, M., Li, N., & Zhang, Y. (2018). Environmental performance evolution of municipal solid waste management by life cycle assessment in Hangzhou, China. *Journal of Environmental Management*, *227*, 23–33.

10 Biogas (Methane, Biohydrogen)

Production, Optimization, and Disposal Guidelines

Gülbahar Bilgiç Tüzemen

10.1 INTRODUCTION

Because of the fluctuating expenses and environmental impacts of traditional energy sources, especially crude oil, interest in renewable energy system (RES) is growing. Sustainability concerns and the environmental damage caused by greenhouse gas emissions are making the use of renewable energy increasingly important. Compared to others, traditional sources have a limited lifespan and will eventually be depleted. As a result of ongoing research, common renewable energy sources include wind, biomass, solar, geothermal, and hydro. While traditional hydrogen (H_2) production methods are widely used, they have drawbacks such as high costs, environmental damage, and unsustainability. Hydrogen production from biogas has recently attracted significant interest due to its low cost, rapid reaction time, and simple operating conditions. Additionally, sludge, solid municipal waste, agricultural, and forestry residues are examples of waste biomass that microorganisms can utilize as raw materials to produce H_2. Biofuels such as H_2, ethanol, methanol, dimethyl ether, synthetic natural gas, and others can be produced from lignocellulose biomass, which can prevent resource waste and environmental degradation. Numerous countries and academics have launched initiatives and studies to support biofuel utilize, with the aim of comprehensively exploring the utilize of biomass in energy production (Sawyerr et al., 2019) For example, the European Union aims to utilize 10% biofuel as an energy contribution in the transportation sector by 2020. Additionally, the United States is projected to produce approximately 36 billion gallons of biofuel annually by 2022 (Molino et al., 2018; Sawyerr et al., 2019). Currently, biogas production for energy generation and biomethane upgrading for grid injection are being adopted by industrial units. A key area of development in organic waste management is the energy recovery of organic waste. Upgraded biogas can be utilized to produce electricity and heat, similar to natural gas, instead of fossil fuels (Balagurusamy & Chandel 2021; Molino et al., 2018; Nie et al., 2024; Sawyerr et al., 2019).

The first topic covered in this chapter is the manufacturing of biogas. Anaerobic digestion (AD), a promising RES that can turn organic waste into biogas that is high in methane (CH_4). The four key biochemical stages of AD, a complex biochemical process: hydrolysis, acidogenesis, acetogenesis, and methanogenesis, will

 DOI: 10.1201/9781003642589-10

be explained in detail, followed by an evaluation of optimization parameters. Finally, biogas disposal proposals, first mentioned in the literature, will be discussed. This review chapter will serve as a precursor to future biogas studies.

10.1.1 Biogas (Methane) Production

Biogas, a gas mainly composed of carbon dioxide (CO_2) and CH_4, is produced by the biological process of AD of various organic materials. Bacteria that thrive during the anaerobic biodegradation of organic compounds produce this gas. Depending on the process parameters and the type of organic matter being fermented, the CH_4 content in biogas can range from 55% to 75% (Ruggeri et al., 2015). Biogas also contains trace quantity of other gases such as nitrogen, hydrogen sulfide, oxygen, water vapor, and so on. The various gases that make up biogas are primarily CO_2 and CH_4, but they also contain 1% to 5% of other gases such as H_2. The amount of CH_4, in biogas specifies its energy value (Heiker et al., 2021). Due to its high CH_4 content, biogas is a desirable energy source. Because of the energy it releases, biogas can be utilized as a fuel for cooking and heating. Biogas can also be used, where a gas engine converts the gas's energy into heat and power (Heiker et al., 2021; Nie et al., 2024; Ruggeri et al., 2015). In many countries, biogas is an important component of an increasingly RES. Biogas facilities differ from other renewable energy plants in a few ways, such as the ability to store gas and provide heat and power when needed or the potential to contribute energy to the transportation sector. Figure 10.1 shows the system boundaries of biogas plants in RES.

Natural AD occurs in anaerobic habitats, including the digestive systems of ruminants at the bottom of wetlands, ponds, and marshes, and some insect species. AD

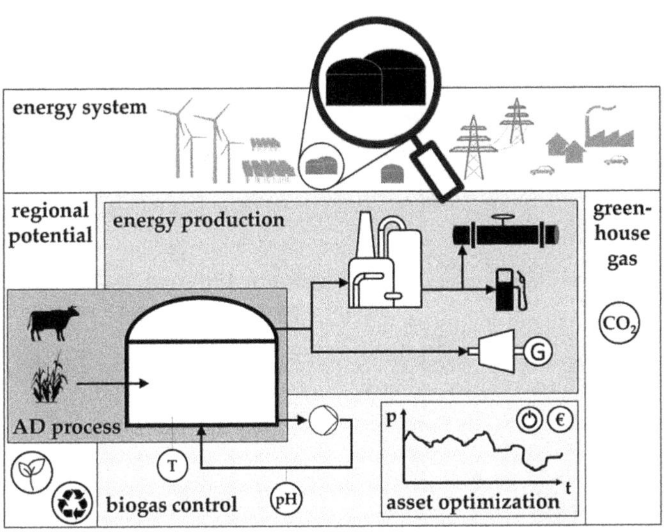

FIGURE 10.1 System boundaries of biogas plants in energy systems (Heiker et al., 2021).

utilizes a variety of organic feedstocks, including energy crops, animal waste, algae, solid municipal waste, and sludge from wastewater treatment plants. While some of these are always available, others are limited by seasonal and regional factors. Biogas plants form the basis of an economic concept that aims to reduce greenhouse gas emissions, recycle nutrients, and utilize them for biorefinery purposes. The reaction of producing biogas from the different organic biomass used in AD involves four basic steps: hydrolysis, acidogenesis, acetogenesis, and methanogenesis (Figure 10.2). The AD reactions are completed by anaerobically decomposing organic materials into CH_4, CO_2, compost, and inorganic nutrients.

The process of hydrolysis: The first stage in AD is hydrolysis. This occurs by breaking down and solubilizing biopolymer particles and colloidal waste into soluble oligomeric or monomeric compounds. During this synthesis, complex organic polymeric compounds, including lipids, carbohydrates, and proteins, are broken down. Enzymes produced by fermentative microorganisms hydrolyze these complex matters into smaller, water-soluble compounds such as amino acids, sugars, and long-chain fatty acids.

Large organic compounds such as carbohydrates, proteins, cellulose, and lipids are broken down into smaller molecules by anaerobic bacteria using enzymes in the first stage. During this stage, proteins are hydrolyzed into amino acids, lipids into fatty acids and glycerol, and polysaccharides into monosaccharides. By oxidizing organic materials through a process known as aerobic biodegradation, enzymatic catalysis boosts up the hydrolysis process. The biogas produced is converted from oxygen to carbohydrates through rapid hydrolysis and aerobic degradation (Hashim, 2021; Sachdeva et al., 2022; Sawyerr et al., 2019).

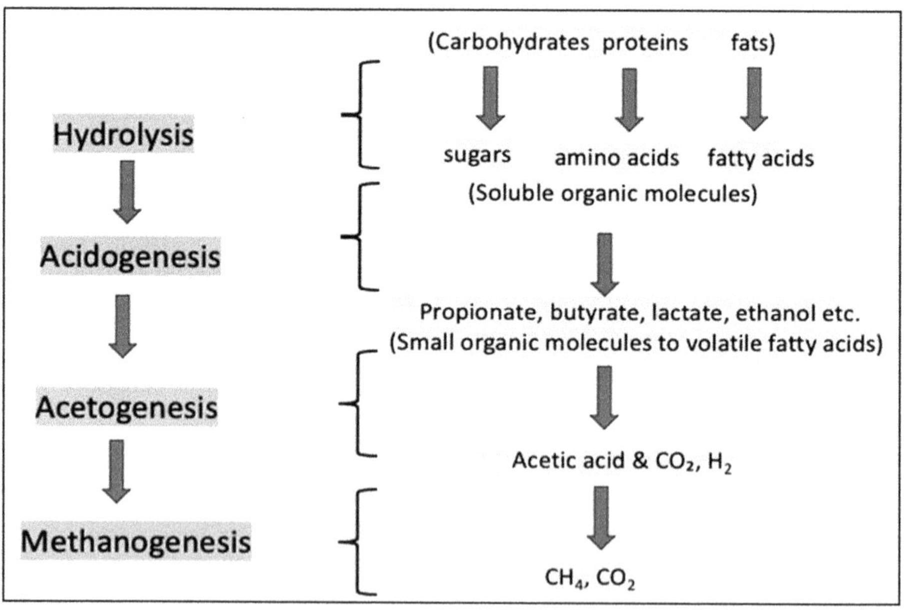

FIGURE 10.2 AD stages scheme.

Final of this stage, a simple organic chemical is formed. Following hydrolysis, the substrate is ready for cellular transport, and fermentative microorganisms can degrade it during the acidogenesis stage. To prevent inefficient breakdown of macromolecules, which can negatively impact digestion rates or other biological processes and, consequently, biogas output, it is imperative to optimize the hydrolysis process.

For the second step to occur, it is very essential to ensure that the microbial culture is actively functioning (Sawyerr et al., 2019). It is also possible to promote the dissolution of organic substances using physicochemical processes. However, air must not enter the system, because if air is present, the biomass cannot function as an anaerobic unit. The fermentation reaction begins at this point, where fermentable organic matter is prepared, combined with water, and deposited in the digestion reservoir after all contaminants have been removed.

Acidogenesis: A number of obligatory and facultative fermentative bacteria break down the first step's products further in the acidogenic stage, producing weak acids like acetic, lactic, butyric and propionic acid, alcohols, H_2, and CO_2. Basically, these bacteria decomposition complex organic substances into volatile fatty acids (VFA), producing acetic and propionic acids, along with some ammonia and CO_2. A group of bacteria called aquatic bacteria is responsible for transforming heavy molecules into smaller ones made of elements like H_2, CO_2, fatty acids, and amino acids (Sawyerr et al., 2019). The acidogenesis stage, often the fastest part of a balanced anaerobic process, includes the generation of high concentrations of H_2 by acid-producing bacteria. A low H_2 level is necessary for acidogenic processes to be thermodynamically favorable. Acidogenesis is the bacterial response to a high concentration of H_2 in the system, which causes acetogenic microorganisms to produce acetate. Acid (acidogens) and CH_4 producing bacteria (methanogens) are two main types of microorganisms that cooperate in an intricate way to break down organic matter and produce biogas. For any anaerobic digester to operate successfully, a symbiotic interaction between acidogenic and methanogenic bacteria must be maintained. This step is crucial as it connects the fermentation and CH_4 production stages. Additional acid must be generated to generate CH_4-producing methanogens.

Acetogenesis: Acetate-forming bacteria convert alcohols with two or more carbon atoms or VFA to acetate during the acetogenesis stage; the primary byproducts are CO_2 and H_2. This step is crucial because homoacetogenic microorganisms continuously convert CO_2 and H_2 to acetate. This reduces H_2 accumulation, which can impair the activity of acetogenic bacteria. Low partial H_2 pressures are required for the acetogenic process to occur. This is due to the fact that acetogenic bacteria may endure in conditions with incredibly low H_2 concentrations. However, if the partial H_2 pressure concentration increases further, acetogenic bacteria may lose their ability to produce acetate. A mutually symbiotic relationship between hydrogenotrophic methanogens and acetogens is necessary to maintain low pressures during the acetogenesis stage of the AD process (Hashim, 2021; Sawyerr et al., 2019). This is because acetogenic bacteria produce acetate, which methanogenic bacteria can use as a substrate. This is the last step of fermentation before methanogenesis.

Methanogenesis: This stage is a key step in the AD process. Because this step generates approximately 70% of the CH_4 used in AD, it has a significant impact on the reactions (Hashim, 2021). During this step, methanogens use acetate to form

CH_4, while CO_4-reducing, and H_2 oxidizing methanogens convert H_4 and CO_4 to CH_4. Methanogens produce CH_4 and CO_2 from substrates such as acetate, H_2, and, to a lesser extent, methylamines, formate, and methanol. These methanogenic microorganisms utilize the end products of the acetogenesis stage as their primary substrates to generate biogas, which typically contains 50–75% CH_4, 25–50% CO_2, and trace amounts of H_2, nitrogen, and hydrogen sulfide (Kurade et al., 2019). Methanogenesis provides information about the level of digestion and biological activity in an anaerobic system. The more CH_4 the system generates, the more stable and efficient it is (Thanarasu et al., 2022).

10.2 OPTIMIZATION AD PROCESS

Biogas-producing bacteria have the unique characteristic of being highly sensitive to environmental factors such as temperature, water content, acidity, concentration, and so on, in addition to their ability to survive without oxygen. According to the AD procedure, microorganisms degrade organic molecules in an oxygen-free environment during the AD (Khan, 2021). The breakdown of organic matter during biochemical transformations and modifications results in the breakdown of large molecules into smaller ones, which is essentially part of the biogas generation process. The ability of methanogenic bacteria to degrade and decompose both simple and complex organic matter ultimately results in biogas production through anaerobic reactions and decomposition. The two categories of these bacteria are mesophilic and thermophilic. They can withstand temperatures of 37–45°C and 50–52°C, respectively. This temperature can increase to 70°C in the presence of thermophilic microorganisms (Mahajan & Balachandran, 2017; Sawyerr et al., 2019). Because their enzyme activity is at its highest, the bacteria are able to degrade organic matter and produce biogas at this temperature (Kurade et al., 2019). Digestion reactions in a biogas plant involve a series of chemical and biological activities that occur in the presence of anaerobic bacteria and in the absence of oxygen. In solids digestion, particularly where feedstocks are concentrated and high levels of VFA are produced, inhibiting methanogen growth, it might not be feasible to use a single-phase system due to the different growth requirements of methanogens and hydrolytic/acid-forming bacteria. Therefore, separating methanogenesis from hydrolysis/acidogenesis could potentially improve the process. Methanogenesis could be optimized in the second stage, while growth of hydrolytic and acid-forming bacteria could be optimized in the first step (Balat & Balat, 2009). In parallel, micro-aeration could be used to maximize the preliminary stage reaction rate. Typical AD reactions include methanogenesis and hydrolysis/acidification, which are considered rate-limiting steps in the complex process.

10.2.1 Key Optimization Parameters Affecting Biogas Production

For a biogas plant to operate efficiently, changes in parameters such as feedstock pH, nutrients, temperature, loading rate, carbon/nitrogen ratio, and retention time must be monitored and controlled. Sudden changes in these parameters can negatively impact biogas production. Therefore, for the biogas plant to operate efficiently, these parameters must be adjusted within the desired range. This section presents some of these factors.

The feedstock's pH: One of the mostcrucialvariables in the ADis the pH of the material. Different ideal pHs are achieved at various steps. This is due to the biological transformations that occur during the various stages of the AD. Acidic conditions can affect methanogenic microorganisms, which can negatively impact CH_4 formation and bacterial growth. During the acetogenesis, when organic acids are generated, the pH can drop below 5 (Liu et al., 2018, Qian et al., 2025). A low pH can inhibit digestion and is a sign of system malfunction or insufficient buffering capacity. The methanogenesis process can also be restricted by high pH. The concentration of VFAs the bicarbonate rate, the alkalinity of the system, and the CO_2 content in the digester gas all influence the pH. According to literature studies, a pH range of 6.5 to 7.5 is ideal for AD biogas yield (Isha et al., 2021; Liu et al., 2018; Qian et al., 2025). However, the ideal pH can vary depending on a number of factors, including the digestion method and the substrate used. Therefore, maintaining a certain pH is crucial, and to achieve this, the system is provided with equilibrium buffers such as lime or calcium carbonate. The pH is acrucial indicator in anaerobic reactions (Sawyerr et al., 2019). It gives a clear picture of how well the system is working, especially how stable the digestion is.

Ratio of C/N: Maintaining the correct feedstock composition is crucial for the plant's efficient operation and to maintain the C/N ratio of feed within the desired range. Microorganisms utilize carbon 25-30 times faster than nitrogen during AD (Bidiko et al., 2025). Equation 3 is used to determine the carbon content from the volatile solids (VS) data.

$$\text{Carbon\%} = \text{Volatile Solids}\,(\%)/1.8 \tag{10.1}$$

Therefore, microbes need a C/N ratio of 20-30:1 to meet this requirement; most carbon is readily degradable (Bidiko et al., 2025). Methanogens are using nitrogen quickly when the C/N ratio is high, which lowers the amount of gas produced. Conversely, a low C/N ratio leads to ammonia accumulation and pH above 8.5, which are detrimental to methanogenic microorganisms (Puyuelo et al., 2011). The best C/N ratios for digestate matters can be achieved by combining materials with high and low C/N ratios, such as sewage or animal waste with organic solid waste.

OLR (organic loading rates): The OLR, usually represented as chemical oxygen demand (COD) (kg/m^3 day) and VS in total solids (TS)/L/day/VS/m^3 day, is the amount of biomass fed to the digester (Dhar et al., 2016). The amount of raw material is commonly expressed as Eqn. (2):

$$\text{OLR} = \text{C/HRT} \tag{10.2}$$

where C is the feed concentration in g·VS/L.

It has been noted that methanogens do not consume H_2 and VFA produced by acidogenic bacteria at the same rate, and that problems can arise if the AD of dry wastes in a single step exceeds the OLR system capacity. This is because, at higher OLRs, the VFA intermediates produced during the acidogenic and acid-forming stages cause arise in acidogenic bacteria, which slows the expansion of the methanogenic population. The increase in OLR and acidogenic activity can lead to the accumulation of organic acids and a decline in pH and biogas generations. The biological activity of methanogens producing CH_4, the primary byproduct of biogas, is inhibited at pH 6.6, resulting in less production (Qian et al., 2025). Therefore, increasing CH_4 production and optimizing reactor performance depends on finding the appropriate OLR for a given substrate. The CH_4 yield is typically calculated as the quantity of gas that can be produced per unit volume of VS contained in the feedstock following exposure to sufficient anaerobic reactions at a certain temperature and set of circumstances. Yield is also a measure of the biodegradability of the organic matters, as feedstocks with low VS/TS are difficult to degrade by AD. Consequently, the volume of gas generated also depends heavily on the substrate.

Duration of retention: There are two types of reactors used in AD technology: batch and continuousreactors (Balat and Balat, 2009). In the batch method, the biomass is sealed during digestion and placed in the reactor at the beginning of the decomposition period. Biogas production follows a bell-shaped curve, with all reaction stages occurring roughly sequentially. Only about one-third of the reactorcontent is used for active digestion, and the retention period ranges from 30 to 60 days. For AD to compete with alternative municipal solid waste disposal options, the retention period must be shorter than the current 20-day norm (Kariyama et al., 2018). Based on the COD and biological oxygen demand of the effluent, the retention period is the average time required for complete digestion of organic material. The process will be more efficient if it is accelerated. The digestion rate, which determines how long the biomass should stay in the digestion chamber and therefore the size and cost of the digester, is regulated by the microorganisms that digest the organic material. Reducing the retention time results in a smaller digester, which in turn reduces costs. Therefore, the design of methods that can achieve complete digestion in a shorter time is encouraged. The overall degradation will be less, but there will be a higher output rate per unit of reactor volume and a shorter retention time. It is widely accepted that several techniques can help shorten retention times. Low solids handling and continuous mixing are two of these.

Hydraulic retention time: The HRT shows the time it takes for liquid and solid wastes in a reactor to come into contact with microbial substrate.

The digester volume, V, and digester flow rate, Q, can be used to HRT, which is frequently represented as θ in literature (Sarker et al., 2019) (Eqn. 10.3):

$$\theta = V/Q \tag{10.3}$$

The HRT can be thought of as the treatment time for a waste stream passing through the AD. The longer the HRT, the more effectively it removes contaminants from the waste, as the biomass has sufficient time to come into close contact with the waste. The retention time of the microbial biomass or sludge, and the HRT are related to the retention time of the sludge. In situations where large amounts of particulates are present in the influent streams, in order to optimize the production of bioenergy, longer retention periods are required.

Temperature: AD is commonly worked at two different temperatures: 35°C (mesophilic) and 55°C (thermophilic) (Balagurusamy and Chandel, 2021). While mesophilic digestion can operate similarly to thermophilic digestion, the bacteria in mesophilic digestion require fewer nutrients. This is true even though thermophilic digestion offers some advantages over mesophilic digestion, such as increased reaction temperature and reduced pathogens. Temperature is a good indicator of biological reaction rates. Regular monitoring is necessary for this precise measurement, especially during weather changes. The optimal temperature, thermophilic or mesophilic, will depend on the desired outcome (Sawyerr et al.,2019). However, the temperature must be appropriate for the type of microorganism utilized in waste treatment.

Nutrients: In AD, due to issues like, acidification, poor CH_4 yields and process instability caused by the insufficient availability of nutrient concentration in biomass, OLRs and extended HRTs have been applied (Wang et al., 2023). They affect how well the AD process works and how stable it is. According to the aforementioned setbacks, for the biogas process to proceed, enough amounts of macro and micronutrients are necessary. Phosphorus (P), potassium (K), and sulfur (S), which are necessary nutrients for the formation of biogas, must be present in sufficient amounts. These nutrients can be supplied naturally or, in the event of a deficiency, can be added externally, depending on the type of feedstock being utilized.

Table 10.1 shows the optimum levels of parameters affecting plant performance. Therefore, regular monitoring of parameters affecting plant performance through optimization allows for early detection of potential problems. The optimal C/N ratio and nutrient balance can be achieved by combining various resource components. It is best to gradually increase the OLR of the biomass rather than suddenly. The bacterial community can adapt to the changing environment. Furthermore, when pH levels begin to drop, balance should be restored immediately using alkaline substances approved for use in biogas plants.

10.3 BIOHYDROGEN PRODUCTION

Due to the increasing demand for H_2 energy in the last years, significant attention has been paid to the development of cost-effective and efficient H_2 production technologies. Traditional H_2 generation methods include water electrolysis, steam reforming of CH_4 and other hydrocarbons, non-catalytic partial oxidation of fossil fuels, and

TABLE 10.1
Optimizing Parameters Affecting Biomass Plant Performance

Parameter	Ideal Range	Optimization and Its Effect
pH	6.8–7.5	The pH range where methanogenic bacteria function best is this one. CH_4 production may stop if the pH falls below 6.5 because of the acidic environment. Unbalanced feedstock quality or excessive organic loading are common causes of a pH drop. The pH can be raised by using alkaline materials like sodium bicarbonate or limestone.
C/N Ratio (Carbon/Nitrogen Ratio)	20–30:1	The carbon and nitrogen balance required for the growth of bacteria and the production of CH_4 is provided by this ratio. Bacterial growth is slowed by nitrogen deficit caused by a high C/N ratio (higher than 30). Ammonia buildup from a low C/N ratio (less than 20) may have harmful consequences. Blending various source materials will result in the optimal C/N ratio. For instance, nitrogen-rich animal dung can be combined with carbon-rich straw or sawdust.
OLR	1–5 kg COD/m³ day	The amount of substrate added to the reactor each day is demonstrate by the OLR. Although a high OLR can boost the production of biogas, it can also cause acid accumulation and system failure. Variations in reactor volume, feedstock type, and plant design affect the ideal OLR. Increasing the OLR gradually while keeping an eye on the system's reaction is crucial.
Temperature	Mesophilic: 30–40°C Thermophilic: 50–60°C	The ideal temperatures for various bacterial communities are represented by mesophilic and thermophilic conditions. Although thermophilic systems produce more biogas due to their rapid digestion, they are more energy-intensive and more brittle. Variations in temperature upset the balance of bacteria, which has an adverse effect on the production of biogas.
HRT	20–40 days (Mesophilic) 10–20 days (Thermophilic)	The average amount of time biomass stays in the reactor is known as HRT. For microbes to break down all organic materials and produce biogas, there must be an adequate HRT. While an excessively long HRT might lower the plant's economic efficiency, a short HRT can result in inefficient digestion and low biogas generation.
Nutrients	Balanced	P, K, and S, which are necessary nutrients for the formation of biogas, must be present in enough amounts. These nutrients can be supplied naturally or, in the event of a deficiency, can be added externally, depending on the type of feedstock being utilized.

autothermal reforming, which combines both methods (Bilgiç et al., 2023). Among these techniques, water electrolysis may be the most environmentally friendly way to produce H_2 gas. However, since electricity accounts for a significant portion of the operating costs of H_2 production, electrolysis should be used in regions with low energy prices. Furthermore, the feedwater must be demineralized to prevent electrode corrosion (Bhuiyan & Siddique, 2025). An alternative method, such as biological biohydrogen generation from renewable sources, has received significant attention as a "green technology" in last decades, following concerns about sustainable development and waste reduction. Biohydrogen can be produced from non-toxic, high-carbohydrate feedstocks by anaerobic and photosynthetic microorganisms. Under AD, H_2 is created as a byproduct during the conversion of organic waste into organic acids, which is then used in the production of CH_4.

The raw materials for biohydrogen production can be obtained from different sources, including energy crops (rapeseed meal and cottonseed), agricultural residues (corn straw), forest waste, and industrial and municipal wastes (food industry residues, kitchen waste, and sewage sludge) (Ashfaq et al., 2024; Bhuiyan & Siddique, 2025). All of these can be sustainably sourced from nature. Figure 10.3 shows the sustainable raw materials used in biohydrogen production. Organic wastes containing cellulose and starch from the food and agricultural sectors can be used to produce biohydrogen. Wastewater treatment plants also produce wastewater sludge rich in protein and carbohydrates. Additionally, starch, cellulose-based products, and crude glycerol are found in large quantities in wastewater from the food and agricultural industries. However, the inclusion of other components can limit the biodegradability of these organic wastes due to their complex structure.

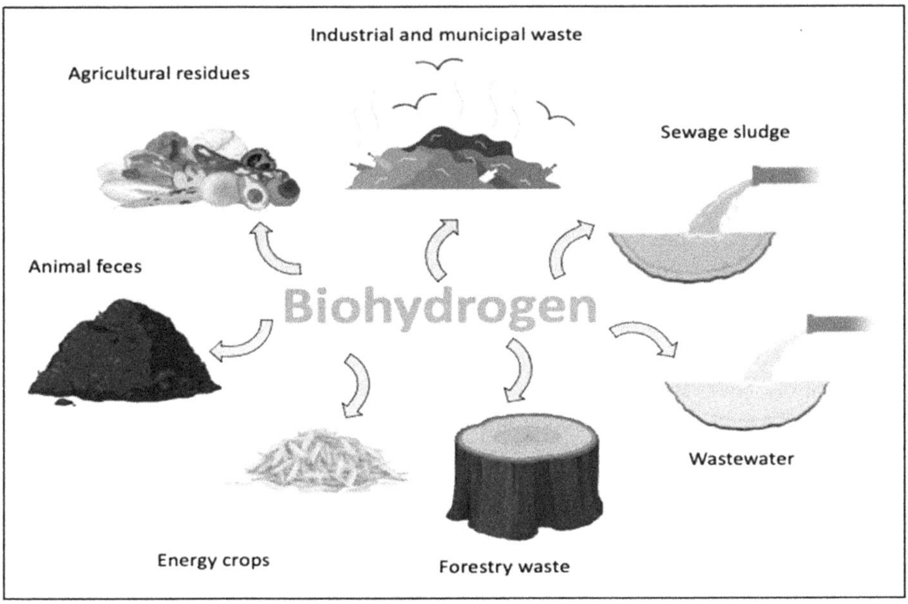

FIGURE 10.3 Sustainable raw materials used in biohydrogen production.

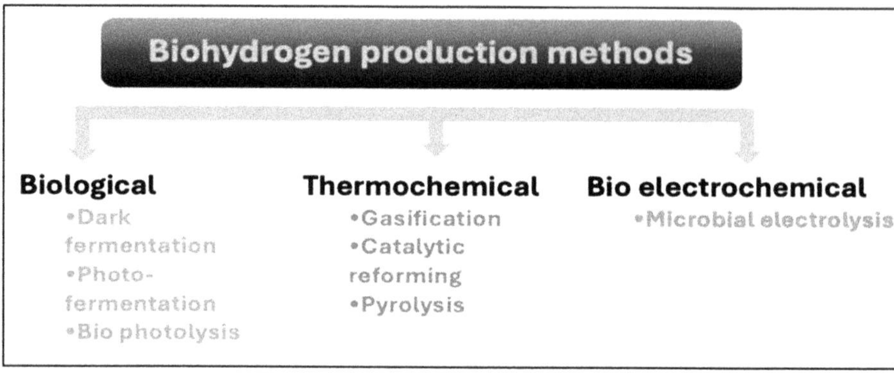

FIGURE 10.4 Schematic diagram of biohydrogen production technologies.

Biohydrogen can be produced as an intermediate product in the acidogenesis stage of AD. However, specific processes and microorganisms are used for specific biohydrogen production. These include: the generation of H_2 through anaerobic fermentation of organic materials by certain bacteria is called dark fermentation, while the use of photosynthetic bacteria to generate H_2 from organic acids in the presence of light is called photofermentation.

Biophotolysis also refers to the direct breakdown of water using sunlight and microalgae or cyanobacteria. Photosynthetic processes include algae that use CO_2 and H_2O to produce H_2 gas. Some photoheterotrophic bacteria use organic acids such as lactic, acetic, and butyric acid to create H_2 and CO_2. It is possible to utilize waste materials for higher H_2 gas production. However, the H_2 production rate is low, and the technology used for this process requires further development (Kapdan & Kargi, 2006). Figure 10.4 shows the schematic diagram of biohydrogen production technologies.

10.4 BIOGAS DISPOSAL GUIDELINES

Biogas disposal refers to the safe and sustainable processing of gases released by biogas-producing facilities. CH_4 and CO_2 make up the majority of biogas. Methane, in particular, is a potent greenhouse gas that contributes 28 times more to global warming than CO_2 (Kumar & Kanwar, 2025; Mundra & Lockley, 2023). Therefore, unregulated release of biogas into the atmosphere can pose significant environmental problems. The primary goal of biogas disposal is to utilize the energy from biogas while reducing its harmful effects on the environment. The main techniques used for this purpose are:

Energy recovery: This strategy, which relies on the utilization of the CH_4 contained in biogas, is the most economically and environmentally beneficial. The most popular approach to energy recovery is cogeneration, or heat and electricity production. Gas engines or microturbines are used to burn biogas, producing heat (hot water) and power. The electricity generated can be used to power the plant or fed into the grid. The generated heat can be used for other industrial operations or for the facility's heating needs, particularly in thermophilic and mesophilic reactors.

Direct heating: Plant buildings or adjacent facilities can be heated by burning biogas in boilers. This approach is quite simple.

Biomethane production ımprovement: Biogas is purified to over 95% by removing CO_2 and other pollutants (such as hydrogen sulfide). Biomethane is the name given to this refined gas. Biomethane can be used as a natural gas alternative in industrial operations, as a vehicle fuel (bio-CNG or bio-LNG), or as a feedstock for the natural gas grid. This technique significantly increases the value and diversity of biogas applications.

Controlled combustion: Stored biogas is burned rather than released into the atmosphere, especially when the gas cannot be transferred to energy recovery systems or when the system is undergoing repairs. CH_4 is converted to CO_2 through this mechanism, significantly reducing its potential to cause global warming. While better for the environment than CH_4 being released into the atmosphere, it also wastes energy.

Direct atmospheric release: CH_4 is a potent greenhouse gas that contributes to global warming, so it is strictly regulated. In modern biogas plants, this is only done in emergencies.

10.5 PROSPECTS AND CHALLENGES

For biomass to be used on a large scale in AD, several significant obstacles must be overcome to ensure its functionality and affordability, and to expand its use on a large scale.

Below is a summary of future research and viewpoints based on existing knowledge.

- To provide a cost-effective, environmentally friendly, and practical biogas production method, the synthesis process can be simplified by selecting specific biomass types, such as nitrogen and iron-rich biomass. Green economy and production techniques are also needed to minimize secondary contamination brought on by the process of preparation.
- Currently investigated modification techniques are environmentally risky and quite complex. Firstly, post-process modification processes are often complex due to technical complexity. On the other hand, the pre-process modification procedure is simpler.
- Given the sensitivity of ADs imbalances resulting from an excess of certain intermediates in a reactor can quickly lead to process failures or inhibition (Anukam et al., 2019). To ensure efficient and comprehensive digestion, the design of any single-chamber reactor whether continuous or not, must account for these factors. Furthermore, there must be ways to detect and correct potential malfunctions in the digesters. These potential malfunctions can be identified through optimization studies.
- Numerous tests are available to measure AD. A digester design should consider co-digestion as a way to balance the C/N ratio. For example, an

organic matter used for AD can also be described in terms of the C/N ratio, which can provide information about its suitability for digestion and the likelihood of VFA acidification or ammonia inhibition (Wang et al., 2023). Factors affecting AD efficiency should also be considered when designing a digester.

- Theoretically, shorter digestion times should increase the amount of biogas produced by allowing batch digesters to process more waste in a given time.

AD has seen an unusually high level of interest in AD research in recent years, partly due to the growing interest in sustainable and renewable energy sources. However, there are still gaps. We believe this study will pave the way for further research.

10.6 CONCLUSIONS

This chapter present of biogas production and optimization guidelines. The study demonstrated that the technologies and procedures used to produce biogas via AD are an effective way to generate sustainable energy alternatives. Several critical parameters in the anaerobic zone were optimized, including retention times, pH, temperature, and the availability of nutrients and OLRs at various steps of the AD process. One of the crucial steps in achieving the best possible biogas production from the AD process is the evaluation and optimization of the AD. This study offers a new perspective for researchers.

REFERENCES

Anukam, A., Mohammadi, A., Naqvi, M., & Granström, K. (2019). A review of the chemistry of anaerobic digestion: Methods of accelerating and optimizing process efficiency. *Processes*, *7*(8), 504.

Ashfaq, M. M., Tüzemen, G. B., & Noor, A. (2024). Exploiting agricultural biomass via thermochemical processes for sustainable hydrogen and bioenergy: A critical review. *International Journal of Hydrogen Energy*, *84*, 1068–1084.

Balagurusamy, N., & Chandel, A. K. (Eds.). (2021). *Biogas production: From anaerobic digestion to a sustainable bioenergy industry*. Springer Nature.

Balat, M., & Balat, H. (2009). Biogas as a renewable energy source—A review. *Energy Sources, Part A*, *31*(14), 1280–1293.

Bhuiyan, M. M. H., & Siddique, Z. (2025). Hydrogen as an alternative fuel: A comprehensive review of challenges and opportunities in production, storage, and transportation. *International Journal of Hydrogen Energy*, *102*, 1026–1044.

Bidiko, G. B., Sangib, E. B., & Gnaro, M. A. (2025). Optimization of biogas production through co-digestion of cafeteria food waste and cow dung using response surface methodology. *Frontiers in Energy Research*, *13*, 1568478.

Bilgiç, G., Bendeş, E., Öztürk, B., & Atasever, S. (2023). Recent advances in artificial neural network research for modeling hydrogen production processes. *International Journal of Hydrogen Energy*, *48*(50), 18947–18977.

Dhar, H., Kumar, P., Kumar, S., Mukherjee, S., & Vaidya, A. N. (2016). Effect of organic loading rate during anaerobic digestion of municipal solid waste. *Bioresource Technology*, *217*, 56–61.

Hashim, A. F. (2021). *Biomethane production from food waste mixture by anaerobic digestion reactor* (Master's thesis). Universiti Malaysia Pahang Al-Sultan Abdullah.

Heiker, M., Kraume, M., Mertins, A., Wawer, T., & Rosenberger, S. (2021). Biogas plants in renewable energy systems: A systematic review of modeling approaches of biogas production. *Applied Sciences*, *11*(8), 3361.

Isha, A., D'Silva, T. C., Subbarao, P. M., Chandra, R., & Vijay, V. K. (2021). Stabilization of anaerobic digestion of kitchen wastes using protein-rich additives: Study of process performance, kinetic modelling and energy balance. *Bioresource Technology*, *337*, 125331.

Kapdan, I. K., & Kargi, F. (2006). Bio-hydrogen production from waste materials. *Enzyme and Microbial Technology*, *38*(5), 569–582.

Kariyama, I. D., Zhai, X., & Wu, B. (2018). Influence of mixing on anaerobic digestion efficiency in stirred tank digesters: A review. *Water Research*, *143*, 503–517.

Khan, A. (Ed.). (2021). *Sustainable bioconversion of waste to value-added products*. Springer Nature.

Kumar, R., & Kanwar, S. S. (Eds.). (2025). *Waste management: From trash to treasure*. CRC Press.

Kurade, M. B., Saha, S., Salama, E. S., Patil, S. M., Govindwar, S. P., & Jeon, B. H. (2019). Acetoclastic methanogenesis led by *Methanosarcina* in anaerobic co-digestion of fats, oil, and grease for enhanced methane production. *Bioresource Technology*, *272*, 351–359.

Liu, C., Luo, G., Wang, W., He, Y., Zhang, R., & Liu, G. (2018). The effects of pH and temperature on acetate production and microbial community compositions by syngas fermentation. *Fuel*, *224*, 537–544.

Mahajan, G. B., & Balachandran, L. (2017). Sources of antibiotics: Hot springs. *Biochemical Pharmacology*, *134*, 35–41.

Molino, A., Larocca, V., Chianese, S., & Musmarra, D. (2018). Biofuels production by biomass gasification: A review. *Energies*, *11*(4), 811.

Mundra, I., & Lockley, A. (2023). Emergent methane mitigation and removal approaches: A review. *Atmospheric Environment: X*, *21*, 100223.

Nie, W., He, S., Lin, Y., Cheng, J. J., & Yang, C. (2024). Functional biochar in enhanced anaerobic digestion: Synthesis, performances, and mechanisms. *Science of the Total Environment*, *906*, 167681.

Puyuelo, B., Ponsá, S., Gea, T., & Sánchez, A. (2011). Determining C/N ratios for typical organic wastes using biodegradable fractions. *Chemosphere*, *85*(4), 653–659.

Qian, S., Chen, L., Xu, S., Zeng, C., Lian, X., Xia, Z., & Zou, J. (2025). Research on methane-rich biogas production technology by anaerobic digestion under carbon neutrality: A review. *Sustainability*, *17*(4), 1425.

Ruggeri, B., Tommasi, T., & Sanfilippo, S. (2015). *$BioH_2$ & $BioCH_4$ through anaerobic digestion: From research to full-scale applications*. Springer.

Sachdeva, S., Garg, V. K., Labhsetwar, N. K., Singh, A., & Yogalakshmi, K. N. (2022). *Zero waste biorefinery* (Y. K. Nandabalan, V. K. Garg, N. K. Labhsetwar, & A. Singh, Eds.). Springer.

Sarker, S., Lamb, J. J., Hjelme, D. R., & Lien, K. M. (2019). A review of the role of critical parameters in the design and operation of biogas production plants. *Applied Sciences*, *9*(9), 1915.

Sawyerr, N., Trois, C., Workneh, T., & Okudoh, V. (2019). An overview of biogas production: Fundamentals, applications and future research. *International Journal of Energy Economics and Policy*, *9*(2), 105–116. https://doi.org/10.32479/ijeep.7375.

Thanarasu, A., Periyasamy, K., & Subramanian, S. (2022). An integrated anaerobic digestion and microbial electrolysis system for enhanced methane production from organic waste: Fundamentals, innovative design, and scale-up deliberation. *Chemosphere*, *287*, 131886.

Wang, Z., Hu, Y., Wang, S., Wu, G., & Zhan, X. (2023). A critical review on dry anaerobic digestion of organic waste: Characteristics, operational conditions, and improvement strategies. *Renewable and Sustainable Energy Reviews*, *176*, 113208.

11 Financial and Techno-economic Aspects of Green Chemistry

Valentin Romanovski

11.1 INTRODUCTION

Green chemistry is formed at the intersection of chemistry, ecology and economics, responding to the challenges of global change and growth of production systems in the context of limited natural resources and the need to minimize the negative impact on the environment. The concept of "green" chemistry (green chemistry), first systematically formulated by Paul Anstas and John Warner (Anastas & Warner, 2000), is aimed at the development and implementation of processes and products that prevent the formation of hazardous substances, minimize the consumption of energy and raw materials, and reduce the toxicity of the final product. At the same time, issues of economic feasibility and financial efficiency are becoming an integral part of strategies for the transition to "green" technologies. With the development of sustainable development principles, green chemistry is considered not only as a tool for reducing environmental risks, but also as a strategic resource for increasing the competitiveness of companies, forming new markets and attracting investment. According to modern concepts, sustainable development is based on three equally important components – economic, environmental, and social. Green chemistry integrates all these aspects, allowing the creation of highly efficient technological solutions taking into account the product life cycle (life cycle assessment, LCA) and complex technical and economic indicators. An important factor stimulating the development of green chemistry is economic and legal instruments that support the greening of industry. As noted in modern environmental management practices, without clearly structured economic incentives, the transition to waste-free or low-waste technologies (LWT) is difficult. In this context, the principles of paid nature management, environmental taxation, and compensation mechanisms for environmental pollution play a key role. Environmental payments should not only be a source of budget formation but also an effective tool for motivating businesses to modernize production processes. The development of green chemistry is closely related to the concept of waste-free and LWT. As the experience of industrialized countries shows, the transition to closed production cycles, the use of secondary resources, and the reduction of waste and energy consumption are not only environmental but also economically feasible. The concept of zero-waste technology (ZNWT) focuses on the full use of raw materials and energy, creating a closed cycle in which the waste from one process becomes the raw material for another. These ideas are reflected in

 DOI: 10.1201/9781003642589-11

modern "green" chemistry practices, from developing new catalysts and biodegradable polymers to using renewable energy sources in chemical production.

International initiatives also contribute to the development of green chemistry. Cooperation with the United Nations Environment Programme (UNEP), the World Bank, the United Nations Economic Commission, European Bank for Reconstruction and Development for Europe, and other international organizations allows us to attract funding for the implementation of environmentally friendly technologies, develop uniform standards and regulations.

From an economic point of view, green chemistry opens up new opportunities for the formation of sustainable business models. More and more companies are implementing ESG (environmental, social, governance) principles, which are becoming an important criterion for investment attractiveness. Green bonds, climate funds, and government support programs make it possible to finance projects related to next-generation chemical processes – energy efficient, safe for ecosystems, and focused on the circular economy.

The issue of assessing the economic efficiency of "green" technologies deserves special attention. The use of LCA, calculation of return on investment (ROI), reduction of waste disposal costs, and environmental fines demonstrate that environmentally oriented solutions are often more profitable than traditional ones.

Modern green chemistry is based on such principles as replacing hazardous reagents with safe analogues (Jayalekshmi et al., 2025), increasing the atomic efficiency of reactions, using renewable raw materials and biocatalysts, and reducing the number of stages and energy-consuming processes. These approaches intersect with the concept of "clean" technologies: reducing the amount of materials used, switching to recyclable and reusable raw materials, minimizing packaging, eliminating toxic solvents and organic compounds, and reducing energy consumption. Thus, the role of green chemistry in sustainable development lies not only in solving environmental problems but also in the formation of an innovative economy capable of ensuring a balance between production interests and environmental protection. The integration of financial mechanisms, international standards, and technological innovations makes green chemistry a key element of modern industrial progress. The following sections of the chapter will consider the financial aspects of green chemistry, methods of technical and economic assessment (TEA), as well as specific business models and examples of the implementation of environmentally friendly technologies.

11.2 PRINCIPLES OF RATIONAL USE OF NATURAL RESOURCES AND CONCEPTS OF WASTE-FREE TECHNOLOGIES

Rational nature management is not just an economic activity with an environmental component, but a system of approaches aimed at the optimal use and reproduction of natural resources with minimal impact on the environment. Its goal is to harmonize human economic activity with natural processes, eliminate irreversible damage to ecosystems, and preserve resources for future generations. The main objective of rational nature management is to meet the needs of society without violating the stability of the biosphere. Modern production should be structured in such a way as to reproduce the cycles of nature as much as possible, eliminating the overconsumption of raw materials and the accumulation of hazardous waste. Any technological chain must take into account three most important factors: the state of natural resources, environmental safety, and long-term economic efficiency. The key principles of rational nature management are as follows:

1. Systems approach. Any natural resource is considered as a part of a larger ecosystem. Economic activity should take into account the interrelationship of all natural components, since isolated protection of one resource can lead to the degradation of others. Systems thinking involves taking into account natural, social, and economic factors as elements of a single complex.
2. Optimization of nature management. Optimization means preventing negative consequences of economic activity by designing technologies with a minimum level of environmental risk. It involves forecasting possible impacts, taking into account environmental limits and priority use of resources within their reproducibility.
3. The principle of production concentration. We are talking about the creation of territorial production complexes, where enterprises are united for the purpose of joint use of raw materials, energy and infrastructure. This approach allows for the comprehensive use of regional resources, reducing waste volumes through cooperation and increasing production efficiency. For example, one enterprise generates waste, but in close proximity there is another enterprise for which this waste is raw material for its production.
4. Greening/ecologization of production. Greening implies the maximum approximation of production cycles to natural biogeochemical cycles. Although it is impossible to completely close all technological processes, the desire to do so is a key area of development. The main tools are resource-saving and LWT, recycling, and increased energy efficiency.

Zero waste production is a concept that involves the use of raw materials, energy, and materials in a closed cycle "raw materials – production – consumption – secondary resources," so that the impact on the environment does not disrupt its normal functioning. The Commission of the European Communities uses a different definition: Clean technology is the production of products with the most rational use of raw materials and energy, which allows simultaneously reducing the volume of substances emitted into the environment and reducing the amount of waste generated during the production and consumption of products. Although the term "Zero Waste Technologies" (ZWT) was first officially used in Geneva in 1979, it appeared back in 1973 thanks to the American chemist Paul Palmer (Yale). In 2000, at the first international Conference in Cathay (New Zealand), the zero waste concept received global recognition and became the basis for the modern movement to transition to "zero waste" and a circular economy. At the international level, after the Geneva Conference (1979), it was recognized that the future of industry lies in low-waste (LWT) and clean technologies. The definition of LWT was clarified at the UNECE Seminar on LWT in Tashkent in 1984. The main goal of LWT is to eliminate the causes of waste generation, rather than combat its consequences. The ideal option is production in which the by-products of one process become raw materials for another. At the same time, the final product should have a long service life and the ability to be repaired, reused, and easily recycled after wear. Three features of LWT can be distinguished: (1) Comprehensive use of raw materials and energy, i.e., there is a principle of mandatory inclusion of all components of raw materials in production, as well as maximum use of the potential of energy resources. (2) Cyclicity, i.e., use of resources in a cycle (Figure 11.1). (3) Ensuring normal functioning of the environment. To fulfill this feature, it is necessary to comply with quality standards in the field of environmental protection. LWT are an intermediate step on the way to LWT. The LWT allows waste generation, but within sanitary and hygienic standards, with priority given to its recycling or safe storage. The LWT focuses on minimizing harmful impacts on the environment through the integrated use of raw materials and the introduction of resource-saving processes.

In international practice, the term "clean technology" is used as a synonym for ZWT. Such technologies are characterized by a low level of waste (liquid, solid, gaseous), minimal emissions into the atmosphere, and efficient use of resources.

Characteristic features of clean production:

- minimization of raw material and energy consumption;
- use of materials suitable for recycling;
- reduction of types of materials to simplify recycling;
- refusal of hazardous chemicals (e.g., chlorinated solvents);
- transition to renewable energy sources;
- reduction of noise and odor levels;
- recirculation of water and heat exchange flows.

The strategy of clean production includes designing processes that minimize hazardous waste, refusal of technologies requiring toxic reagents, optimization of logistics, and replacement of hazardous catalysts with environmentally friendly ones. There is no single criterion for assessing the degree of waste-free production. However, the following indicators are widely used:

- the share of raw materials used that go into finished products;
- the amount of waste per unit of output;
- the ratio of recycled waste to its total volume;

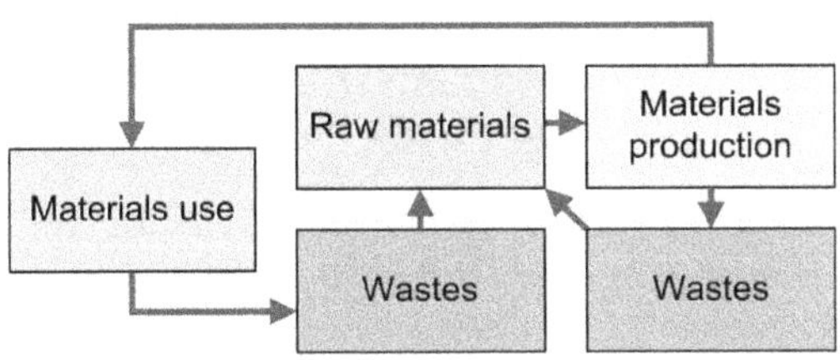

FIGURE 11.1 The principle of cyclicity of ZWT.

- the level of waste toxicity and its impact on ecosystems;
- energy costs for production and processing.

A waste-free coefficient (K) has been developed for the chemical industry, taking into account the completeness of raw material use, energy efficiency, and the degree of impact on the environment. The higher the K, the closer the production is to the ZWT category. The standard is the production in which the technological stages form a closed cycle, while the waste of one process becomes a resource for another, and energy losses are minimal.

The development of ZWT is impossible without deep greening of various industries. Energy: transition to renewable sources (solar, wind, hydropower), minimization of the use of fossil fuels, and development of closed carbon cycle technologies. Transport: reduction of toxic emissions and use of environmentally friendly fuels (e.g., replacing fossil fuels with water/hydrogen and obtaining energy from natural processes) and development of public transport. It should be noted that the introduction of electric transport in itself is not an environmentally friendly approach, in many cases even less environmentally friendly than traditional ones. The use of electric transport is more associated with the transfer of pollutant emissions from the city beyond its boundaries. Agriculture: biological methods of pest control, minimization of chemical fertilizers and pesticides, introduction of closed soil technologies. Industry: equipment modernization, introduction of continuous processes, increase in the unit capacity of units, reduction in the number of processing stages.

Economic methods play a key role in stimulating the transition to clean technologies. The system of paid nature management is an important instrument of environmental policy, including: i) payment for the use of natural resources; ii) payments for environmental pollution; iii) compensation contributions for damage caused. This system encourages enterprises to invest in technologies that reduce their environmental load and, accordingly, reduce the amount of payments. The normative approach allows the state to set prices for environmental services that take into account the costs of resource reproduction and compensation for damage. At the same time, a mechanism of tax incentives is used for companies implementing ZWT and LWT. Institutional measures play a significant role: control over compliance with environmental standards, development of insurance mechanisms, and public involvement in monitoring.

The concepts of ZWT and LWT for green chemistry are of strategic importance. They allow minimizing emissions of hazardous chemicals, reducing waste disposal costs and increasing the use of secondary resources.

11.3 TECHNICAL, ECONOMIC, AND ENVIRONMENTAL ASPECTS OF GREEN TECHNOLOGY ANALYSIS

A comprehensive analysis of green technologies is a key tool for substantiating decisions on their implementation and scaling. Such an analysis should take into account technical feasibility, economic feasibility, and environmental sustainability at all stages of the life cycle (Figure 11.2). The choice of life cycle boundaries depends on the need. It can include either the entire cycle (Figure 11.2) from the extraction

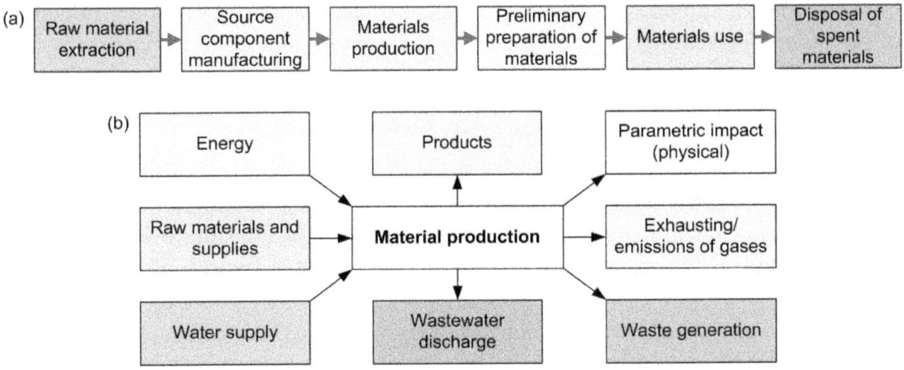

FIGURE 11.2 (a) Stages of the product life cycle, (b) types of environmental impact at stages of the life cycle.

of natural resources to the final disposal of the spent product (more often used to assess the environmental impact), or from the purchase of equipment and its operation, modernization, and final disposal (assessment of life cycle costs). Considering one aspect in isolation from the others inevitably leads to erroneous conclusions: a technology with high energy efficiency may be economically unprofitable or not comply with environmental standards. For this reason, feasibility studies usually include all three components for analysis: technical, environmental, and economic (Figure 11.3). Modern approaches to green chemistry are increasingly based on the integration of TEA, LCA, and life cycle costing (LCC) methodologies, allowing for a comprehensive assessment of the viability and sustainability of innovations.

Technical analysis within the framework of green technologies is aimed at determining the efficiency of using raw materials, water, energy and other resources, as well as analyzing key production characteristics. It includes building material and energy balances, calculating yield factors for target products and identifying bottlenecks in the process flow diagram. Of great importance is determining the process readiness level, which shows how close the technology is to industrial

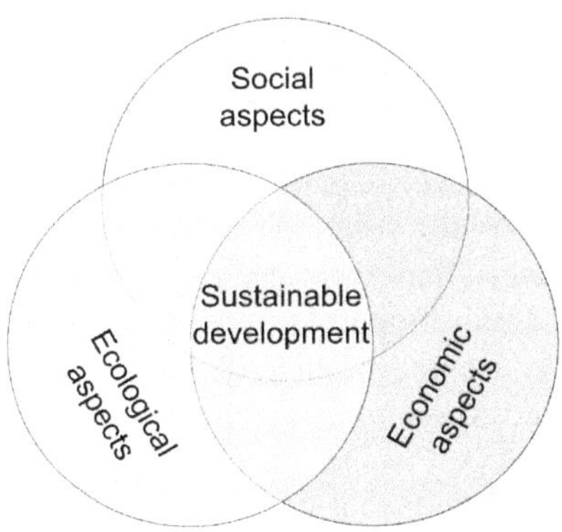

FIGURE 11.3 Three main components of feasibility studies.

implementation. In addition, the operating experience of pilot plants, the flexibility of processes for scaling and resistance to fluctuations in raw material quality are taken into account. Professional software packages are used to model technological processes – Aspen Plus, Aspen HYSYS, CHEMCAD, SuperPro Designer. Aspen Energy Analyzer tools enable energy audits and pinch analysis, while COMSOL Multiphysics and ANSYS Fluent are used for CFD (Computational Fluid Dynamics) modeling of fluid dynamics, heat and mass transfer, and chemical reactions. The main results of technical analysis are specific energy consumption, conversion and product yield factors, energy efficiency, and complex indices of technological complexity.

Economic analysis considers the technology from the perspective of costs, profitability, and investment attractiveness, with the possibility of such an assessment at all stages of the life cycle. There are static and dynamic performance indicators. Static indicators, such as simple profitability, simple payback period, and specific cost of production, are based on fixed calculations without taking into account the time value of money (discount rate). They are easy to use, but do not reflect the risks associated with price changes or project delays. Dynamic indicators, on the other hand, take into account the time factor and discounting of cash flows: these include net present value (NPV), internal rate of return (IRR) and profitability index. These indicators allow you to evaluate a project taking into account future income, inflation, cost of capital, and market fluctuations. A special place in economic analysis is occupied by the LCC method, which takes into account all costs, starting from the design and procurement stage of technology, including capital investments (CAPEX), operating expenses (OPEX). To calculate these indicators, software tools such as Aspen Process Economic Analyzer (APEA), SuperPro Designer, @RISK or Oracle Crystal Ball are used, which provide scenario modeling and sensitivity analysis to changes in external factors. Hybrid models linking TEA with LCA allow you to simultaneously analyze economic and environmental parameters, which is becoming a standard in the development of sustainable industrial processes.

Environmental analysis is aimed at assessing the environmental impact of technology at all stages of its life cycle. The main tool is the LCA methodology, which allows for a quantitative assessment of greenhouse gas emissions, toxic compounds, water pollution, solid and hazardous waste generation, as well as energy and water consumption. The key indicators of environmental analysis are the carbon footprint (in terms of CO_2 equivalent), water footprint, energy consumption, global warming potential, as well as integral impact indices, such as eutrophication, acid rainfall, or ozone layer depletion. SimaPro, GaBi, OpenLCA, Umberto LCA+, and Brightway2 software solutions are used to implement LCA. They support international databases (Ecoinvent, ELCD, GaBi Database) and allow for an assessment of the entire environmental profile of the technology.

The ongoing trend is to use integrated TEA + LCA + LCC approaches that allow the economic, technical and environmental indicators to be presented simultaneously. The use of Aspen Plus or SuperPro Designer packages alongside LCA tools such as SimaPro or GaBi, allows to produce a single model that incorporates both material balances, economics and environmental impacts. A more comprehensive life cycle sustainability assessment methodology extends this integration by including social factors (S-LCA), such as impacts on public health, quality of working conditions or social benefits for regions. These approaches are used in the analysis of technologies for biofuels, biopolymers, CO_2 capture and use, waste recycling and renewable energy, where it is important to find a balance between economic profitability, technical feasibility and minimization of environmental damage.

TEA of "green" processes is a systems assessment of technological solutions from both economic and environmental efficiency perspectives over the life cycle of the product. Similar to defining costs in TEA, the environmental footprint assessment is complex in defining the costs of the process as it includes not only the costs of raw materials, energy and equipment, but also the indirect costs of disposal, environmental payments, reputation risks and regulatory costs. Such a comprehensive analysis provides an understanding of not only the technological, but also the strategic value of new solutions. For a comprehensive assessment of the selected technical, environmental and economic indicators, expert assessment methods and other systems analysis approaches can be used (Hurynovich et al., 2021). The main objective of the feasibility study of green processes is to justify their effectiveness in comparison with traditional technologies. New technologies should demonstrate an acceptable level of profitability and a justified payback period for invested funds. For example, the criterion may be a simple payback period of no more than 5 years, and a dynamic payback period of no more than 8 years.

To illustrate affordances of green technologies, let us consider a contrast between the traditional production of polyethylene and polylactic acid (PLA) bioplastics. Recent research indicates PLA-based polymers have a carbon footprint that is 60–70% lower than polyethylene with similar strength and durability. Although the capital costs of implementing a PLA line are 15–20% higher, life cycle cost analysis (LCCA) shows that reduced waste disposal costs and lower environmental fees reduce the payback period of the project from 8 to 5 years (Bala et al., 2022; Rezvani Ghomi et al., 2021). This approach demonstrates how the integration of LCA and financial calculations allows for the formation of more sustainable business models.

The traditional chemical industry is focused on maximizing output while minimizing costs, often ignoring long-term environmental impacts. Green chemistry changes this trend. For example, switching from organochlorine solvents to aqueous or biodegradable substitutes may still involve some investment in new equipment, but it minimizes the potential for toxic chemicals entering the environment and cuts the costs of their neutralization. Another option is to use catalysts based on natural enzymes. Despite the high cost of developing such catalysts, they provide higher selectivity of reactions and reduce the formation of by-products, which reduces both material costs and waste treatment costs.

When comparing technologies, it is important to consider all possible indirect benefits: reduced fines, freeing up areas previously occupied by waste, increased investment attractiveness, simplified receipt of environmental certificates, social factors, etc. Such factors are often decisive when choosing in favor of green technologies (especially social factors, which can influence the choice in favor of implementing a project even if its implementation is not economically efficient).

The regulatory environment has a significant impact on the TEA of green processes. Strict environmental standards and tax mechanisms encourage companies to switch to new technologies, as the cost of inaction becomes too high. However, such a stimulating role of environmental taxes and charges is sometimes limited by the state so as not to close enterprises, instead of building effective treatment facilities. In this regard, in some countries with developing economies, the system of calculating environmental charges has not developed, but is gradually degrading. But if we look globally, the demand for environmentally certified products is growing in the world market, which also increases the economic attractiveness of green solutions. In the context of globalization, access to foreign markets is increasingly associated with ISO 14001 certificates and other sustainability standards. Closed cycles of water supply and secondary processing of raw materials are being introduced at enterprises. For example, the integration of wastewater treatment technologies with subsequent recycling allows for a simultaneous reduction in water costs and pollution charges. Other examples include the processing of by-product gases (e.g., CO_2) into chemical products with high added value. Such projects show that "green" processes can be not only environmentally friendly but also highly profitable.

The future of TEA is associated with digitalization and the use of forecasting models based on big data. Such approaches allow taking into account dynamic factors – changes in raw material prices, tariffs for environmental payments, and market demand. The integration of LCA with financial models (LCCA) is becoming a key tool for strategic planning. In addition, the development of the concept of carbon reporting and carbon trading markets will enhance the role of "green" technologies as a competitive advantage. Companies that can calculate the economic benefits of reducing their carbon footprint in advance will receive tangible advantages in accessing investments and government grants.

11.4 FINANCIAL ASPECTS OF GREEN CHEMISTRY

Financial sustainability and cost-effectiveness are the essential boundaries to successfully implementing the principles of green chemistry. It is impossible to ensure a mass transition to environmentally friendly technologies without clear development of associated economic incentives/financing sources and confidence in long-term financial benefits. Unlike traditional chemical processes, which have been optimized historically by primarily criteria of cost and productivity, green chemistry achieves economy through a very different spectrum of product life cycle, including the costs of requisite raw materials, energy resources, waste sink, payments for environmental degradation, and risk of regulatory restrictions. Today, green chemistry is considered one of the priority areas of "green" investment. World practice shows that companies implementing sustainable technologies are becoming more attractive to international funds, climate initiatives, and government support programs. For example, the issue of "green" bonds allows financing projects to modernize chemical production, reduce emissions, or switch to renewable raw materials. ESG factors have now become a part of long term investment consideration. So, firms developing through green chemistry have more opportunity for capital because the investors who are considering ESG will favor these companies. Government cash and tax incentives are very important too: Anyway, just money in the form of tax breaks, grants for innovative projects, interest rate compensation programs which incentivize the introduction of energy efficient non-waste producing technologies will be stimulants.

International financial instruments occupy a special place. Of particular importance is cooperation with organizations such as the World Bank, the UNEP, and the United Nations Economic Commission for Europe. These structures not only provide investments but also help to form standards for assessing the effectiveness of projects, which is especially important for the chemical industry, where the risks of environmental accidents or fines can be critical.

"Green" bonds (Green Bonds) (Axious 2025; Reuters 2025b) are an important mechanism to support sustainable projects. Specifically, it allows companies to raise targeted capital for projects to reduce emissions, to initiate intangible renewables, or to improve energy efficiency. For example, in 2023 BASF issued €2 billion in green bonds to fund projects directed towards low-carbon production and plastics recycling. Public development banks such as the European Investment Bank and the World Bank support common initiatives likely to be financed via 'green' bonds. Therefore, green bonds are one of the most useful financing instruments in the chemical industry.

When implementing new technologies, the analysis of their payback and long-term benefits comes to the fore. Traditional indicators such as ROI, NPV, and payback period are supplemented by environmental criteria – reduction of environmental payments, reduction of waste disposal costs, and reduction of the risk of lawsuits for environmental damage. Example: the introduction of new generation catalysts that increase the selectivity of chemical reactions may require significant investments at the start, but due to the reduction in the amount of waste, saving raw materials, and reducing the cost of wastewater treatment, the project pays off faster than traditional process schemes.

An important tool is the calculation of "hidden" economic effects. For example, the use of biodegradable polymers may be more expensive at the production stage but it reduces waste management costs and improves the company's reputation, which in the long term increases its market capitalization. Likewise, using renewable energy sources in chemical manufacturing means it no longer relies on the volatile pricing of fossil fuels, making it more financially stable overtime. Alongside traditional payback measures (NPV, ROI, IRR), the ESG assessment framework is also finding its way into green chemistry as it simultaneously considers the financial outcomes a long with the project's impact on climate, biodiversity, and social impact. Most recently, these types of integrated financial-environmental measures have become obligatory for gaining funding from large institutional investors.

The environmental tax system is one of the most effective tools that encourages enterprises to modernize. Payment for natural resource use as well as compensation payments for pollution are becoming an important factor in the economic pressure on non-ecological production. The higher the polluting effect of the technology, the higher is its indirect cost.

Green chemistry, on the contrary, allows you to reduce these costs. For example, the transition to low-waste processes reduces the volume of landfilled waste and, accordingly, the amount of payment for its placement. In EU countries, the "polluter pays" principle is widely used, which encourages companies to invest in environmentally friendly technologies in order to reduce the financial burden. Economic incentives include not only fines, but also incentives. Companies that use green processes can receive tax breaks, access to preferential loans and government grants. Such policies are already in place in a number of countries that support the transition to a circular economy. As a result, companies view investments in green chemistry as a strategic step to reduce risks and improve competitiveness.

Despite the obvious advantages, green chemistry projects often face high start-up costs and a long payback period. For small and medium-sized enterprises, this becomes a serious barrier. In such cases, the role of consortiums, clusters, and partnerships that unite industrial and scientific organizations is important. Joint investments and attracting international grants and subsidies help reduce the financial burden on individual companies. Another tool is the use of voluntary certification mechanisms and the implementation of international standards (e.g., ISO 14001). Although this requires expenses for auditing and implementing environmental management systems, such steps open up access to new markets and strengthen investor confidence, which in the long term brings more financial benefits than costs.

World practice shows that the financial benefits of green chemistry can be illustrated by specific cases. For example, the introduction of closed water supply technologies at chemical plants allows for a reduction in water consumption and wastewater treatment costs several times. Another example is the use of by-products as secondary raw materials: the processing of carbon dioxide into raw materials for the synthesis of polymers not only reduces emissions but also opens up new market niches. For large corporations, green chemistry is becoming part of the corporate strategy. They use it as a tool to increase brand value and consumer confidence. A reduced carbon footprint, confirmed by financial reports and certification, helps attract international investors focused on ESG projects.

In the coming years, we can expect an increased role of financial mechanisms in promoting green chemistry. The introduction of carbon taxes, the development of carbon quota markets, and the creation of climate funds will stimulate companies to switch to new technologies. Modern financial policy is gradually being restructured in such a way as to take into account environmental risks in assessing the cost of capital. Thus, the financial aspects of green chemistry are not only a matter of costs but also a source of strategic advantages. Companies willing to invest in green technologies benefit from reduced operational risks, increased consumer confidence, and sustainable growth against the backdrop of the global transition to a low-carbon economy.

11.5 INDICATORS AND INTERNATIONAL STANDARDS

Without developing a system of clear indicators and international standards that allow for an objective assessment of the efficiency and sustainability of the technologies deployed, the evolution of green chemistry is not possible. These indicators are a necessary tool for all stakeholders: both manufacturers and investors, regulators, and consumers. They are key to monitoring progress along the path to greening our industry and moving toward the sustainable development agenda. Indicators are quantitative and qualitative metrics that characterize the impact of technologies on the environment, the level of resource conservation, and economic efficiency. They allow for a comprehensive assessment of processes not only in terms of their financial return but also taking into account the impact on ecosystems, public health and climate.

The major groups of indicators that are commonly used are as follows: (1) Environmental and resource indicators. This refers to energy, water and raw materials consumed; the emissions and waste produced per unit of output. The indicators help us understand how well the resource base is being utilized and potential risks of natural resource depletion. (2) Economic indicators. Including profitability of green technologies, savings in environmental payments, reduced waste disposal costs and costs of complying with environmental standards. (3) Social indicators. Level of environmental and safety for employees and the population; decreased toxicity in products; improved safety; and a heal their work environment. (4) Comprehensive sustainability indices. For example, an integral carbon footprint indicator or an energy efficiency index, which includes several aspects of environmental and economic assessment.

The use of indicators allows you to justify decisions on the choice of technologies and demonstrate their advantages in the long term. These indicators are useful for establishing the relationship between environmental efficiency and economic efficiency, which is critical to investment decision-making. In this context of globalization, environmental reporting and management standardization is a useful tool to integrate into global markets. The main standards are as follows:

- ISO 14001 – an environmental management system that serves as a framework for organizations to put in place a comprehensive set of practices to reduce organizational impacts on the environment.
- ISO 14040/44 – LCA standards for developing standardized analysis of product environmental footprints.
- Responsible Care – a voluntary initiative of the chemical industry at the international level that promotes the safe production, transportation, and use of chemical products.
- GRI (Global Reporting Initiative) standards are a non-financial reporting system that includes environmental, social, and governance impact indicators. Using such standards helps companies structure internal processes, increase transparency for investors and partners, and meet the requirements of international markets. Moreover, ISO certification or participation in ESG initiatives may be a prerequisite for obtaining green financing or participating in international tenders.

Chemical industry companies actively use international standards not only to comply with regulations but also to improve investment attractiveness. For example, BASF publishes key KPIs in its annual ESG reports compiled in accordance with GRI standards: a 25% reduction in CO_2 emissions over the past 10 years, a 20% increase in energy efficiency, and a share of recycled materials in production of up to 15% (BASF 2025a, 2025b). These indicators have a direct impact for rating agencies on a company's rating and increase the confidence of investors. One of the key parts of analyzing green technologies is assessing the energy intensity and material intensity. These indicators tell us how well a technology uses natural resources and what savings may be realized from scaling. For example, reducing the energy intensity of a process by 10–15% can significantly reduce energy costs and, in the long term, reduce the company's carbon footprint.

Material intensity (the volume of raw materials used per unit of production) is directly related to resource savings and waste reduction. Integrated use of raw materials is one of the key criteria for classifying production as low-waste or waste-free. In relation to green chemistry, this means developing processes in which almost the entire mass of reagents is converted into useful products, and by-products are reused.

For a global assessment of sustainable development, international indices are used, for example the Environmental Performance Index (EPI), Dow Jones Sustainability Index (DJSI), and Carbon Disclosure Project (CDP). These determine a company's environmental approaches, carbon footprint, resource efficiency, and reporting standard compliance. Chemical companies that demonstrate progress on these metrics gain competitive advantages in the market, including access to investment and preferential support programs. For the effective use of indicators, it is important not only to collect data but to use it to make management decisions. In terms of green chemistry, this indicates the amount of environmental monitoring systems, audits and analysis of production processes. For example, a company can monitor the reduction of greenhouse gas exposure after a process line has been upgraded and correlate the amounts of reduced emissions against savings on carbon taxes or pollution levies. In this way, environmental and financial indicators are combined into a single strategy for increasing business sustainability.

In the coming years, we can expect the development of more integrated assessment systems that take into account not only the carbon footprint but also the water footprint, biodiversity, and social effects. In the context of increasing pressure from regulators and society, companies will be forced to expand the range of reporting data and move to more detailed analysis systems. The development of digital platforms and artificial intelligence technologies will automate data collection and analysis, which will significantly improve the accuracy and efficiency of decision-making.

11.6 EXAMPLES OF BUSINESS MODELS AND INNOVATIONS IN GREEN CHEMISTRY

The development of green chemistry is not possible without the development of new business models based on circular economy, resource saving, and new forms of organizing production. Unlike the classic model of the chemical industry, where the key success criterion is the cost and quantity of manufactured products, modern businesses are shifting their focus towards reducing ecological footprint, resource efficiency, and long-term sustainability.

The circular economy is the maximum closure of material flows and reuse of resources. In the chemical industry, this is reflected in the creation of closed technological cycles where the waste of one process becomes raw material for another process. For example, the production of biodegradable polymers not only helps reduce the volume of plastics in landfills but also creates secondary raw materials compostable or recyclable. Companies working in this area are building business models focused on an extended product life cycle: from the design and selection of environmentally friendly materials to the organization of the return and disposal of used products.

Waste-free technologies are the basis for rational use of natural resources, and their development is associated with economic feasibility. The implementation of the "waste = resource" concept allows us to reduce not only the environmental impact but also production costs. This is well demonstrated by chemical clusters, where enterprises of different profiles are united into a single system of by-product exchange, creating synergy and increasing profitability. Technological innovations are the driving force of green chemistry. One of the promising areas is the use of renewable raw materials – vegetable oils, starch, and cellulose, as well as agricultural waste for the synthesis of chemical products. Such solutions not only reduce dependence on oil and gas, but also reduce the carbon footprint of production. These solutions demonstrate not only an ecological but also an economic effect, as they allow the creation of products with high added value using what was previously considered waste. The world's largest chemical corporations, such as BASF, Dow Chemical, and DuPont, are actively implementing the principles of green chemistry in their strategy. Much attention is also paid to local initiatives. In a number of countries, including the European Union, eco-industrial parks are actively developing, where enterprises unite to exchange resources and waste. This allows for a reduction in production costs and an increase in environmental efficiency.

Green chemistry is strongly connected to other aspects of sustainable development – renewable energies, recycling waste, green building. For example, it requires the creation of chemicals to make the solar panels or biodegradable building products which can lead to new markets and potential synergies within and between industries. An interesting example is projects where the chemical industry cooperates with agriculture, creating biotechnological complexes. Here, agricultural waste is processed into chemical raw materials, and by-products of chemical processes are used for fertilizers or feed additives. Modern companies are moving from a product sales model to a service provision model. In the chemical industry, this is expressed in the so-called chemical leasing concept, when the customer pays not for the chemical substance but for its function (e.g., for the number of cleaned parts and not for the volume of cleaning solution). This model encourages manufacturers to produce less chemicals and safer solutions because their profit is based on decreasing costs and decreasing chemicals of use. This simultaneously reduces the environmental load, while creating new financial opportunities. The chemical leasing concept, developed with support from UNIDO, assumes that the manufacturer does not sell the chemical but offers a service for its use (Schwager et al., 2016, 2017). This should encourage companies to lower the volume of chemicals and develop more efficient and safer solutions. In an example from one of the pilots in Serbia, undertaken with UNIDO support, they were able to switch to chemical leasing, which resulted in decreasing solvent use by 20%, reducing their purchase costs by a total of 15%, and reducing emissions of volatile organic compounds by almost 1/3 (Lu et al., 2023). Large chemical companies, such as BASF, consider such models as part of ESG strategies, where reducing material intensity and switching to service solutions directly increase the investment attractiveness of the business. Despite the obvious advantages, innovative projects face a number of challenges. The high cost of developing and certifying new materials, limited investment, and low consumer awareness can all hinder implementation. The state has an important role to play here; it should create an environment for positive action, provide tax credits, support scientific research, encourage natural resource users to adopt green technologies and environmental stewardship actions, and promote partnerships between businesses and universities.

The most important components of Green technologies are nanotechnology, resource conservation, bio-oriented production, and processing. Green technologies also affect all areas of human activity. Below we will present a brief analysis of scientific publications confirming the wide range of applicability of green technologies, from management to engineering and technical solutions. Water treatment and wastewater treatment technologies occupy one of the key places in the green transformation. Contamination of water sources necessitates the development of new designs of water treatment system equipment (Gurgenidze & Romanovski, 2023; Hurynovich & Romanovski, 2018; Prapolski & Romanovski, 2025; Yushchenko et al., 2023, 2024), purification of water bodies (Matsukevich et al., 2025), use of generated waste (ion-exchange resins (Romanovski & Gruzinova, 2025), lime mud (Kamarou et al., 2023, p. 2025), coagulation sediments (Kamarou et al., 2025b), disinfection of water supply systems (Ramanouski et al., 2015; Romanovski et al., 2024c; Rymovskaia & Ramanouski, 2016), combined purification methods (Romanovski et al., 2025b), as well as new approaches to the management of water treatment facilities (Khmel et al., 2025). Water treatment waste can be processed into sorption (Ramanouski & Andreeva, 2012) and catalytic materials, as well as a wide range of building materials (gypsum, anhydrite (Liu et al., 2024; Wang et al., 2024), gypsum and anhydrite binders (Romanovski et al., 2022). In wastewater treatment systems for organic and inorganic substances, new effective sorbents (Gruzinova and Romanovski, 2024, 2025) and catalytic materials (Glinskaya et al., 2023; Matsukevich et al., 2021, 2023), including those based on waste (Zeng et al., 2023; Zhou et al., 2024), including nanostructured and polyfunctional materials (Romanovski et al., 2025a), are also being developed. The integration of green chemistry into energy solutions has been made possible by the development of highly efficient and reliable energy sources (Kamenskii et al., 2025), new approaches to the synthesis of these materials (Beletskii & Romanovski, 2024; Beletskii et al., 2024; 2025; Romanovski et al., 2024a), fuel combustion regimes (Amrutsagar et al., 2025) and the search for alternative fuel sources (Hurynovich et al., 2021). New methods are being developed for processing consumer waste to obtain a number of materials or use them as part of a batch or mixtures (Burduladze et al., 2025; Yagubkin et al., 2024; Zalyhina et al., 2022, 2024), waste from galvanic production as a source of d-elements (Kong et al., 2024), polymer composite materials (Kalaimagal et al., 2025; Kumar et al., 2025a, 2025b, 2025c). A huge number of solutions are proposed for the secondary use (re-use) of industrial waste: the extraction of rare earth elements from spent catalysts (Kozlovskaya et al., 2024), important metals from mining waste (Smorokov et al., 2022, 2023a, 2023b), increasing the efficiency of ore processing (Osipova et al., 2022), obtaining pigments (Zalyhina et al., 2021), recycling electrode and battery waste (Zhou et al., 2023). Nanotechnology in agriculture not only provides increased crop yields, but is also used in all areas of agriculture (Periakaruppan et al., 2023). Metal oxide nanoparticles have shown advantages over other nanomaterials in agronomic applications (Romanovski et al., 2024a; Selvaraj et al., 2025). They act as seed germination stimulants, enhance root nutrition, and increase tolerance to extreme climatic conditions (Romanovski, 2024; Romanovski and Periakaruppan, 2023). For this some wastes or wastewater can be considered as a source of vital elements for plants (Romanovski et al., 2023a; 2023b). Also, the development of new nanosensors, which have also found wide application in all aspects of human economic activity (Romanovski et al., 2023c, 2025b). One of the long-known industries in which waste can be used is the construction industry. Eco-friendly composites have been developed from wood, paper, bamboo, and glass fibers using waste from the cement and paper industries. In the automotive and aviation industries, aluminum composites doped with nano-additives from TiO_2 (Kumar et al., 2023; Ogunsanya et al., 2023) to nanotubes (Abedi et al., 2023, 2024), HEA (Akinwande et al., 2023a, 2023b, 2024), and various dispersed wastes (Abdulwahid et al., 2023; Olukayode et al., 2023), as well as composite building materials (Akinwande et al., 2025; Vasumathi et al., 2024), are being actively studied. One aspect of green technologies is the synthesis of nano- and meso-sized materials using plant extracts, microorganisms, etc. as reducing agents (Palanimuthu et al., 2024). In recent years, new approaches to modeling the synthesis (Bogdanov et al., 2022; Samsonov et al., 2025) and degradation (Savina et al., 2023) of nanomaterials, additive technologies (Bhowmik et al., 2025; Hematibahar et al., 2025a, 2025b), the use of artificial intelligence, and machine learning to improve efficiency and better understand the mechanisms of ongoing processes have been increasingly used. All this clearly demonstrates how green technologies are integrated into real production and resource management, increasing their efficiency, environmental friendliness and reducing the risk of failure. The integration of these solutions, based on deep scientific research, makes it possible to make the future more sustainable and the economy circular and energy efficient.

11.7 PROBLEMS AND PROSPECTS OF FINANCING AND IMPLEMENTATION

Despite the obvious environmental and economic advantages of green chemistry, its widespread implementation faces a number of barriers – both financial and technological, organizational and regulatory. In order for green technologies to become the standard of the chemical industry, it is necessary not only to improve production processes but also to create effective mechanisms for financing and supporting innovation. The major challenge continues to be the high cost of developing and implementing new technologies (Matus et al., 2007). Green processes often require the purchase of upgraded equipment, more expensive new materials, or even complex integrated processing systems. The costs are prohibitive for small and medium-sized enterprises, especially when long-term, cheap financing is unavailable (Palma & Hodgett, 2025).Another factor is the high uncertainty of economic returns. New technologies may take longer to pay off than traditional ones, which makes them less attractive to investors focused on quick capital returns (De Haas et al., 2025). This is particularly noticeable in countries with underdeveloped green finance systems, where access to specialized loans, grants, or subsidies is limited. Moreover, in some cases the market price for green materials or products continues to be higher than its standard analogue which also lowers its competitive appeal. Bioplastics or biodegradable packaging materials, for example, are often pricedhigher than conventional polyethylene, requiring more policy support measures as compensation for the price differential. As stated in Reuters (2025a, b), the global chemical industry has some significant risk elements around decarbonizing because of fossil feedstock processing and the incredibly high capital associated with some of the newer e-methanol and e-cracking processes. Asa result, many companies in Europe and across the globe are delaying the transition to carbon-free processes because of the lack of profitability in an environment where energy prices are phenomenally volatile. It is significant that only about 15% of new projects in the chemical industry receive financing through green bond mechanisms, while the rest are forced to rely on government subsidies or private venture funds (Reuters 2025a, b). These examples confirm the need for comprehensive measures: strengthening tax incentives, introducing carbon tariffs, and the active participation of international financial institutions in creating a stable infrastructure for financing green technologies.

The regulatory environment is critical in promoting the uptake of green chemistry. In countries where environmental standards are low, or where the enforcement of standards is weak, companies tend to adopt older, less environmentally friendly technologies that are cheaper. The legislation on paid nature management and environmental charges is a strong incentive for modernization, though in many cases the implementation of this legislation needs work. The lack of coordination between government bodies, investors, and businesses can also delay the uptake of innovations. A lack of clear strategy to organize the transition to sustainable production by aligning scientific, financial, and industrial assets means the process of implementation remains fragmented. Although the use of green chemistry is developing rapidly, not all technologies undertake research and development for mass implementation. The development of new catalysts, biodegradable products, or CO_2 capture processes can have a substantial amount of money invested in scientific research and development, as many of these products do not make business sense until testing is complete. These research and development projects bring the possibility of technology risks: a new technology may not be as effective as initially designed, or it does not maintain performance over a long period of time, which means companies are often cautious when selecting technology. Integrating green processes into production chains also comes with challenges. In many cases, the introduction of an environmentally friendly or sustainable solution requires the company to fundamentally alter existing logistics and supply chains, including retraining the workforce and installing environmental monitoring systems. Despite the existing barriers, there are several promising areas that can accelerate the financing of green chemistry: (1) Development of green financial instruments. Green bonds, ESG investment funds, and climate loans are becoming an important source of capital for projects in the field of environmentally friendly technologies. Investors increasingly consider a company's sustainability as a factor in long-term reliability, which increases the attractiveness of green projects. (2) Government assistance and benefits. Many countries already have mechanisms to subsidize modernization of enterprises, tax incentives, and grants for

R&D on green chemistry. Expanding these types of measures wouldhelplessenthe-costor shorten the payback period of these projects. (3) Partnerships and consortia. Combining the efforts of businesses, research centers, and the state within clusters and technology platforms allows for the distribution of risks and the acceleration of innovation.

Global environmental challenges such as climate change and environmental pollution require collective action. International organizations – UNEP, the World Bank, the UN Economic Commission for Europe – actively support projects aimed at developing green technologies. Such initiatives provide not only funding, but also exchange of experience, standardization of approaches, and development of uniform requirements for environmental reporting. An example of effective international cooperation is the Montreal Protocol, which stimulated the transition to environmentally friendly refrigerants and new chemical processes. Similar mechanisms can be applied to other sectors of the chemical industry, especially in the area of reducing carbon emissions.

In the years to come, there will likely be increased interest in carbon financing mechanisms and carbon trading. Companies that can show they have reduced greenhouse gas emissions can acquire additional funding by trading allowances on specialist markets. Another trend is the digitization of project evaluation and monitoring. The utilization of digital twins or forecasting systems based upon big data and AI will lead to improved evaluation of the efficacy of green technologies and their risk of application. Increased consumer demand for environmentally friendly products will also have a significant impact. Companies that are the first to adapt to these trends will receive not only financial benefits, but also a long-term competitive advantage. To accelerate the implementation of green chemistry, it is necessary to: (i) form transparent mechanisms to support innovation and access to preferential loans for modernization; (ii) develop partnerships between industry, research centers and the state; (iii) introduce mandatory environmental reporting standards and sustainability indicators; and (iv) raise the level of environmental culture and competencies among specialists and managers.

11.8 CONCLUSION

This chapter has captured the intricacies surrounding the development of green chemistry practice: encompassing environmental principles, economic rationality, financial mechanisms, and modern technologies, thus laying the groundwork for the circular economy. The analysis has shown that green chemistry fits naturally into a new production model where energy efficiency, waste minimization, and product life cycle enablement are emerging priorities. Energy efficiency is one area that allows modern enterprises to combine what they seek to do as companies to survive in the long run and apply ESG principles in their short- and longer-term decisions. Similarly, LCA standards allow modern enterprises to incorporate multiple sustainability criteria into their respective change agendas. This is no longer a question of image or "social responsibility," but a factor that directly affects financial results and investment attractiveness. The ways in which issues of financing green chemistry today happen through a variety of instruments of green bonds, funds for climate change, tax incentives, and grants from government programs, but the key determiner of success is not just because a company could access various funding sources, but the ability of accompany that is processed through their arguments, the corporations can effectively justify the economic efficiency of their environmental solutions.

TEA of green processes requires a systematic approach. A simple calculation of profitability no longer reflects the whole picture. Integration of LCA and LCCA allows identifying hidden economic benefits – reduced disposal costs, reduced fines for environmental violations, improved reputation, and increased resilience to regulatory risks. This is especially relevant in the context of tightening environmental regulations in Europe and North America, as well as against the backdrop of global initiatives to reduce the carbon footprint.

The system of indicators and international standards plays the role of the "language" of sustainable development. ISO 14001, GRI, CDP, as well as specialized ratings (EPI, DJSI) form uniform criteria by which companies, projects, and industries are assessed. This allows investors, governments, and society to objectively compare results and support those initiatives that truly contribute to the transition to a green economy.

Green chemistry business models are going beyond the usual schemes. Service approaches such as chemical leasing, extended producer responsibility (EPR) models, or the integration of CO_2 capture technologies into value chains are becoming drivers of innovation. These solutions open up opportunities not only to reduce environmental damage but also to create new market niches where value is determined not by the volume of the product produced but by its functionality and sustainability.

However, the transition to green chemistry is accompanied by significant challenges. Among them are the high capital intensity of innovative projects, the lack of long-term financial mechanisms, the imperfection of the regulatory framework, and technological risks. World practice shows that without the support of the state and international institutions, most green projects do not reach the commercialization stage. This is especially noticeable in developing countries, where environmental requirements are not yet sufficiently integrated into economic policy. To accelerate the implementation of green chemistry, it is necessary to (i) more actively develop national and regional support programs (tax incentives, grants, and state guarantees); (ii) create sustainable financial instruments (green bonds and climate funds) available to both large corporations and small businesses; (iii) invest in research centers and technology clusters that unite business and science; and (iv) introduce mandatory ESG reporting and transparent metrics for all enterprises in the chemical industry.

REFERENCES

Abdulwahid, M. Y., Akinwande, A. A., Kamarou, M., Romanovski, V., & Al-Qasem, I. A. (2023). The production of environmentally friendly building materials out of recycling walnut shell waste: A brief review. *Biomass Conversion and Biorefinery*, *1*, 10. https://doi.org/10.1007/s13399-023-04760-2

Abedi, M., Moskovskikh, D., Nepapushev, A., Suvorova, V., Wang, H., & Romanovski, V. (2023). Advancements in laser powder bed fusion of carbon nanotubes-reinforced AlSi10Mg alloy: A comprehensive analysis of microstructure evolution, properties, and future prospects. *Metals*, *13*(9), 1619. https://doi.org/10.3390/met13091619

Abedi, M., Moskovskikh, D., Romanovski, V., Ozherelkov, D., & Gromov, A. Unlocking the potential of graphene-reinforced AlSi10Mg nanocomposites in laser powder bed fusion: A comprehensive review. *Journal of Alloys and Compounds.* 2024, 173441. https://doi.org/10.1016/j.jallcom.2024.173441

Akinwande, A. A., Balogun, O. A., Adediran, A. A., Adesina, O. S., Romanovski, V., & Jen, T. C. (2023b). Experimental analysis, statistical modeling, and parametric optimization of quinary-(CoCrFeMnNi)100 –x/TiCx high-entropy-alloy (HEA) manufactured by laser additive manufacturing. *Results in Engineering, 17,* 100802. https://doi.org/10.1016/j.rineng.2022.100802

Akinwande, A. A., Kumar, M. S., & Romanovski, V. (2023a). Mechanical strengthening and micro-mechanical modelling of newly fabricated 7068-aluminium/lightweight-high-entropy-alloy as aero-material. *Materials Chemistry and Physics, 309,* 128344. https://doi.org/10.1016/j.matchemphys.2023.128344

Akinwande, A. A., Moskovskikh, D., & Romanovski, V. (2024). Enhancing mechanical and damping performance of 3D-printed aluminium-7075 with shape memory high-entropy alloy (SMHEA): Parametric optimization and mathematical modeling. *The International Journal of Advanced Manufacturing Technology, 130*(1), 369–383. https://doi.org/10.1007/s00170-023-12651-z

Akinwande, A. A., Ogunsanya, O. A., Balogun, O. A., Bhowmik, A., & Romanovski, V. (2025). On the tribo-mechanical and damping properties of aluminum composite modified with ecofriendly fillers. *Particulate Science and Technology,* 1–16. https://doi.org/10.1080/02726351.2025.2462941

Amrutsagar, L. V., Kumar, R., Kanabar, B., Beemkumar, N., Sangwan, G., AlHazaa, A., Romanovski, V., & Bhowmik, A. (2025). Exploring methanol and CNG as low-reactivity fuels in reactivity controlled compression ignition for single-cylinder diesel engines: An experimental based comparative analysis at full load condition. *Energy, 135910.* https://doi.org/10.1016/j.energy.2025.135910

Anastas, P. T., & Warner, J. C. (2000). *Green chemistry: Theory and practice.* Oxford University Press.

Axious. (2025). https://www.axios.com/2024/07/24/green-bond-boom

Bala, A., Arfelis, S., Oliver-Ortega, H., & Méndez, J. A. (2022). Life cycle assessment of PE and PP multi film compared with PLA and PLA reinforced with nanoclays film. *Journal of Cleaner Production, 380,* 134891.

BASF. (2025a). https://report.basf.com/2025/en/

BASF. (2025b). https://www.basf.com/global/en/who-we-are/sustainability/our-contributions-to-enabling-the-green-transformation/carbon-footprint/corporate-carbon-footprint

Beletskii, E., Pinchuk, M., Snetov, V., Dyachenko, A., Volkov, A., Savelev, E., & Romanovski, V. (2024). Simple solution plasma synthesis of Ni@NiO as high-performance anode material for lithium-ion batteries application. *ChemPlusChem.* https://doi.org/10.1002/cplu.202400427

Beletskii, E., & Romanovski, V. (2024). Direct plasma solution recycling of cathode materials for lithium-ion batteries with simultaneous removal of contaminants and relithiation. *Journal of Power Sources, 624,* 235576. https://doi.org/10.1016/j.jpowsour.2024.235576

Beletskii, E., Volkov, A., Evshchik, E., Kolmakov, V., Shikhovtseva, A., & Romanovski, V. (2025) A comprehensive evaluation framework for lithium iron phosphate cathode relithiation techniques: Balancing production costs, electrochemical performance, and environmental impact. *Energy & Environmental Materials,* e12850. https://doi.org/10.1002/EEM2.12850

Bhowmik, A., Kumar, Kumar, Y. K., Samal, P., Mahapatro, A., Alhazaa, A. N., Romanovski, V., & Santhosh, A. J. (2025). Predictive modeling of the mechanical behavior of 3D-printedpolylacticacid/wood composite: Comparison of GEP and ANN methods. *AIP Advances.* https://doi.org/10.1063/5.0268653https://doi.org/10.1063/5.0268653

Bogdanov, S., Samsonov, V., Sdobnyakov, N., Myasnichenko, V., Talyzin, I., Savina, K., Romanovski, V., & Kolosov, A. (2022). Molecular dynamics simulation of the formation of bimetallic core-shell nanostructures with binary Ni-Al nanoparticle quenching. *Journal of Materials Science*, *57*(28), 13467–13480. https://doi.org/10.1007/s10853-022-07476-2

Burduladze, A., Papushvili, T., Rurua, N., Khetsuriani, B., & Romanovski, V. (2025). Stabilization of highway pavement base using coal ash, experience of Georgia: Laboratory characterization and field evaluation. *Engineering Reports*. https://doi.org/10.1002/eng2.13063

De Haas, R., Martin, R., Muûls, M., & Schweiger, H. (2025). Managerial and financial barriers to the green transition. *Management Science*, *71*(4), 2890–2921.

Glinskaya, A., Petrov, G., Vialikanava, I., & Romanovski, V. (2023). Crystal structure, magnetic and photocatalytic properties of solid solutions Bi2-xLaxFe4O9 (x = 0.05, 0.1). *Chemistry Select*, *8*(8), e202204285. https://doi.org/10.1002/slct.202204285

Gruzinova, V., & Romanovski, V. (2024) Optimizing oil-contaminated wastewater purification with polypropylene thread waste. *Waste and Biomass Valorization*. https://doi.org/10.1007/s12649-024-02806-8

Gruzinova, V., & Romanovski, V. (2025) Optimizing oil-contaminated wastewater purification with aluminum coagulants. *Water, Air, & Soil Pollution*. 236, 263.https://doi.org/10.1007/s11270-025-07928-w

Gurgenidze, D., & Romanovski, V. (2023). The pharmaceutical pollution of water resources using the example of the Kura River (Tbilisi, Georgia). *Water*, *15*(14), 2574. https://doi.org/10.3390/w15142574

Hematibahar, M., Fediuk, R., Momeni, K., Kharun, M., Bhowmik, A., & Romanovski, V. (2025a). Strategic roadmap for 3D-printed reinforcement using fused deposition modeling: A state-of-the-art review. *Engineering Reports*, *7*(6), e70232. https://doi.org/10.1002/eng2.70232

Hematibahar, M., Kharun, M., Bhowmik, A., & Romanovski, V. (2025b). The SADRA algorithm as a framework for sustainable innovation: A review of case studies on 3D-printed reinforced concrete beam and fiber-reinforced concrete. *Engineering Reports*. https://doi.org/10.1002/eng2.70208

Hurynovich, A., Kwietniewski, M., & Romanovski, V. (2021). Evaluation of the possibility of utilization of sewage sludge from wastewater treatment plant – Case study. *Desalination and Water Treatment*, *227*, 16–25. https://doi.org/10.5004/dwt.2021.27199

Hurynovich, A., & Romanovski, V. (2018). Artificial recharge of deep aquifers/ VI MiędzynarodowaKonferencjaNaukowo-Techniczna nt. "Nowoczesnemiasta. Infrastrukturaiśrodowisko" INFRAEKO 2018, 07–08.06.2018, Kraków/E3S Web of Conferences 45, 00025, 2018. https://doi.org/10.1051/e3sconf/20184500025

Jayalekshmi, C., Periakaruppan, R., Bharathi, A., Selvaraj, K. S. V., Al-Dayan, N., & Romanovski, V. (2025). Latex of *Calotropis gigantea* based biogenicmagnesium hydroxide nanoparticles: Synthesis, characterization, antioxidant and antibacterial activity. *Journal of Chemistry*. https://doi.org/10.1155/joch/2054584

Kalaimagal, S., Vasumathi, M., Rashia Begum, S., Saravana Kumar, M., Romanovski, V., & Salah, B. (2025). Evaluation of impact resistance of sustainable Sandwich structures with optimized FDM coregeometries using low-velocity drop impact tests. *Polymer Composites*. https://doi.org/10.1002/pc.29659

Kamarou, M., Moskovskikh, D., Chan, H. L., Wang, H., Li, T., Akinwande, A. A., & Romanovski, V. (2023). Low energy synthesis of anhydrite cement from waste lime mud. *Journal of Chemical Technology and Biotechnology*, *98*, 789–796. https://doi.org/10.1002/JCTB.7284

Kamarou, M., Moskovskikh, D., Kuskov, K., Yudin, S., Akinwande, A. A., Smorokov, A., Özkılıç, Y. O., Abdulwahid, M. Y., Bhowmik, A., Romanovskaia, E., Korob, N., Paspelau, A., & Romanovski, V. (2025). High-strength gypsum binder with improved water resistance coefficient derived from industrial wastes. *Waste Management Research*. https://doi.org/10.1177/0734242X241240042

Kamenskii, M., Volkov, A., Romanovski, V., & Beletskii, E. (2025). Electrochemical properties of $LiCoO_2$ as anode material of lithium-ion batteries: Implementation for recycling and reuse of spent material. *Engineering Reports*. https://doi.org/10.1002/eng2.70204

Khmel, E., Hurynovich, A., Holubava, V., & Romanovski, V. (2025). Organizational models for managing the operation of water supply facilities of the agro-industrial complex. *World Water Policy*. https://doi.org/10.1002/wwp2.70012

Kong, D., Wu, J., Yan, X., Zhang, H., Iqbal, A., Ivanets, A., Romanovski, V., Zhang, L., & Su, X. (2024). Electroplating sludge derived $CuFe_2O_4/MgFe_2O_4$ metal oxide composites for highly efficient removal of Congo red. *Environmental Science and Pollution Research*, *31*, 58109–58118. https://doi.org/10.1007/s11356-024-34974-4

Kozlovskaya, I., Martsul, V., & Romanovski, V. (2024). Value-added lanthanum-containing products recovered from spent cracking catalyst. *ACS Sustainable Resource Management*. https://doi.org/10.1021/acssusresmgt.4c00364

Kumar, M. S., Akinwande, A. A., Yang, C. H., Vignesh, M., & Romanovski, V. (2023). Investigation on mechanical behaviour of Al–Mg–Si alloy hybridized with calcined eggshell and TiO_2 particulates. *Biomass Conversion and Biorefinery*, 1–12. https://doi.org/10.1007/s13399-023-04215-8

Kumar, M. S., Pulikkalparambil, H., Ishfaq, K., Ramasamy, S., Romanovski, V., & Salah, B. (2025a). Elevating green materials sustainability through the integration of ceramic particles. *Polymer Composites*. https://doi.org/10.1002/pc.29894

Kumar, M. S., Yang, C. H., Pulikkalparambil, H., Edoziuno, F. O., Romanovski, V., & Salah, B. (2025c). Improving thermal stability and dynamic mechanical properties of eco-friendly bamboo fiber-reinforced epoxy composites with SiC particle integration. *Fibers and Polymers*. https://doi.org/10.1007/s12221-025-00990-5

Kumar, M. S., Yang, C.-H., Vignesh, M., Edoziuno, F. O., Romanovski, V., & Salah, B. (2025b). Optimizing the filler distribution to enhance the interlaminar shear strength and impact resistance of fiber reinforced epoxy composites. *Polymer Composites*. https://doi.org/10.1002/pc.29885

Liu, D., Chen, J., & Ma, X.*et al* (2024). Effect of calcination temperature and superplasticizer on the properties of anhydrite II from phosphogypsum. *Journal of Thermal Analysis and Calorimetry*. https://doi.org/10.1007/s10973-024-13608-4

Lu, C. F., Cheng, C. Y., & Hung, Y. L. (2023). Assessing consumer perceptions of chemical leasing business models and possibilities for adoption by Taiwanese manufacturers. *Sustainable Production and Consumption*, *35*, 325–337.

Matsukevich, I., Kulinich, N., Kulbitskaya, L., Kuznetsova, T., Popkov, V., Chebanenko, M., Moskovskikh, D., Kuskov, K., & Romanovski, V. (2023). Mesoporous nanocomposites based on CeO_2 and MgO: Preparation, structure and photocatalytic activity. *Journal of Chemical Technology and Biotechnology*, *98*(10), 2497–2505. https://doi.org/10.1002/jctb.7476

Matsukevich, I., Lipai, Y., & Romanovski, V. (2021). Cu/MgO and Ni/MgO composite nanoparticles for fast, high-efficiency adsorption of aqueous lead (II) and chromium (III) ions. *Journal of Materials Science*, *56*(8), 5031–5040. https://doi.org/10.1007/s10853-020-05593-4

Matsukevich, I. V., Beljin, J., Kulinich, N. V., Apostolović, T., Maletić, S., & Romanovski, V. (2025). Photocatalytic degradation of polycyclic aromatic hydrocarbons under visible light irradiation in water using TiO_2/MgO nanocomposites. *Environmental Science and Pollution Research*. https://doi.org/10.1007/s11356-025-36055-6

Matus, K., Anastas, P. T., Clark, W., & Itameri-Kinter, K. (2007). Overcoming the challenges to the implementation of green chemistry. https://dash.harvard.edu/server/api/core/bitstreams/7312037e-20dc-6bd4-e053-0100007fdf3b/content

Ogunsanya, O., Akinwande, A., Balogun, O. A., Romanovski, V., & Kumar, S. (2023). Mechanical and damping behavior of artificially agedal 6061/TiO_2 reinforced composites for aerospace applications. *Particulate Science and Technology*, *41*(2), 196–208. https://doi.org/10.1080/02726351.2022.2065652

Olukayode, O. M., Akinwande, A. A., Balogun, O. A., Oyediran, A. O., Mosuru, S. G., Abdulwahid, M. Y., & Romanovski, V. (2023). Applicability of the response surface method in the optimization of compressive strengths of banana-fiber-reinforced concrete using waste glass as partial cement replacement. *Innovative Infrastructure Solutions*, *8*(8), 207. https://doi.org/10.1007/s41062-023-01178-7

Osipova, E., Shevchuk, V., Stromski, A., & Romanovski, V. (2022). Intensification of potash ore flotation by the introduction of industrial oils. *Journal of Chemical Technology and Biotechnology*, *97*(1), 312–318. https://doi.org/10.1002/jctb.6945

Palanimuthu, V., Periakaruppan, R., Romanovski, V., Bharathi, A., Selvaraj, K. S. V., Anukeerthana, S., Nishanthi, R., & Vanajadevi, G. (2024). Synthesis and structural characterization of SiO_2 nanoparticles using extract of *Gracilaria crassa* via green chemistry approach. *Chemistry Open*. https://doi.org/10.1002/open.202400356

Palma, D., & Hodgett, R. (2025). Barriers to sustainable manufacturing in the chemical industry: A qualitative study. *Sustainability*, *17*(7), 3241.

Periakaruppan, R., Romanovski, V., Thirumalaisamy, S. K., Palanimuthu, V., Sampath, M. P., Anilkumar, A., Sivaraj, D. K., Ahamed, N. A. N., Murugesan, S., Chandrasekar, D., & Selvaraj, K. S. V. (2023). Innovations in modern nano technology for the sustainable production of agriculture. *ChemEngineering*, *7*(4), 61. https://doi.org/10.3390/chemengineering7040061

Prapolski, D., & Romanovski, V. (2025). Resent advances in underground water deironing and demanganization: Comprehensive review. *Journal of Water Process Engineering*, *70*, 107089. https://doi.org/10.1016/j.jwpe.2025.107089

Ramanouski, V. I., & Andreeva, N. A. (2012). Purification of washing waters of iron removal stations. *Proceedings of BSTU. Chemistry and Technology of Inorganic Substances*, *3*, 62–65.

Ramanouski, V. I., Likhavitski, V. V., Rymovskaya, M. V., & Hurynovich, A. D. (2015). Determination of the main parameters of disinfection of drinking water supply facilities by ozone. *Proceedings of BSTU. Chemistry and Technology of Inorganic Substances*, *3*, 64–67.

Reuters. (2025a). https://www.reuters.com/sustainability/decarbonizing-industries/chemicals-industry-struggles-kick-its-fossil-fuel-habit-2025-07-28/

Reuters. (2025b). https://www.reuters.com/sustainability/cop/green-bond-issuance-dives-almost-third-amid-climate-backtracking-2025-07-23/

Rezvani Ghomi, E., Khosravi, F., Saedi Ardahaei, A., Dai, Y., Neisiany, R. E., Foroughi, F., … & Ramakrishna, S. (2021). The life cycle assessment for polylactic acid (PLA) to make it a low-carbon material. *Polymers*, *13*(11), 1854.

Romanovski, V. (2024). Positive and negative effects of different metal-based nanoparticles on seed germination and plant growth. In *Nanofertilizer delivery, effects and application methods* (pp. 251–270). Elsevier. https://doi.org/10.1016/B978-0-443-13332-9.00006-X

Romanovski, V., Dubina, A., Sehat, A. A., Su, X., & Moskovskikh, D. (2024a) Green and bio-waste-based materials for energy production, conversion, storage, and hybrid technologies. In *Materials for energy production, conversion, and storage* (pp. 118–135). CRC Press. https://doi.org/10.1201/9781003318859-8

Romanovski, V., & Gruzinova, V. (2025). Aggregate from spention-exchange resins for petroleum products removal from wastewater. *Waste and Biomass Valorization*. https://doi.org/10.1007/s12649-025-03158-7

Romanovski, V., Kulichik, D., Periakaruppan, R., Palanimuthu, V., Kulinich, N. V., Popkov, V., Dubina, A., Matsukevich, I. V., Shtepa, V., & Romanovskaia, E. (2025a). Multifunctional ZnFeO-based composite from iron sludge for photocatalytic degradation of azo, thiazine, anthraquinonedyes, formaldehyde and oil sorption from aqueous media. *Journal of Water Process Engineering*, *77*, 108355. https://doi.org/10.1016/j.jwpe.2025.108355

Romanovski, V., Matsukevich, I., Romanovskaia, E., & Periakaruppan, R. (2023c). Nano metal oxide as nanosensors in agriculture and environment. In *Nanometal oxides in horticulture and agronomy* (pp. 321–352). Academic Press. ISBN: 9780323918091. https://doi.org/10.1016/B978-0-323-91809-1.00016-0

Romanovski, V., Paspelau, A., Kamarou, M., Likhavitski, V., Korob, N., & Romanovskaia, E. (2024c). Comparative analysis of the disinfection efficiency of steel and polymer surfaces with aqueous solutions of ozone and sodium hypochlorite. *Water, 16*, 793. https://doi.org/10.3390/w16050793

Romanovski, V., & Periakaruppan, R. (2023). Why metal oxide nanoparticles are superior to other nanomaterials for agricultural application? In *Nanometal oxides in horticulture and agronomy* (pp. 7–18). Academic Press. ISBN: 9780323918091. https://doi.org/10.1016/B978-0-323-91809-1.00014-7

Romanovski, V., Pilipenko, M., Dubina, A., Likhavitski, V., Volodko, S., Moskovskikh, D., & Romanovskaia, E. (2025b). Optimizing dye wastewater purification: Ultrasonic and flotation with ozonation synergy. *Engineering Reports, e13044.* https://doi.org/10.1002/eng2.13044

Romanovski, V., Roslyakov, S., Trusov, G., Periakaruppan, R., Romanovskaia, E., Chan, H. L., & Moskovskikh, D. (2023a). Synthesis and effect of CoCuFeNi high entropy alloy nanoparticles on seed germination, plant growth, and microorganisms inactivation activity. *Environmental Science and Pollution Research, 30*(2), 23363–23371. https://doi.org/10.1007/s11356-022-23918-5

Romanovski, V., Su, X., Zhang, L., Paspelau, A., Smorokov, A., Sehat, A. A., Akinwande, A. A., Korob, N., & Kamarou, M. (2023b). Approaches for filtrate utilization from synthetic gypsum production. *Environmental Science and Pollution Research.* https://doi.org/10.1007/s11356-022-24584-3

Romanovski, V., Zhang, L., Su, X., Smorokov, A., & Kamarou, M. (2022). Gypsum and high quality binders derived from water treatment sediments and spent sulfuric acid: Chemical engineering and environmental aspects. *Chemical Engineering Research and Design, 184*, 224–232. https://doi.org/10.1016/j.cherd.2022.06.008

Romanovski, V., Zhang, Z., & Akbarisehat, A. (2025b) Nanobiosensors for precision farming and sustainable agriculture. In E. K. Radhakrishnan (Ed.), *Agriculture sustainability through nanotechnology* (pp. 156–174). CRC Press. https://doi.org/10.1201/9781003354253-10

Rymovskaia, M. V., & Ramanouski, V. I. (2016). Effect of used solutions for drinking water supply facilities disinfection to soil. *Proceedings of BSTU. Chemistry and Technology of Inorganic Substances, 4*, 151–155.

Samsonov, V., Nepsha, N., Sdobnyakov, N., Talyzin, I., Kolosov, A., Puitov, V., Savina, K., Zhigunov, D., & Romanovski, V. (2025) Chemical and structural segregation in Pt-Pd-Ni ternarynano systems: Molecular dynamics simulation. *Materials Chemistry and Physics.* 130827. https://doi.org/10.1016/j.matchemphys.2025.130827

Savina, K., Grigoryev, R., Myasnichenko, V., Sokolov, D., Bogdanov, S., Kolosov, A., … & Samsonov, V. (2023). Dealloying in Pt-based nanoalloys as a way to synthesize bimetallic nanoparticles: Atomistic simulations. *Nano-Structures & Nano-Objects, 34*, 100977. https://doi.org/10.1016/j.nanoso.2023.100977

Schwager, P., Decker, N., & Kaltenegger, I. (2016). Exploring green chemistry, sustainable chemistry and innovative business models such as chemical leasing in the context of international policy discussions. *Current Opinion in Green and Sustainable Chemistry, 1*, 18–21.

Schwager, P., Dunjic, B., & Kaltenegger, I. (2017). Success and failure of the chemical leasing model in addressing sustainability challenges: Evidence from practice. *Current Opinion in Green and Sustainable Chemistry, 8*, 14–17.

Selvaraj, K. S. V., Periakaruppan, R., Palanimuthu, V., Romanovski, V., Bharathi, A., & Mohan, M. (2025). Analysis of heavy metal accumulation in *Ulva rigida* and its effects on seed germination of *Pennisetum glaucum. Chemistry Open.* https://doi.org/10.1002/open.202400386

Smorokov, A., Kantaev, A., Bryankin, D., Miklashevich, A., Kamarou, M., & Romanovski, V. (2022). Low-temperature desiliconization of activated zircon concentrate by NH_4HF_2 solution. *Minerals Engineering*, *189*, 107909. https://doi.org/10.1016/j.mineng.2022.107909

Smorokov, A., Kantaev, A., Bryankin, D., Miklashevich, A., Kamarou, M., & Romanovski, V. (2023a). Low-temperature method for desiliconization of polymetallic slags by ammonium bifluoride solution. *Environmental Science and Pollution Research*, *30*(5), 30271–30280. https://doi.org/10.1007/s11356-022-24230-y

Smorokov, A., Kantaev, A., Bryankin, D., Miklashevich, A., Kamarou, M., & Romanovski, V. (2023b). A novel low-energy approach of leucoxene concentrate desiliconization by ammonium bifluoride solutions. *Journal of Chemical Technology and Biotechnology*, *98*(3), 726–733. https://doi.org/10.1002/jctb.7277

Vasumathi, M., Rashia Begum, S., Gokul Nath, P., Kavimani, V., Romanovski, V., & Salah, B. (2024). Insights on the bearing response of glass fiber-reinforced stainless steel laminates: Investigating the influence of natural fiber and specimen geometry. *Polymer Composites*. https://doi.org/10.1002/pc.28455

Wang, C., Chen, J., Liu, D., Gong, H., Tan, H., Yang, F., … & Ramanovski, V. (2024). Preparation of anhydrite whiskers from red gypsum and application in polyvinyl chloride. *Journal of Material Cycles and Waste Management*, 1–11. https://doi.org/10.1007/s10163-024-02150-y

Yagubkin, A., Shabanov, D., Niyakovskii, A., & Romanovski, V. (2024). Maximizing strength and durability in wood concrete (arbolite) via innovative additive control and consumption. *Biomass Conversion and Biorefinery*. https://doi.org/10.1007/s13399-024-06071-6

Yushchenko, V., Velyugo, E., & Romanovski, V. (2023). Influence of ammonium nitrogen on the treatment efficiency of underground water at iron removal stations. *Groundwater for Sustainable Development*, *22*, 100943. https://doi.org/10.1016/j.gsd.2023.100943

Yushchenko, V., Velyugo, E., & Romanovski, V. (2024). Development of a new design of deironing granulated filter for joint removal of iron and ammonium nitrogen from underground water. *Environmental Technology*, *45*(14), 2735–2742. https://doi.org/10.1080/09593330.2023.2185820

Zalyhina, V., Cheprasova, V., & Romanovski, V. (2021). Pigments from spent ammonium chloride zinc plating electrolytes. *Journal of Chemical Technology and Biotechnology*, *96*(10), 2767–2774. https://doi.org/10.1002/jctb.6822

Zalyhina, V., Cheprasova, V., & Romanovski, V. (2022). Paper industry slag for the production of building ceramics. *Journal of Chemical Technology and Biotechnology*, *97*(11), 3091–3099. https://doi.org/10.1002/JCTB.7174

Zalyhina, V., Cheprasova, V., & Romanovski, V. (2024). Recycling of fine fraction of spent foundry sands into fireclay bricks. *Journal of Material Cycles and Waste Management*. https://doi.org/10.1007/s10163-023-01825-2

Zeng, J., Xu, R., El-Kady, A. A., Oranj, B. T., Ahmed, R., Valentin, R., Hu, X., Wu, W., Wang, D., Mao, J., Wu, H., Gu, X., Li, P., Xu, W., & Zhang, Z. (2023). Nanomaterials enabled photoelectron catalysis for removing pollutants in the environment and food. *TrAC Trends in Analytical Chemistry*, 117187. https://doi.org/10.1016/j.trac.2023.117187

Zhou, H., Zhang, D., Jiang, Y., Zeng, B., Zhao, C., Zhang, M., Zeng, B., Zhu, X., Su, X., Romanovski, V., & Bi, R. (2023). Recovery of carbon from spent carbon cathode by alkaline and acid leaching and thermal treatment and exploration of its application in lithium-ion batteries. *Environmental Science and Pollution Research*, *30*(53), 114327–114335. https://doi.org/10.1007/s11356-023-30404-z

Zhou, Z., Zhang, L., Yan, B., Wu, J., Kong, D., Romanovski, V., Ivanets, A., Li, H., Chu, S., & Su, X. (2024). Removal of chromium from electroplating sludge by roasting-acid leaching and catalytic degradation of antibiotics by its residue. *Journal of Environmental Chemical Engineering*, *12*(1), 111754. https://doi.org/10.1016/j.jece.2023.111754

12 Economic and Technological Drivers of Green Chemistry

Sadia Abdullah, Mamoona Munir, Mushtaq Ahmed, Bisma Javed, Shazia Sultana, and Nigar Mursal

12.1 INTRODUCTION

Green chemistry has become an important scientific and industrial practice geared toward coming up with chemical products and processes that minimize or completely avoid production of dangerous substances. In contrast to the conventional chemistry, in which the functional and performance aspects are often the only ones that are considered, the green chemistry incorporates the environmental, economical, and technological aspects into the molecular design process (Idoko et al., 2024). The adoption of green chemistry has been on an upward trend over the last several decades all over the world because of the growing anxieties over the destruction of the environment, the threat to human health and the loss of resources. With industries around the world trying to adopt greener solutions, the economic and technological factors that led to the development of green chemistry have been discussed more critically, especially on the aspects of being viable, affordable, and profitable to the investment (Yadav & Dubey, 2024).

There is undoubtedly a necessity for chemistry in our civilization to ensure a thriving and healthy existence. In this context, the "12 Principles of Green Chemistry" were indeed transformative. As professionals, we discovered that through careful initial design, we can deliver new chemicals, products, and technologies that are safe for both people and the environment, enabling us to responsibly care for our planet (Patel et al., 2022). Professor Anja-VerenaMudring, the chief editor of Green Chemistry Letters and Reviews, which is one of the most popular journals in chemistry with its main aim at improving the sustainability in chemistry, made these comments. In the next text, she claims that such a green chemistry embodied in the 12 principles will be able to confront the upcoming challenges, including climate change and the COVID pandemic (Krasnodębski, 2024). Figure 12.1 represents the Principles of green chemistry.

Green chemistry financial aspect is multi-faceted which involves initial cost of implementing green technology, maintenance cost of using more sustainable approach and the economic benefits in the long term such as reduction of waste, increased energy efficiency, and lower regulatory costs Souza et al., 2024). Adopting

DOI: 10.1201/9781003642589-12

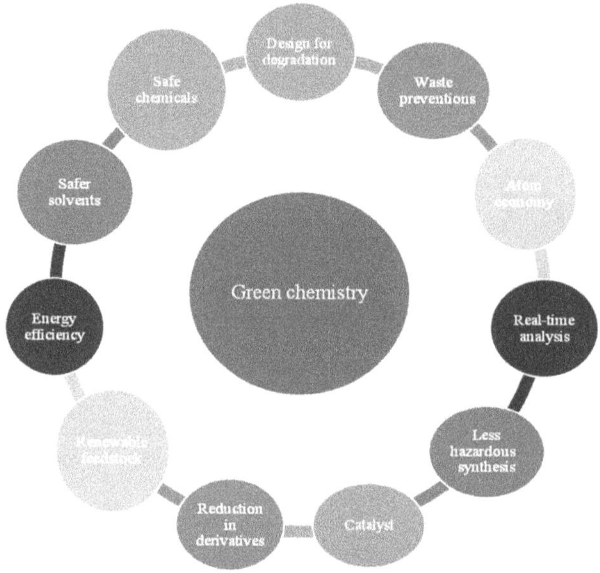

FIGURE 12.1 Principles of green chemistry.

the idea of green chemistry can also necessitate high upfront expenditures, the expenses are often offset by higher efficiency levels in the processes and reduction in waste disposal costs. Tax breaks, state grants, and environmental green certifications have become the important financial instruments who are promoting the deployment of green chemistry in developed and developing economies (Chen et al., 2020).

Besides economic factors, technological development is also important in establishing the viability of green chemistry. Increased design of more sustainable synthesis procedures, catalytic platforms, solvent-free technologies and renewable feedstock depict the technological variables that make green chemistry not only feasible but also competitive (Ferrari & Buscemi, 2023). The tools and technologies that can be used at the present stage to decide the economically and environmentally advantageous paths include a high-throughput screening, computational chemistry, and lifecycle assessment (LCA) tools. These tools not only contribute to the cost reduction of production but also enhance the selectivity and yield of chemical processes which are important factors in the competitiveness of industries (Zuin et al., 2021).

The trend that favors the use of green chemistry is that it aligns with corporate sustainability goals and the circular economy system. It is increasingly becoming recognized by industries that integrating the concept of green chemistry in their production operations enhances brand image, regulatory compliance, and sustainable profitability (Chen et al., 2020). The agrochemical, pharmaceutical, and consumer products industries have their companies spending a lot on green chemistry in their efforts to achieve their environmental, social, and governance (ESG) objectives (Ncube et al., 2023). These changes denote an even wider shift of a linear system of take-make-dispose to a more circular and regenerative model of the economy, where waste is eliminated, and value is reused and maintained within the system as long as possible (Abdussalam-Mohammed et al., 2020).

Regardless of these developments, there are still a number of techno-economic challenges that still exist in the popular practice of green chemistry (Abdussalam-Mohammed et al., 2020). These are large start-up capital, insecurity in the regulatory framework, inaccessibility to skilled employees, and little knowledge among the small and medium enterprises (SMEs). The disconnection between short term business cycles and the long term benefits of green chemistry still stands as a hindrance in the decision making in cost sensitive industries. To overcome these issues, it is necessary to have a unified system of policies that promotes research and development (R&D), collaboration between the government and business, and entry into sustainable funding options (Venkatesan et al., 2024).

Green chemistry has opportunities and challenges in the techno-economic application in developing countries such as Pakistan and other Global South countries. These regions are rich in the resources of bio-diversity and biomass which can be exploited in the application of green chemistry. Conversely, technological diffusion is more likely to be prevented by infrastructural constraints, insufficient foreign investment in R&D, and ineffective enforcement of policies (Tahir et al., 2025). On-site inventions, local entrepreneurship, and cross-national ventures have portended positive platform regarding the development of green chemistry practices that are cost effective and context relevant (Wijaya, 2025).

This chapter is going to offer a comprehensive treatment of both financial and techno-economic aspects of green chemistry with a particular emphasis on the combined influence of economic incentives and technological advances. It reviews the current economic systems and cost-benefit analyses which demonstrate the feasibility of green chemistry in industrial application. The future of green chemistry, according to Ogbuagu et al. (2024) is definite as it is going to be determined by not only technological innovations or regulatory measures but also by thorough financial and techno-economic assessments. Financial analysis provides important information on the costs-effectiveness, scalability, and profitability of green chemistry innovations, and the techno-economic assessment (TEA) assists in the process of filling the gap between laboratory feasibility and commercial feasibility. These analyses offer decision-makers such as policymakers, and investors or leaders in the industry, some measurable metrics to assess green options compared to the traditional methods. They also recognize financial constraints, risk assessment and direct strategic investments. Thus, a profound knowledge of financial and techno-economic factors is important to speed up the implementation and spread of green chemistry in global value chains.

Empirical studies published recently indicate a growing dependence between the development of finance, the growth of green technology, and the level of environmental performance. Analysis of both panel information of OECD and developing nations has revealed that improved financial systems including green finance, credit, insurance, venture capital, and capital market will enhance financing of energy efficient and low carbon innovations, which will eventually lower the CO_2 emission and other environmental pollution. Moreover, econometric models using difference-in-differences (DID) and system GMM approaches confirm that green finance policies (such as those implemented in China since 2017) significantly enhance firms' environmental investments, thereby improving their environmental performance. In such cases, technology improvement (particularly green patent and clean technology R&D) can be considered as a critical mediator: a better green technology innovation (GTI) not only increases green total factor productivity, but also acts as a financial buffer between the compliance and investment costs associated with the environmental regulations (i.e., the Porter hypothesis). Research using CDM frameworks indicates that non-green patents, on the one hand, tend to increase productivity, on the other hand, green innovation in highly polluting or industrial sectors are providing similar or superior results, and the cost-benefit curve tends to form inverted U-shaped curves where productive impacts are concentrated in mid to high percentile.

To the techno-economic assessment perspective, the case studies that are process-based demonstrate a high correlation of economic viability and environmental performance in both green chemistry and bio-valorization. An estimated capital investment of approximately PS100 million is the pyrolysis system in the United Kingdom that uses the help of a microwave to convert the agricultural pea waste into activated carbon. It anticipates a rate of return in the billions and internal rate of return (IRR) of about 55% and a net present value (NPV) recouping expenditure by the third year and retaining significant returns over two decades, thus encouraging the use of resources in a circular economy. Likewise, an agro-industrial-based production of biodiesel with oil-rich yeasts using agro-industrial byproducts, the techno-economic analysis of the process has shown that recycling glycerol feedstock makes it possible to produce close to zero waste CO_2 emissions, as well as reducing costs of operation, especially through the elimination of energy-intensive heating and sterilization processes; this makes cost efficiency and environmental advantages complement each other. Recent studies on the extraction of rare earth elements in phosphogypsum waste take LCA and techno-economic analysis (TEA) to a wider level of analysis. A probabilistic analysis of 87 scenarios above a 15% IRR of economic feasibility is found, but it also indicates trade-offs: reduced effects on ecosystem depletion, but more severe effects on human health in some instances, dilemmas on how to juggle the concerns of various kinds of sustainability.

12.2 PRINCIPLES OF GREEN CHEMISTRY AND ECONOMIC RELEVANCE

The intersection of sustainable operations and economic profitability has become one of the key areas of industrial and scholarly study. The principles of green chemistry have emerged as a foundational framework for designing chemical products and processes that are not only environmentally benign but also economically advantageous. Professor Mudring refers to the 12 principles of green chemistry (Table 12.1), which were first introduced by two American academics, Paul Anastas and John Warner, in their influential 1998 publication *Green Chemistry*. Anastas replied in 2007 that: "Since the 1990s, when the Twelve Principles of Green Chemistry were established, there have been significant achievements in creating new products and processes that align better with human health, environmental standards, and sustainability objectives" (Devierno Kreuder et al., 2017). The Twelve Principles of Green

TABLE 12.1
Twelve Principles of Green Chemistry

	12 Principles of Green Chemistry (Ivanković et al., 2017)
Prevention	Preventing waste is more effective than managing or cleaning it once it has already been created.
Atom Economy	Synthetic techniques must be developed to enhance the integration of every material involved in the process into the end product.
Less Hazardous Chemical Syntheses	Synthetic methods should be formulated to utilize and produce materials that have minimal or no toxicity to human health and the ecosystem wherever feasible.
Designing Safer Chemicals	Chemical products ought to be developed to maintain functional efficacy while minimizing toxicity.
Safer Solvents and Auxiliaries	The application of auxiliary materials (such as solvents, separation agents, etc.) should be avoided whenever feasible and remain harmless when employed.
Design for Energy Efficiency	Energy needs must be acknowledged for their environmental and economic effects and should be reduced. Synthetic procedures ought to be performed at room temperature and atmospheric pressure.
Use of Renewable Feedstocks	A feedstock for raw materials should be renewable instead of exhausting whenever it is feasible both technically and economically.
Reduce Derivatives	Unwarranted derivatization (blocking agents, protection/deprotection, temporary alteration of physical/chemical processes) should be evaded whenever feasible.
Catalysis	Catalytic agents (as selective as feasible) are preferable to stoichiometric agents.
Design for Degradation	Chemical substances ought to be formulated in a way that, once they have served their purpose, they do not remain in the environment and decompose into harmless byproducts.
Real-time Analysis for Pollution Prevention	Analytical approaches must be advanced to facilitate real-time, in-process monitoring and control before hazardous substances are generated.
Inherently Safer Chemistry for Accident Prevention	The substances and their forms selected for a chemical process should be chosen to reduce the risk of chemical accidents, such as releases, explosions, and fires.

Chemistry offered chemists guidelines to create clean, eco-friendly methods that are sustainable over the long haul are described in Table 12.1.

12.3 ECONOMIC DRIVERS OF GREEN CHEMISTRY

Green chemistry is being spurred more and more by economic forces that link environmental returns to commercial interests. As customers are more and more demanding, companies that operate green are discovering that green and profitable are not an oxymoron.

12.3.1 Market Demand and Sustainability Trends

The increased level of environmental awareness and stringent rules with high standards have forced industries to opt for greener manufacturing processes. The economic impacts of green production processes are critically examined in this research work. Short-run and long-run economic benefits associated with green manufacturing, including cost reductions, generation of revenues, and increased market competitiveness, are the topics of discussion based on a range of empirical studies in the study. The research by Zhang (2024) shows that although the initial costs of green methods and technologies may be high, the long-term benefits often outweigh these costs many times over through waste minimization, improved efficiency, and better environmental compliance. Additionally, businesses that adopt green manufacturing practices can enhance their brand value and customer loyalty, thereby strengthening their overall economic performance.

The high demand of the market on sustainable chemical products is triggered by the increasing customer awareness and pressure to adhere to regulatory requirements (Ogbuagu et al., 2024)). In 2016, BCC Research estimated that the U.S. market for end-use products in the alternative (sustainable) chemicals sector was valued at \$150 billion and projected it to reach \$884 billion by 2026, with a compound annual growth rate (CAGR) of 19.4%. Similarly, companies such as Accenture project that demand for bio-based chemicals will increase 2.3-fold between 2023 and 2028, reaching \$570 billion—approximately 4.5 times faster growth than that of traditional chemicals—driven by rising consumer preference for sustainable products. Sustainability trends are also tracking changing investor expectations and business models. Higher transparency and circularity of supply chains are being expected by stakeholders, with most willing to pay extra for products responsibly made: surveys show that around 70% of consumers will pay a 5% premium for environmentally sourced goods (de Castro Ribeiro, 2023). Meanwhile, institutional investors and regulators require companies to adopt green chemistry and sustainable procurement, driving company policy in the direction of low-carbon feedstock and recycling (Veleva & Cue, 2019).

12.3.2 Cost-benefit Analysis of Green vs. Conventional Processes

Research consistently proves that less waste, less energy consumption, and reduced operations save dollars. For example, application of green chemistry principles such as atom economy and low E-factor processes reduces inputs on material and waste disposal directly. Pharmaceutical companies have reduced E-factors significantly over the years with a decrease in environmental footprint and manufacturing cost (William et al., 2016). Green chemistry is embraced by minimizing the off-site recovery and health costs incurred by administrations and businesses. Wetlands constructed for wastewater treatment in Italy had a BCR > 1 in Sicily, but BCR = 10 and ROI ≈ 9 in Emilia Romagna when ecosystem services were taken into consideration. Also, lab and industrial scales comparisons of green oxidation reactions through solar-driven microgel catalysis revealed that energy OPEX per volume decreased from around 12–18 kWh/m³ pilot scale to around 0.1–0.2 kWh/m³ industrial scales,

TABLE 12.2
The Economic Factors and Effects of Sustainable Practices (Hami et al., 2015)

Economic Driver	Key Economic Implication
Market demand and sustainability trends	Rising consumer willingness to pay premiums and investments from sustainable sourcing
Cost-benefit comparison	Lower material/energy use, reduced waste disposal, avoided health/environmental costs
ROI on green innovations	Productivity gains, regulatory-driven innovation benefits, better long-term survival

proving cost competitiveness by scale and integration of renewable energy (Sgroi et al., 2015).

12.3.3 Return on Investment (ROI) in Green Innovations

The payback of green innovations is industry-variant but generally superior to traditional techniques, especially in high-pollution and high-energy industries. A 2024 green vs. non-green innovation sample study concluded that in manufacturing firms and high-pollution businesses, green patents cover equally or even higher middle levels of productivity (50th–80th percentile) with economic success accompanied by environmental advantages (Xie et al., 2022).

The Porter hypothesis suggests that it is possible for environmental controls to actually stimulate innovation, which can repay the cost of compliance through increased efficiency and competitiveness. A study of Italian green technology start-ups shows that firms using environmentally friendly practices have survival rates more than two times those of non-green firms, suggesting an association between long-term financial performance and green business models (Farza et al., 2021). Table 12.2 points out the economic impetus and impacts of sustainable measures, showing how cost-benefit analysis and market demand can increase the willingness to pay for sustainably produced products and reduce environmental cost. It also points out the ROI in green technology to bring about productivity enhancement and long-term sustainability (Tan et al., 2019).

12.4 TECHNO-ECONOMIC ASSESSMENT (TEA) METHODOLOGIES

12.4.1 Life Cycle Cost Analysis

Whole life cost analysis or life cycle cost analysis (LCCA) is a technique used for calculating all costs associated with a chemical process or product over its entire life cycle from design and construction to operation, maintenance, and waste disposal. LCCA in green chemistry helps decision-makers determine whether a sustainable route saves money in the long term even though it can be costlier in the beginning (Giacomella, 2021). For example, during wastewater treatment involving graphene oxide-based adsorbents, researchers compared several options and found graphene oxide foam (GOFa) to have the lowest life cycle cost (~\$1.73 ± 0.09/m³) because of its highest material usage and resistance to raw material price volatility. LCCA often incorporates environmental externalities estimating CO_2 emissions' or energy usage's embodied costs where there are no evident exposures through methods like activity-based costing of the environment (Barahmand & Eikeland, 2022). Life cycle costing usually happens in stages:

a. Set the parameters for the system,
b. List all capital outlays, operating costs, utility consumption, maintenance, and disposal expenses,
c. Discount (e.g., NPV) over an allotted lifecycle, and
a. Conduct sensitivity or uncertainty analysis (e.g., Monte Carlo or worst case/best case).

The new computational systems support parametric and dynamic life cycle costing embedding within LCA, with provision for time-variable inputs, uncertainty, and optimization phases.

12.4.2 Techno-Economic Modeling Tools

Techno-economic analysis (TEA) brings together process simulation, cost estimation, and financial evaluation in order to choose the economic viability of novel or existing technologies. TEA software typically includes:

- Process flow sheet and mass/energy balance modeling (e.g., using Aspen Plus, SuperPro Designer, or BioSTEAM).
- Equipment sizing and cost estimation via cost curves or databases (e.g., Perry's, Marshall & Swift indices).
- CAPEX evaluation by direct equipment factors and indirect cost factors.
- Calculation of OPEX from raw materials, labor costs, utility costs, waste disposal, overheads, and maintenance.
- Financial appraisal by use of discounted cash flows to determine NPV, IRR, payback period, or minimum selling price (MSP).
- Sensitivity analysis or uncertainty analysis (for instance, tornado diagrams, Monte Carlo simulation) to identify variables that affect economic outcomes.

Atsbha et al. (2024) conducted a combined TEA and LCA for an electrochemical CO_2 capture and conversion technology (CO_2RR vs. traditional CO_2RR) and concluded that while CO_2RR is currently some 3.5 times more expensive, it could be perhaps 6% less expensive and more sustainable using technology advancement by employing renewable energy. A 2023 RSC report of catalytic fast pyrolysis of mixed plastic waste proposed a CAPEX of \$77.6 million to \$107 million for a 240 TPD unit and an OPEX annually that would be as much as \$84 million, an MSP of around \$2.18/kg naphtha, roughly four times the petroleum norm. Forward-looking ("prospective") TEA models for novel carbon capture and utilization technologies use hybrid costing (bottom-up for first plants) along with learning curves to estimate reductions in CAPEX and OPEX with advancing technology.

12.4.3 Capital and Operational Expenditures (CAPEX and OPEX)

CAPEX encompasses all initial expenditures required to design, construct, install, and commission an installation. This includes equipment acquisition, piping, instrumentation, civil works, land, engineering, contingency, and working capital (Nikookaran et al., 2017). In TEA, CAPEX can be estimated by major equipment factored methods: begin with purchase cost of equipment (by scaling principles or vendor quotations), and append auxiliary and indirect costs by typical multipliers. Case studies demonstrate how selection of technology has a direct influence on CAPEX—for instance, hydrogen production by photocatalytic systems can be between \$4.15 and \$9.80/kg H_2, where capital and labor contribute ~75% of the total cost (Clavé & Estécahandy, 2023).

12.4.4 Operating Expenses (OPEX)

OPEX also includes direct, repeating costs: raw materials and feedstock, utilities (electricity, heat, solvents), labor, maintenance, waste treatment, and overhead. Sensitivity analyses will typically show raw material cost or feedstock price to be among the biggest OPEX drivers, for example, mixed plastics pyrolysis facilities have feedstock alone accounting for ~\$52 million/year of ~\$84 million OPEX (Akbar et al., 2021). Labor depreciation or feedstock fluctuations can account for 39–42% of total OPEX in bioenergy and bio-products. Uncertainty and variability are essential: recent reviews indicate Monte Carlo simulation used in both CAPEX and OPEX segments to study intervals of MSP or IRR and determine sensitive cost variables. Learning curves in CAPEX and variable OPEX are used independently in future TEA models to project future cost savings with increasing deployment (Vartiainen et al., 2020).

12.5 TECHNOLOGICAL ADVANCEMENTS SUPPORTING GREEN CHEMISTRY

12.5.1 Process Intensification and Automation

Over the past decade, developments in continuous flow chemistry, micro-reactor systems, and modular automated platforms have greatly enhanced the capacity to carry out more sustainable chemical processes. Flow chemistry enables precise control over reaction temperature, time, and mixing, which greatly reduces waste and energy consumption, improves safety, and provides greater scalability compared to traditional batch reactors. Slug flow systems and microreactors possess better heat and mass transfer and hence reactions can be performed more safely and efficiently benefiting for rapid biodiesel production, fine chemicals, and pharmaceuticals. The use in drug production (e.g., Eli Lilly's continuous asymmetric hydrogenation of a lead clinical candidate) serves as an example for commercial application for intensified flow-based processes. Process analytical technology (PAT) and automation have additionally promoted process intensification (Gude & Martinez-Guerra, 2018).

12.5.2 Use of Bio-Based and Renewable Feedstock

Current research focused particularly on sustainable biodiesel and bio-based chemicals indicates that feedstock diversification to waste oils, non-edible oils, algae, and lignocellulosic biomass reduces fossil carbon dependency considerably and is lower in life cycle greenhouse gas emissions. For example, multifunctional nano catalysts (such as ZnO and CaO) have succeeded in yielding over 95% biodiesel from waste cooking oil at relatively low temperatures and short reaction times using improved reactor geometries to achieve feedstock conversion efficiency (Kohli et al., 2019). Biomass-derived platform molecules (levulinic acid, furfural, γ valerolactone) are utilized more and more in catalytic processes of bioplastics, biofuels, and chemicals with the support of advancements in biochemical and thermochemical conversion technologies. Parallel development in solvents has created green bio-based solvents such as ethyl lactate, 2-methyltetrahydrofuran, γ-valerolactone, and dihydrolevoglucosenone (Cyrene) that are low in toxicity, highly biodegradable, and also compatible with catalytic and asymmetric reactions. These solvents and raw materials are green and abide by the tenets of green chemistry by reducing the use of toxic chemicals, maximizing biodegradability, and enabling closed-loop material cycles (Iglesias et al., 2020).

12.5.3 Green Catalysts and Solvent Alternatives

Green catalysis has made significant strides over the last few years, with advances in heterogeneous catalysts, enzymes (biocatalysis), photocatalysts, electrocatalysts, and MOF-derived catalysts that can function under milder and greener conditions (Nanda et al., 2021). Supported Pt, Pd, and Ru nanoparticles as heterogeneous catalysts have been utilized with good efficiency in supercritical CO_2 systems for hydrogenation or hydrodeoxygenation reactions with high conversion and selectivity and enabling catalyst recycling over many cycles. Enzymes and biocatalysts still occupy a place of their own, especially for fine chemicals and asymmetric synthesis, due to their excellent selectivity, moderate reaction conditions, and minimal waste generation. Meanwhile, more benign solvent replacements like supercritical CO_2, ionic liquids (originating from bionome), deep eutectic solvents (DES), and natural deep eutectic solvents (NADES) gained more attention as even greener media for catalysis and extraction operations (Hessel et al., 2022).

DES have been successful in asymmetric catalysis and extraction operations, with solvent tunability and biodegradability; natural feedstock-derived ionic liquids preclude the volatility problem; and supercritical CO_2 is a recyclable, nonflammable reaction medium in hydrogenation and polymer processing. Such technology advances process intensification and automation, renewable bio-based feedstocks, and environmentally benign catalyst/solvent alternatives have revolutionized the practice of sustainable chemistry. Robotic and continuous systems reduce the amount of energy consumed, waste created, and improve consistency and scalability (Sheldon, 2016a).

Use of renewable feedstocks (biomass-based solvents and feedstocks) decreases petrochemical dependence and carbon emissions. Concurrently, sophisticated catalysts, especially recyclable heterogeneous systems, enzymes, and photothermal/electro catalysts enable high selectivity under gentle, green conditions (Winterton, 2021).

12.6 SUCCESSFUL GREEN CHEMISTRY IMPLEMENTATIONS

12.6.1 Industrial Examples with Economic Analysis

Since 2016, more and more industries have applied green chemistry tactics successfully, proving that profitability and sustainability are not opposing forces. Such initiatives are reporting common gains in profitability and sustainability, providing measurable experience of return on investment, waste reduction, energy conservation, and longevity in international supply chains (Sheldon, 2016b). An example is the production of bio-based 1,3-propanediol (bio-PDO) through renewable corn sugars by fermentation using engineered microorganisms instead of using the more traditional petrochemical feedstocks (Ivanović et al., 2021).

Since the company traditionally relied on formaldehyde resins, which were associated with the release of VOC and potential cancer risks, the company transitioned to lignin-based adhesive technology developed together with the agrochemical giants. This helped improve indoor air quality and reduced workplace exposure, reducing regulatory threat and minimizing long-term healthcare costs for factory workers. The new adhesive technology is derived from lignin too, a renewable wood preservative waste product, and has cost benefits via the use of in-plant waste streams. IKEA has utilized this marketing innovation, which helps its sustainability image and customer loyalty, hence ultimately leading to enhanced revenues and brand value (Pratama et al., 2025).

Bio-derived plasticizers are yet another game-changing trendsetter in the market. Its phthalate-free plasticizers, which are compounded with epoxidase soybean oil and other vegetable oils, have acquired a significant market share in flexible PVC applications such as cable, flooring, and medical devices (Plass et al., 2021). They not only meet EU REACH and RoHS compliance but also outperform conventional phthalates in lifecycle toxicity, carbon footprint, and processability. Market research studies conducted by independent market research firms have shown ECOLIBRIUM™ manufacturers have been saving via reduced regulatory fines, better waste management, and simplified logistics for global markets. In addition, such firms have reduced reputational risks and are in better sustainability rankings and will receive more solid streams of investments and more investor confidence. Moreover, BASF's biomass balance approach, since 2018 launched in Europe and Asia, has enabled the replacement of fossil feedstocks with sustainable ones (e.g., bio-naphtha) in traditional petrochemical facilities (Robinson & Dow, 2021).

This strategy avoids costly new installations by using mass balance certification to authenticate bio-content throughout entire value chains. As a result, BASF has minimized Scope 1 and Scope 3 emissions, achieved eco-labels like EU Ecolabel and Blue Angel, and achieved market access to premium markets for sustainable materials. Analysis indicates that this approach enables break-even in 2–4 years depending on the industry and extended payback from carbon credits and lower environmental taxes (Franz et al., 2024). Pfizer in the pharmaceutical sector has been at the forefront of green chemistry innovations such as continuous flow synthesis and resin recovery loops and has made significant strides in material efficiency while making medicines. One study asserts that such operations reduce solvent use by 85% and reduced the amount of wastes by 50% without deteriorating product quality, thereby reducing the overall cost of pilot-scale production by nearly 20% (Zhang et al., 2024).

12.6.2 Break-Even and Payback Period Comparisons

Green chemistry measures effectiveness of sustainable projects both in environmental terms but also in principal economic studies particularly highlighting break-even point and payback periods. Both are important financial metrics used in facilitation of industrial adoption as well as long-term investment planning. One of the best examples is the GlaxoSmithKline (GSK) carbon-neutral chemistry lab at the University of Nottingham, opened in 2017 (Van Fan et al., 2019). The building has been optimized using a range of linked green technologies, including solar photovoltaic systems, passive natural ventilation, and a biofuel-powered CHP plant. Collectively, the technologies have enabled the lab to conserve more than 60% energy compared to conventional laboratory design. The 25-year payback time of carbon however is indicative of the fact that the changes in this case are long-term and permanent. Short-term investment returns may take time to materialize; however, the cumulative energy and carbon savings over the facility's lifetime provide substantial and consistent value, particularly within a climate-conscious regulatory environment (Padulles et al., 2025).

Although it is true that batch operations sometimes require increased usage of solvents, which may affect some green chemistry indicators like E-factor or mass intensity, some numerous reports demonstrate definite economic advantages. Constant production improves the efficiency of the process since it allows achieving greater throughput, minimal downtime, and a lower batch-to-batch variation (Prokhorenko & Kokoshynska, 2024). These advantages actualize reduced operation costs, reduced inventory holding cost, and reduced waste production. Even though the original capital outlay in equipment and rearrangement of facilities would typically be higher, the payback period is much lower due to the higher process mass intensity (PMI), improved space-time yield, and optimal utilization of resources (Prajapati et al., 2023). In an API synthesis scenario, a flow process cut production time by more than 40%, achieving a PMI gain of almost 30% relative to its batch counterpart. The advantages allowed companies to recover their capital costs in a couple of years and to reach the green chemistry objectives such as waste reduction and saving of energy (Baumann & Baxendale, 2015).

12.6.3 Adoption in Pharmaceuticals, Agrochemicals, etc.

Green chemistry has fully transformed the drug industry, whereby the new sustainable methodologies have replaced the old, wasteful production processes (Fenibo et al., 2022). The most notable example here is the landmark biocatalytic synthesis of sitagliptin (type 2 diabetes drug) in Merck that was done enzymatically. The sustainable route avoided the need for asymmetric hydrogenation catalyzed by heavy metals, which used to yield ca. 100 kilograms of waste per kilogram of product. The novel process eliminated toxic aqueous waste streams while also significantly improving yield, process safety, and energy efficiency (Alcántara & Alcántara, 2018). The innovation was rewarded to Merck with the prestigious U.S. EPA Presidential Green Chemistry Challenge Award, highlighting its economic and environmental importance in reducing raw material and purification expenses as well as improving scalability and operating reliability (Das et al., 2019). Similarly, Pfizer made great strides toward eco-friendly production of the widely used depression drug sertraline (Sharma et al., 2024).

During second-generation manufacturing, Pfizer replaced less toxic reagents for the toxic reagents benzoyl chloride and titanium tetrachloride and improved the process by removing many steps, increasing the yield by 80%. The reconstructed synthesis removed 20–60% of raw material consumption, reduced toxic solvent use, and reduced energy and water consumption by a significant amount (Sharma et al., 2017). These innovations contributed to a considerable decrease in total environmental burden and treatment cost of waste without compromising product efficiency and safety still driving the pharmaceutical sector toward process intensification and green-conscious drug development (Agnihotri & Aruoma, 2020). GlaxoSmithKline (GSK) has demonstrated ongoing commitment to green chemistry by applying a green solvent selection guide throughout the company and biocatalysis for most active pharmaceutical ingredients (APIs). Notably, in the production of lapatinib, a cancer medicine, GSK employed continuous flow chemistry with enzyme-catalyzed catalysis to precisely manage reaction conditions and employ the minimum quantities of solvent and reagents. These developments led to significant reductions in poisonous waste, greenhouses decreased, and workplace safety increased. GSK set out to reduce its API waste production by 10% by 2030 as part of its overall sustainability program (Ncube et al., 2023).

Aside from drugs, the development of agrochemicals has also embraced the tenets of green chemistry. One such excellent example is the 2022 U.S. EPA Green Chemistry Award given to Provivi for developing a pheromone system to control fall armyworm infestations a debilitating crop pest. This solution replaces traditional species-specific toxic insecticides with eco-friendly, non-toxic, and biologically derived species-specific pheromones. This technology minimizes environmental contamination and resistance to pesticides by providing a scalable and cost-effective approach to sustainable agriculture particularly to developing countries where environmental conservation and sustaining crop output is at stake (Lane et al., 2023).

12.7 CHALLENGES AND BARRIERS

12.7.1 High Initial Costs

The conversion to green chemistry practices is often associated with a large initial financial cost. This encompasses the acquisition of new typically specialized equipment, modification or replacement of the traditional reactors with reactors that can be compatible with green solvents and catalysts, and adoption of alternative and renewable raw materials that could be costlier and harder to obtain in the first place. Also, organizations tend to need the provision of resources to train employees on how to safely and efficiently implement and manage these new processes (Veleva & Cue, 2019). Though many LCAs and techno-economic analyses suggest that green chemistry can result in significant long-term savings due to the decrease in waste production, decreased energy use, and fewer regulatory burdens, the first transition costs may be a significant obstacle in particular to the small and medium-sized enterprises (SMEs). Such businesses are usually not run with a sufficient margin, and they usually have no capital reserves or access to long-term funding to carry out such changes. Despite the undoubted economic and environmental advantages, it seems that the concern about the financial stress of the moment or the inconvenience of the existing production approaches prevents many of them from changing (Ratti, 2020).

The studies on the investments in sustainable technologies underline the strong influence of behavioral and psychological obstacles. A perception of costs and reduced rate of adoption in comparison to what would happen in standard economic models is compounded by risk aversion, questioning the validity of new technologies, and lack of immediate returns on investment. This is most notable in the businesses that have low profit margin and high competition thus the leaders focus on short-term gains instead of long-term sustainability and environmental outcomes. Research indicates that in the absence of external incentives, such as governmental requirements, financial incentives, or tax benefits, several companies put off or simply do not implement green chemistry-related practices, including those which would be consistent with meeting their larger sustainability objectives. The internal competition toward the investment capital is an added challenge in larger companies where the accessibility of resources might be more enhanced. Green chemistry proposals are usually competing with the traditional projects that are offering higher rate of returns or more reliable financial returns. Lacking regulation demands and obvious incentives, green chemistry projects might struggle to move forward, not because it is tricky technically but because of institutional resistance and the latent expenses involved in getting rid of outdated systems. Consequently, this continues to widen the disparity between the theoretical opportunities of more sustainable, clean methods, and their actual application. The technological advancement, favorable policies, ready funding, and a change in organizational culture that appreciates long-term sustainability and short-term results are the keys to bridging this gap (Sovacool et al., 2022).

12.7.2 Technology Readiness Levels (TRLs)

Despite the fact that a lot of the innovations in green chemistry are the result of scholarly researches, the process of transferring the results to pilot testing and then to the massive production is a complicated, time-consuming, and complicated affair. There are always major challenges to the process of converting scientific developments in the laboratory environment to the industrial one, and the main reason is the low technology readiness levels (TRLs) of innovative procedures and materials.

These issues are especially pronounced when it comes to newly developed catalysts, novel solvents or technologies that use biomass feeds where the initial yields and product purities that we observe in controlled laboratory settings tend to decline when exposed to the realities of industrial operation. With the increase in scale, the reaction kinetics, mass and heat transfer constraints, the life of catalysts, and the accumulation of impurities are increasingly important and difficult to manage (Buchner et al., 2019).

The reasons that often prevent companies to expand include: the combination of engineering risks, absence of confirmed data through the real-life situations and the inability to incorporate sustainable practices within the existing production systems without expensive changes or redesigns. In addition, there is often a great disparity between the academic goals like innovation or the understanding of mechanisms and demand of the industry on credible processes, safety, and cost-effectiveness. These dynamics are embodied in a new model of evolutionary competition, where the adjustment costs are considered as a factor that increases the transition costs, and transition costs are regarded as a factor that increases the adjustment costs: green technologies appear to be beneficial, and they start being used, but the uncertainties regarding the performance indicators, the market appropriateness and the ROI, can push the adoption until systems achieve high TRLs (usually TRL 7–9) when the operational performance will confirm the scalability and reliability (Sandoval, 2024).

This is creating a permanent chicken-and-egg dilemma. Unproven systems scare away investors and other players in the industry to inject their capital and resources because they fear technological failure, regulatory stalling, or financial inefficiency. Conversely, this dedication during small-scale work, wide-scale testing, and continuous improvement is what is essential to push the innovations out of the developmental stage of the so-called valley of death and into the field of profits. Most green chemistry innovations are still confined in the textbooks of academic literature instead of an actual change in practice in the scale-up stage without institutionalized funding systems, public-private partnerships, or regulatory incentives to reduce risks (Roschangar et al., 2015; Sheldon & Brady, 2022).

12.7.3 Regulatory and Policy Limitations

Fragmented, outdated or inconsistently implemented regulatory systems are major obstacles to the mainstream implementation of green chemistry practices. Most nations have found outdated chemical regulations that do not keep up with the development of chemical science or the increasing imperative of environmental health issues (Roschangar et al., 2015). The Toxic Substances Control Act (TSCA) was initially passed in 1976 and has since faced endless criticism in the United States over its form which has led to major data gaps, pre-market safety standards, and lack of innovation promotion. In spite of the fact that the TSCA was revised in 2016 as the Frank R. Lautenberg Chemical Safety for the 21st Century Act, critics believe that the development is slow because of the delay in regulations and the continuity of using weak toxicological data to adopt safer chemical alternatives (Dhayalan et al., 2017; Krimsky, 2017).

These challenges are further compounded by regulatory ambiguity and cross-jurisdictional differences. The lack of coherent definitions for terms such as green chemistry, sustainable chemicals, and non-toxic-by-design creates confusion for all stakeholders, including manufacturers and policymakers. The consequence is a conflict or overlap of competing demands which make it difficult to comply with and costly to companies that seek to innovate. The process of approval of new substances can be accompanied by extensive safety assessment, and the schedule can be significantly postponed depending on the regulator, which hinders the introduction of more sustainable alternatives into the market (Zhao et al., 2021).

12.8 POLICY, INCENTIVES, AND FINANCING OPTIONS

Governments across the globe have been using all kinds of direct fiscal instruments to enhance the advancement of the green chemistry and speed the overall process of transition to sustainable industrial development. Such initiatives represent an understanding that the market alone might not be adequate to create the systemic innovation required to address the damage to the environment and climate change. Markedly, the European Union has been in the forefront with its Green Deal Industrial Plan, and this is supported by the EU Green Chemistry Subsidy Program (Schwager et al., 2016). This system gives much support to the general people through grants, tax incentives, and special R&D funding particularly with regard to environmentally friendly chemical manufacturing, sustainable catalysis, renewable resources, and waste minimization technologies. The EU hopes to gain an edge on clean technology by offering direct incentives on low-carbon innovation and sustainable industrial practices, at the same time equipping its native industry with an advantage over international competition (Zuin et al., 2021).

Meanwhile, China has come up with a groundbreaking approach to green finance integration into regional plans of development. Since 2017, the establishment of green financial zones has made it possible to take a diversified approach, a combination of subsidized loans, favorable interest rates, green bonds, and environmental insurance mechanisms to assist businesses toward switching to solutions that are more sustainable than those that cause pollution (Wang et al., 2021). These areas not only encourage cleaner production methods but also assist in considering the environmental costs in the otherwise resource-heavy industries. Empirical research based on quasi experimental DID designs has revealed statistically significant increases in the level of environmental protection spending at the firm level with the adoption of these fiscal policies. Furthermore, such policy efforts have lessened disparities in SMEs in accessing capital since a significant portion frequently encounters non-equivalent difficulties in shifting in the direction of green chemistry practices (Wang & Wang, 2021).

International monetary institutions like the International Monetary Fund (IMF) have been insisting on the need to redistribute funds toward encouraging green innovation. IMF policy reviews point out to the fact that a step-by-step removal of fossil fuel subsidies particularly those in the Middle East, Africa, and Asia-Pacific has created significant fiscal space. These available resources are slowly being channeled into environmentally friendly infrastructure, clean energy innovation, and climate resistant technology. Notably, some part of this redistribution is aimed at increasing the bankability and creditworthiness of clean chemistry projects, thereby making them more appealing to the private investors and multilateral development banks. The changing trend of the global fiscal policy demonstrates growing agreements that specific economic incentives, institutional changes, and regulatory stability are essential to create a robust and low-carbon industrial foundation that is rooted in the concepts of green chemistry (Yin et al., 2022).

12.8.1 Carbon Pricing and Green Credits

Emission-based pricing mechanisms like Emissions Trading Schemes (ETS), carbon taxes, and Carbon Border Adjustment Mechanisms (CBAM) are necessary to ensure that the market operations in line with the environmental sustainability include the costs of carbon emission, which were not previously considered. Such instruments have been effective to assign prices to carbon, increase the costs of pollution, and enhance the financial viability of clean technologies, including green chemistry. The Emissions Trading System of the European Union (EU ETS) is one of the most popular systems which have become a world standard. The EU has recently introduced CBAM wherein tariffs are being imposed on imports that contain heavy carbon like steel, cement, aluminum, and fertilizers. This unifying strategy aims to lock out the carbon leakage and extend the decarbonization impact of the EU to the rest of the world to encourage the exporting countries to adopt cleaner production technologies, such as catalysts, solvents, and synthetic processes, that can align with green chemistry principles (Ren et al., 2025).

As of early 2023, more than 70 jurisdictions worldwide had implemented carbon pricing systems, covering approximately 23% of global greenhouse gas emissions. These include established systems in countries such as Canada, New Zealand, and Sweden, as well as emerging initiatives and pilot projects in developing nations. (Digitemie & Ekemezie, 2024). As an example, Pakistan has begun establishing early products of carbon pricing, but countries in the Middle East and Africa are considering voluntary carbon credit trade schemes to promote the reduction of emissions. Such mechanisms are critical in areas where an industrial sector is rapidly expanding, and green chemistry might bring considerable payoffs to mitigating the environmental impacts of chemical production (Ahmad et al., 2025).

12.8.2 Role of Green Finance and ESG Investments

The development and the active promotion of green financial tools including green bonds, sustainability-linked bond (SLB), and the ESG-based credit options have provided significant opportunities to make investments in eco-friendly innovations, specifically in the chemical industry. Such financial instruments are more frequently used to fund projects as well as companies that revolve around green chemistry, renewable materials, and low carbon technology. Research conducted in China indicates that, green credit policies are very useful in increasing the ESG performance of the firms as the loans and investments are strategically redirected on projects that have demonstrable environmental benefits. These policies enhance the financial access of sustainability-minded firms and affect the corporate behaviors by imparting greater managerial sensitivity to environmental effects, promoting transparency, and improving funding to R&D in cleaner and more circular chemical reactions (Dovidonis, 2024; Li et al., 2023).

The swift expansion of global green bond markets during the 2022–2023 period, where issuance rates reached record highs in both advanced and developing nations. However, the rapid growth has uncovered pressing challenges mainly the need to harmonize international standards regarding transparency, certification, and accountability to address greenwashing and build lasting market trust. Initiatives such as the European Union's green taxonomy, reform proposals for the Sustainable Finance Disclosure Regulation (SFDR), and mandatory ESG-reporting requirements aim to bridge these gaps by making sure that sustainability claims are backed by reliable metrics and independent verification (Sundqvist, 2022). These regulatory frameworks aim at safeguarding investors against false promises, and also through these regulations to make sure capital is realistically correlated with the decarbonization pathways and environmental limits.

However, recent data indicate a shift in trend because the issuance of green bonds globally has declined by some 32% in 2025 a decline primarily linked to regulatory uncertainty and the reversal of major sustainability pledges of major economies. Analysts posit that falling policy indicators especially in the United States and parts of Asia have undermined investor trust and increased concerns as to the sustainability of green financial markets on the long run. This disparity highlights the precariousness of advancement in the face of a decrease in political ambition to the environmental agenda and emphasizes the importance of the sustained, evidence-based policy support that would ensure the perpetuation of investor attention and effective expansion of the program of sustainable financing initiatives (Alsmadi et al., 2023). In this way, the regulation consistency and transparency is crucial to ensure the integrity and strength of green finance as the force of sustainable change in the chemical industry and other domains (Tiwari et al., 2025).

12.9 BARRIERS AND CHALLENGES IN FINANCIAL VIABILITY

Advancement in technology and particularly the emergence of eco-innovations and clean technologies has sometimes been beneficial as well as problematic. Although innovations are capable of decreasing CO_2 emissions, enhancing energy efficiency, and promoting sustainable development, the innovations are usually expensive due to their initial expense of research and implementation. In 2024, a bibliometric analysis revealed that eco-innovation can be costly to businesses in the short-term and can have a negative impact on short-term performance particularly in resource-abundant industries where companies require swift payoffs that could otherwise not be forthcoming involving (Veleva & Cue, 2019). Although green technology has the potential to reduce the emissions, it is important that the financial growth and innovation are managed: a strong financial market and credible institutions boost the efficiency of environmental innovation, and weak systems can stimulate energy consumption and then increase emissions by allowing them to fund pollutant assets. This situation is an example of a fight. Green (2016) notes that despite the positive prospects of the future with environmental innovation, economic benefits, and expectations of investors, as well as institutional preparedness, tend to be slowed down.

Financial sustainability barriers are still common in various situations especially to the SMEs, clean technology start-ups and infrastructure projects. A number of systemic limitations, inefficiency in affordable and customized financing, incomplete information to evaluate risks, lack of awareness of sustainable business models and fixed short-term returns expectations by investors. Corporate venture capital (CVC) may assist in overcoming the valley of death between R&D and commercialization by contributing to patient capital, but the traditional VCs remain cautious because of the long timelines and risk factors of clean tech projects (Zuin et al., 2021). On even a larger scale even financially viable renewable projects are hindered by non-technical barriers like entrenched energy systems, fragmented regulation, absence of interconnection standards, and a lack of public awareness. Policy instability within the government, cultural barriers to change and institutional stagnation are some of the challenges hindering the funding and development of green technologies. Although the three relationships between innovation, sustainability, and profit are quite clear, achieving financial success does not only presuppose technological advancement but also a significant financial support, clear regulations, strong institutional potential, and long-term perspectives of investors (Lancaster, 2025).

12.9.1 High Initial Investment and Long Payback Periods

Existing academic and industrial literature has repeatedly emphasized the presence of high initial capital and long payback times to be the primary barriers to financial viability of the new low-carbon technologies. An analysis of energy performance contracting in high-energy consuming sectors indicate that complicated EPC projects demand huge initial investment on equipment, development, environmental tests and permits as well as grid connections that normally precede any revenue generation making them inherently less appealing unless accompanied by government incentives or friendly regulatory circumstances. The literature on energy finance in the world highlights how many regions, particularly in developing countries, are lacking in long-term project funding, with short-term loans and equity financiers taking the place (Dai et al., 2022). This situation compels renewable and innovation-led ventures to aim at payback periods that normally surpass reasonable risk levels, therefore, ruling out these ventures as investment opportunities. Moreover, unpredictability of income and unpredictable character of the income of renewable sources (the so-called cannibalization effect) can significantly decrease the projected returns, expected profits may decrease by over 33 to potentially negative figures depending on multiple scenarios of market penetration, extending the actual financial break-even point and increasing the investment need by 13–67% using real-options modeling (Gorshkov et al., 2018).

The payback period is longer which increases investor reluctance despite significant environmental advantages. The ease and popularity of the traditional payback models, which disregard the time value of money and do not take into consideration cash flows outside the payback horizon, reinforce biases against more long lasting green investments, although these may also provide substantial total lifetime benefits or reduction in emissions. The off-putting impacts of policy instability add to the situation: a change in subsidies can slow down the development cost or delay the revenue stream, extending the payback even further (Xia et al., 2021). As a result, a variety of new eco-innovations can only be made financially viable in stable regulatory frameworks, supportive capital or customized financing schemes that allow them to depreciate their assets over a lifespan and not individual short-term ROI expectations (Ruiz-Mercado et al., 2016).

12.9.2 Market Entry Barriers and Technology Readiness Levels (TRLs)

The market barriers have a significant influence on the cost-effectiveness of disruptive clean technologies, particularly in case of mid-TRL technologies (TRLs 4–6) also known as the valley of death. Companies tend to face challenges at TRL 4 to transition between proof-of-concept and demonstration because of limited funds, low investor confidence, and insufficient early adopter customers. Research indicates that preliminary funds may be used to validate the lab, but to go to TRL 6 requires investment by the company and its partners, expertise in business, and pilot validation on the side of customers, which many academic and small research groups cannot afford (Moussa, 2022).

These obstacles augment economic risks: innovations below TRL 7 or more often struggle to create viable commercial proposal, and thus avoid successful integration into project pipelines or supply chains. At the same time, the literature sheds light on inherent constraints of relying on TRL as the only measure of being ready to be introduced into the market. TRL evaluates technical readiness but often does not pay attention to regulatory, market and social aspects needed to determine economic feasibility and environmental effects. When it comes to developing sustainable energy technologies, the practitioners suggest incorporating TRL with the other frameworks, including a Commercial Readiness Index (CRI), which allows policymakers and investors to consider the technical factors as well as the issues, associated with funding, regulatory suitability, supply chain, and acceptance of the new technologies by stakeholders. Besides, over-ambitious or exaggerated TRL claims might cause problems that projects that are considered ready might have difficulties in launching into the market or in obtaining regulatory acceptance or standardization qualifications. In order to address these two interrelated problems, an effective preparedness measure is required: the alignment of technical TRL development and business development objectives (IPR strategy, investor presentations, customer pilot trials), the verification of compliance readiness, and the planning of risk management to reduce the obstacles on the way of market entry and support the economic and environmental sustainability agenda (White et al., 2022).

12.9.3 Price Competitiveness with Petrochemical-Based Processes

Despite the rapid development of bio refining technologies during the past ten years, bio-based chemicals usually cannot compete with prices of petrochemical-based counterparts. The price of feedstock is also a significant obstacle naphtha or ethane feedstock costs are about $0.25/kg whereas biomass feedstock costs are 0.38–0.52/kg, which translates into a 30–100% increase in the cost of producing bio-succinic acid or 1,4-butanediol. In addition, the bioprocess productivity is limited typical fermentations can only produce approximately 65–85% of the theoretical limits which require higher production scales which escalate the capital and operational costs. Downstream processing is another factor of cost: the recovery of target molecules in the aqueous fermentation broth may require 60–70% of the total production costs (Sheldon, 2014).

Increased regulatory and infrastructure barriers also affect competitiveness of price. New bio-ethylene plants require substantial capital investment of above 1.2–1.8 billion (compared to 900million) in new petrochemical plants, which translate into long pay-back periods. Inconsistencies in infrastructure are also another barrier: oil refineries are not prepped to accept biomass need specialized equipment, and the costs of updating this process are 30–50% higher than a new facility, and insufficient storage, logistical challenges, and seasonality of biomass in various areas make the process more expensive by 15–20% of a 50-mile radius (Bauer et al., 2022).

12.9.4 Risks and Uncertainties in Technology Commercialization

The commercialization of new technologies, particularly those that may affect the environment, such as clean energy or circular economy innovations are threatened by uncertainty in terms of adoption rates in the market, regulatory regulations, and economics. The key issues are the so-called valley of death between the R&D and scaling up of the project where a huge capital investment collides with the uncertain market indicators (Clark & Pasternak, 2021). Investors should be cautious once speculations are based on future regulations or behavioral patterns; simultaneously, technologies might seem economically viable in a lab or testing setting and encounter problems of obtaining risk-taking investment to scale. Fluctuations in the exchange rates, fluctuation in commodity prices and political instability lead to financial unpredictability, complicating long-term planning in business, although apparently well-performing lifecycle environmental factors may make business work well (Jessop & Reyes, 2018).

Meanwhile, the uncertainty of technical performance is dangerous, new technology can fail to perform or be more expensive to use in expansion, and the projected environmental benefits and financial sustainability will be at risk. Only at a large scale, unexpected integration issues or supply chain bottlenecks or hidden effects on the environment may arise (Thormann et al., 2021). In addition, organizations are exposed to issues regarding gambler dilemma of the innovator and initial converts may get better options, existing dominant companies may still prevail, leading to delays in market acceptance and sales. Innovations that are environmentallyoriented, the uncertainties in terms of how the environmental impacts can be measured, pricing carbon and the criteria of LCA may lead to doubts and skepticism among investors. The linkage between innovation, economic viability, and sustainability is reduced when benefits to the environment are difficult to quantify or dependent on future possible carbon credits or regulations, which hinders commercialization as well as limiting financial sustainability (Hallett, 2025).

12.10 FUTURE TRENDS AND OPPORTUNITIES

12.10.1 Emerging Funding Models (Crowd Funding, Green Venture Capital)

Equity and reward-based crowd funding have become popular ways of financing green and sustainable projects in the last ten years. A review of 771 sustainability-oriented campaigns (2014–2021) revealed that campaigns that mostly address environmental and ESG objectives achieved more success than others, in particular, institutional investors fostering engagement through making prior investments and thereby causing information cascades among the audience in general. Literature on cleantech projects underlines the fact that accurate and straightforward explanations of projects improve the credibility and greatly increase the chances of success compared to vague narratives, even in the case of small investors through platforms such as Kickstarter (Bonini & Capizzi, 2019).

In addition to crowdfunding, there has been an emergence of a new source of growth finance, which is, green venture capital (VC). In June 2025, 2,003 U.S. funding rounds in 2022 were studied showing that environmentally oriented ventures are more attracted to large amounts of pre-IPO capital, particularly those supported by a high-profile lead VC. Simultaneously, the lists of the most advanced climate tech venture funds in the world (such as Breakthrough Energy Ventures, Lowercarbon Capital, Congruent Ventures) demonstrate a rapid increase in the number of deals in the field of clean technology and green chemistry. It is suggested that in the future, integrating blockchain and stablecoins will help lower the platform fee, improve investor confidence due to transparency, and democratize liquidity through tokenization and secondary markets specifically designed to be used in sustainability-oriented crowdfunding systems (Nguyen et al., 2024).

12.10.2 Carbon Neutral and Net-Zero Targets as Market Drivers

Market trends, as outlined by the corporate and governmental commitment to achieve carbon neutrality and net zero emissions by 2050 (or sooner), are affecting the trend of more sustainable chemical processes and materials. These goals act as policy and supply chain benchmarks, encouraging buyers and regulators to choose lower-emission items in their lifecycle, thus creating an incentive to invest in green chemistry discoveries and environmentally friendly industry processes. Though the search did not specifically identify particularly recent articles (between 2022 and 2025), this causality is highly supported by the current literature on climate technology and sustainability practices (Aziz et al., 2024).

Companies that support net zero objectives are associated with improved business opportunities, higher valuations, and increased attractiveness to investors particularly with limited partners being more inclined toward investment strategies that are ESG-oriented. VCs and institutional investors would now like ventures to show quantifiable environmental impacts, which include Scope 1–3 carbon accounting and plausible net zero strategies (Renné, 2022). As green chemistry progresses, those companies that have set net zero requirements such as carbon capture solutions or bio-based products can access new funding and partnership sources by the government and other stakeholders, making decarbonization a competitive force on the market (Walker, 2020).

12.10.3 Global Green Supply Chain Integration and Export Potential

The industry is having an export potential outside the United States as global trends in sustainable sourcing and low carbon materials are creating international export opportunities in the green chemistry products, particularly those based in areas that can provide certified low-emission chemical inputs. The suppliers that prove to act in line with the principles of sustainability (including certifications of renewable materials, the labels of the circular economy) are becoming more sought after by the multinational purchasers and the international regulators (Singh & Trivedi, 2016). Even though specific documents on the export of green chemistry in 2015–2025 could not be found, the literature related to the circular bioeconomy indicates that Europe generates more than EUR1.5 trillion of circular bioeconomic value with the support of the investment program, like the European Circular Bioeconomy Fund (Raman et al., 2023).

It is in the interest of producers in the developing countries to seek high-value export markets through participation in international green supply chains and to obtain better profit margins, particularly in situations where sustainability certification is obligatory. Companies that apply green chemistry to product development and supply chain enjoy the advantages of demand-related aspects (such as global OEMs that are interested in low carbon materials) and finance-related ones (such as ESG-based trade financing and green conditions provided by export credit agencies). As the world is growing more strict regarding the tracking of carbon footprint and chemical safety, those exporters who prove to be exemplars with proven green chemistry certifications will gain a competitive edge (Galera-Quiles et al., 2021).

12.10.4 Financial Opportunities in Industrial Interdependence and Resource Sharing

Industrial cooperation the mutual sharing of materials, energy, water or infrastructure among neighboring businesses brings great cost reductions and value creation, and has been identified as a promising financial pathway in the realms of sustainable chemistry (Van Berkel, 2006). The successful case of the Kalundborg Eco Industrial Park in Denmark that has existed long enough to demonstrate the benefits of resource sharing, including wasted heat, gypsum, fly ash, and water, with savings of hundreds of millions of USD over several years at a cost of approximately 78.5 million USD, growth in many years. Recent extensive reviews emphasize more than 20 opportunities and 13 actionable lessons (such as developing shared databases, a co-benefit portfolio, and regional technology hubs) that guide stakeholders toward viable symbiosis networks (Jensen et al., 2011).

Recent research examines the financial and organizational dimensions of symbiosis: differences in transaction costs across various companies require fair cost-sharing strategies analyzed through cooperative game theory and multi-agent systems to justly distribute infrastructure and governance expenses and sustain long-term cooperation (Sheldon & Brady, 2022). Efforts by global entities like the World Bank to identify and support industrial symbiosis clusters show that financing strategies and government incentives are increasingly targeting resource-sharing frameworks as scalable, low-carbon bases for the progress of green chemistry (Lozano et al., 2018).

12.11 CONCLUSION

During recent years, techno-economic analyses (TEAs) have emerged as a fundamental analytical instrument for market potential assessment of green chemistry innovation. Recent research emphasizes that combined TEA–life cycle analyses enhance the estimation of capital expense, operational expenses, revenue, and environmental externalities of new processes such as plasma-assisted reforming or future biomass-derived fuels. For example, hybrid plasma reforming of methane to syngas is observed to offer distinct techno-economic and environmental benefits over traditional steam reforming, showing how integration of processes conserves energy but wastes more resources. Recovery of the rare earth elements in an industrial phosphogypsum was also considered regarding a combined probabilistic TEA and LCA wherein internal rates of return exceeded 15% in all cases demonstrating that green chemistry using secondary resources is economical at the commercial scales with the proper feed concentration and capacity factors. These results highlight that the green chemistry technologies could be economically viable in case the concerns of scale, the intensification of the process, and feedstock quality are resolved appropriately.

Investment risks and financial gains of green chemistry are increasingly becoming policy motivated and cost-cutting in nature. Encouragement policies such as tax incentives, financing and carbon pricing are critical in resolving the issues of the valleys of death between laboratory innovation and commercialization as were emphasized in the U.S. and EU agendas on industrial policy. The present values of green hydrogen in green hydrogen paired with hydrogen green chemistry are 3.50–6.00/kg, but it is competitive where renewable energy costs are less than 20–30/MWh, and U.S. incentives, such as IRA tax credits, can offset the costs by up to 3.00/kg, which encourages the investment in electrolyzers and the expansion of production. Behavioral studies have indicated that, despite the cheap green technology, there are still adoption biases at the company level: theory predicts a rate of 81% adoption, yet 20% has been realized of an observed pattern of behavior representing the gap between the finance potential in theory and in practice decisions. This implies that technology economic feasibility does not suffice to cause something to occur with adoption in the absence of investor risk perception and organization mentality.

Technology leaders identify the greatest focal point on the arrangement of R&D efforts and implementation strategies that are technologically superior and those supported by regulatory anxiety. Continuous flow reactors, or combined reaction-separation modules may be used to reduce CAPEX, energy utilizations and wastages significantly with the implementation of process intensification technologies such as membrane reactors (such as MACBETH) and lead to acceptable financial performance and reduce environmental footprints. To be in good business to become part of the value chains, firms should use mass-based metrics of green chemistry (measured by mass) performance (E-factor, atom economy) and impact-oriented LCA metrics to construct their business cases. There is a tendency of risk mitigation through R&D relations with universities, technology incubators, and public-private relations and a possibility of access to capital, co-location of pilots, and possible licensing or trade sale exit.

Strategic policy instruments are important instruments in the governance and civil arena. Policymakers must delineate pricing on carbon, emission caps, and financial assistance programs to support pilot-level enterprises and to proceed with commercialization and thus avoid the valley of death that commercialization currently involves in clean-tech. Regulatory approaches under the Porter hypothesis suggest that high environmental regulation can enhance innovation and competitiveness in the chemical industry through converting compliance costs into drivers of innovation. Further, simple-to-apply rules of chemical disclosure (e.g., digital product passports and supply chain transparency) and incentives for circularity (encouraging recycling of toxic or "forever" chemicals) can draw investment and minimize potential liability risks.

One viable way of accelerating green chemistry deployment is through the development of pilot and demonstration plants through well-coordinated funding-supported projects. Various case studies illustrate that organizations and teams, when scaling from laboratory stage to pre-commercial pilots, can minimize risks and achieve financial break even earlier particularly when augmented by government-supported funding or guarantees. Process innovations like bio-sourced nanoparticle production, membrane reactor use, and hybrid renewable electrolyzer systems, if analyzed using TEA + LCA, always exhibit shorter time to market and greater investor confidence.

A quicker absorption route is through using policy-inspiration-driven demand metrics and investor activism. Institutional investors and ESG funds increasingly urge chemical companies to move away from hazardous legacy chemicals and toward greener alternatives, such as in efforts to regulate PFAS and corporate chemical footprint reporting. In combination with third-party certifications and green chemistry benchmark standards, that investor demand has the power to incent companies to push technology forward. Of no less importance is the use of information framing, performance standards, and peer comparison to align firm investment behavior with normatively optimal levels by modeling (81% take-up vs. approximately 20%) to create a comprehensive package of financial, regulatory, and behavioral tools for economic leverage of green chemistry in industry sectors.

REFERENCES

Abdussalam-Mohammed, W., Ali, A. Q., & Errayes, A. (2020). Green chemistry: Principles, applications, and disadvantages. *Chemistry Methodology*, *4*, 408–423.

Agnihotri, A., & Aruoma, O. I. (2020). Alzheimer's disease and Parkinson's disease: A nutritional toxicology perspective of the impact of oxidative stress, mitochondrial dysfunction, nutrigenomics and environmental chemicals. *Journal of the American College of Nutrition*, *39*, 16–27.

Ahmad, R., Liu, G., Rehman, S. A. U., Fazal, R., Gao, Y., Xu, D., Agostinho, F., Almeida, C. M., & Giannetti, B. F. (2025). Pakistan road towards Paris Agreement: Potential decarbonization pathways and future emissions reduction by a developing country. *Energy, 314*, 134075.

Akbar, A., Majid, M. A. B. A., & Mokhtar, A. A. (2021). Life cycle costing of thermal energy storage system using CAPEX and OPEX. *Journal of Mechanical Science and Technology, 35*, 3223–3229.

Alcántara, C. M., & Alcántara, A. R. (2018). Biocatalyzed synthesis of antidiabetic drugs: A review. *Biocatalysis and Biotransformation, 36*, 12–46.

Alsmadi, A. A., Al-Okaily, M., Alrawashdeh, N., Al-Gasaymeh, A., Moh'd Al-Hazimeh, A., & Zakari, A. (2023). A bibliometric analysis of green bonds and sustainable green energy: Evidence from the last fifteen years (2007–2022). *Sustainability, 15*, 5778.

Atsbha, T. A., Yoon, H.J., Cherif, A., & Lee, C.J. (2024). Feasibility and sustainability of emerging CCU pathways for formic acid production. *Green Chemistry, 26*, 10982–10995.

Aziz, S., Ahmed, I., Khan, K., & Khalid, M. (2024). Emerging trends and approaches for designing net-zero low-carbon integrated energy networks: A review of current practices. *Arabian Journal for Science and Engineering, 49*, 6163–6185.

Barahmand, Z., & Eikeland, M. S. (2022). Techno-economic and life cycle cost analysis through the lens of uncertainty: A scoping review. *Sustainability, 14*, 12191.

Bauer, F., Kulionis, V., Oberschelp, C., Pfister, S., Tilsted, J. P., Finkill, G. D., & Fjäll, S. (2022). *Petrochemicals and climate change: Tracing globally growing emissions and key blind spots in a fossil-based industry.* Lund University.

Baumann, M., & Baxendale, I. R. (2015). The synthesis of active pharmaceutical ingredients (APIs) using continuous flow chemistry. *Beilstein Journal of Organic Chemistry, 11*, 1194–1219.

Bonini, S., & Capizzi, V. (2019). The role of venture capital in the emerging entrepreneurial finance ecosystem: Future threats and opportunities. *Venture Capital, 21*, 137–175.

Buchner, G. A., Stepputat, K. J., Zimmermann, A. W., & Schomäcker, R. (2019). Specifying technology readiness levels for the chemical industry. *Industrial & Engineering Chemistry Research, 58*, 6957–6969.

Chen, T.L., Kim, H., Pan, S.Y., Tseng, P.C., Lin, Y.P., & Chiang, P.C. (2020). Implementation of green chemistry principles in circular economy system towards sustainable development goals: Challenges and perspectives. *Science of the Total Environment, 716*, 136998.

Clark, T. J., & Pasternak, A. S. (2021). *How to commercialize chemical technologies for a sustainable future.* Wiley Online Library.

Clavé, N., & Estécahandy, M. (2023). Optimizing the balance between production availability, CAPEX and OPEX of projects while managing environmental aspects with a simulation tool. In *Abu Dhabi International Petroleum Exhibition and Conference* (SPE, D041S130R001).

Dai, H., Li, N., Wang, Y., & Zhao, X. (2022). The analysis of three main investment criteria: NPV, IRR and payback period. In *2022 7th International Conference on Financial Innovation and Economic Development (ICFIED 2022)* (pp. 185–189). Atlantis Press. Series Editor(s): Andrea Appolloni, Francesco Caracciolo, Zhuoqi Ding, Periklis Gogas, Gordon Huang, Gilbert Nartea, Thanh Ngo, Wadim Striełkowski, Seema Joshi.

Das, A. K., Chakraborty, R., & de la Guardia, M. (2019). Teaching green analytical and synthesis chemistry: Performing laboratory experiments in a greener way. In *Green analytical chemistry: Past, present and perspectives.* Springer.

De Castro Ribeiro, L. F. (2023). *The enhanced UN Global Compact communication on progress: A new tool for sustainability benchmarking*(Master's thesis). Universidade do Porto, Portugal.

Devierno Kreuder, A., House-Knight, T., Whitford, J., Ponnusamy, E., Miller, P., Jesse, N., Rodenborn, R., Sayag, S., Gebel, M., &Aped, I. (2017). A method for assessing greener alternatives between chemical products following the 12 principles of green chemistry. *ACS Sustainable Chemistry & Engineering*, *5*, 2927–2935.

Dhayalan, M., Denison, M. I. J., Jegadeeshwari, A. L., Krishnan, K., & Gandhi, N. N. (2017). In vitro antioxidant, antimicrobial, cytotoxic potential of gold and silver nanoparticles prepared using *Embelia ribes*. *Natural Product Research*, *31*(4), 465–468. https://doi.org/10.1080/14786419.2016.1166499

Digitemie, W. N., & Ekemezie, I. O. (2024). Assessing the role of carbon pricing in global climate change mitigation strategies. *Magna Scientia Advanced Research and Reviews*, *10*, 022–031.

Dovidonis, I. (2024). *Green financing: The role in ensuring sustainable economic growth in Europe*(Master's thesis). Vilniaus Universitetas.

Farza, K., Ftiti, Z., Hlioui, Z., Louhichi, W., & Omri, A. (2021). Does it pay to go green? The environmental innovation effect on corporate financial performance. *Journal of Environmental Management*, *300*, 113695.

Fenibo, E. O., Ijoma, G. N., Nurmahomed, W., & Matambo, T. (2022). The potential and green chemistry attributes of biopesticides for sustainable agriculture. *Sustainability*, *14*, 14417.

Ferrari, A., & Buscemi, A. (2023). Sustainability metrics and green finance for chemicals. *Journal of Applied Finance & Banking*, *13*(6), 243–255.

Franz, A. W., Buchholz, S., Albach, R. W., & Schmid, R. (2024). Towards greener polymers: Trends in the German chemical industry. *Green Carbon*, *2*, 33–44.

Galera-Quiles, M. D. C., Piedra-Muñoz, L., Galdeano-Gómez, E., & Carreño-Ortega, A. (2021). A review of eco-innovations and exports interrelationship, with special reference to international agrifood supply chains. *Sustainability*, *13*, 1378.

Giacomella, L. (2021). Techno-economic assessment (TEA) and life cycle costing analysis (LCCA): Discussing methodological steps and integrability. *Insights into Regional Development*, *3*, 176–197.

Gorshkov, A., Vatin, N., Rymkevich, P., & Kydrevich, O. (2018). Payback period of investments in energy saving. *Magazine of Civil Engineering*, 65–75.

Green, S. A. (2016). Green chemistry: Progress and barriers. *Physical Sciences Reviews*, *1*, 20160072.

Gude, V. G., & Martinez-Guerra, E. (2018). Green chemistry with process intensification for sustainable biodiesel production. *Environmental Chemistry Letters*, *16*, 327–341.

Hallett, J. P. (2025). Green chemistry startups: Some lessons learned. *Green Chemistry*, *27*, 403–412.

Hami, N., Muhamad, M. R., & Ebrahim, Z. (2015). The impact of sustainable manufacturing practices and innovation performance on economic sustainability. *Procedia CIRP*, *26*, 190–195.

Hessel, V., Tran, N. N., Asrami, M. R., Tran, Q. D., Long, N. V. D., Escribà-Gelonch, M., Tejada, J. O., Linke, S., & Sundmacher, K. (2022). Sustainability of green solvents–Review and perspective. *Green Chemistry*, *24*, 410–437.

Idoko, F. A., Ezeamii, G. C., & Ojochogwu, O. J. (2024). Green chemistry in manufacturing: Innovations in reducing environmental impact. *World Journal of Advanced Research and Reviews*, *23*, 2826–2841.

Iglesias, J., Martínez-Salazar, I., Maireles-Torres, P., Alonso, D. M., Mariscal, D. R., & Granados, M. L. (2020). Advances in catalytic routes for the production of carboxylic acids from biomass: A step forward for sustainable polymers. *Chemical Society Reviews*, *49*, 5704–5771.

Ivanković, A., Dronjić, A., Bevanda, A. M., & Talić, S. (2017). Review of 12 principles of green chemistry in practice. *International Journal of Sustainable and Green Energy*, *6*, 39–48.

Ivanović, T., Hischier, R., & Som, C. (2021). Bio-based polyester fiber substitutes: From GWP to a more comprehensive environmental analysis. *Applied Sciences*, *11*, 2993.

Jensen, P. D., Basson, L., Hellawell, E. E., Bailey, M. R., & Leach, M. (2011). Quantifying 'geographic proximity': Experiences from the United Kingdom's national industrial symbiosis programme. *Resources, Conservation and Recycling*, *55*, 703–712.

Jessop, P. G., & Reyes, L. M. (2018). Green Centre Canada: An experimental model for green chemistry commercialization. *Physical Sciences Reviews*, *3*, 20170189.

Kohli, K., Prajapati, R., & Sharma, B. K. (2019). Bio-based chemicals from renewable biomass for integrated biorefineries. *Energies*, *12*, 233.

Krasnodębski, M. (2024). The bumpy road to sustainability: Reassessing the history of the twelve principles of green chemistry. *Studies in History and Philosophy of Science*, *103*, 85–94.

Krimsky, S. (2017). The unsteady state and inertia of chemical regulation under the US Toxic Substances Control Act. *PLOS Biology*, *15*, e2002404.

Lancaster, M. (2025). *Green chemistry: An introductory text.* Royal Society of Chemistry.

Lane, M. K. M., Rudel, H. E., Wilson, J. A., Erythropel, H. C., Backhaus, A., Gilcher, E. B., Ishii, M., Jean, C. F., Lin, F., & Muellers, T. D. (2023). Green chemistry as just chemistry. *Nature Sustainability*, *6*, 502–512.

Li, Y., Zhang, Y., &Solangi, Y. A. (2023). Assessing ESG factors and policies of green finance investment decisions for sustainable development in China using the fuzzy AHP and fuzzy DEMATEL. *Sustainability*, *15*, 15214.

Lozano, F. J., Lozano, R., Freire, P., Jiménez-Gonzalez, C., Sakao, T., Ortiz, M. G., Trianni, A., Carpenter, A., & Viveros, T. (2018). New perspectives for green and sustainable chemistry and engineering: Approaches from sustainable resource and energy use, management, and transformation. *Journal of Cleaner Production*, *172*, 227–232.

Moussa, A. (2022). Technology readiness level framework for the chemical industry. In *Systematic process development: From idea to value via technology readiness levels.* ACS Publications.

Nanda, B., Sailaja, M., Mohapatra, P., Pradhan, R., & Nanda, B. B. (2021). Green solvents: A suitable alternative for sustainable chemistry. *Materials Today: Proceedings*, *47*, 1234–1240.

Ncube, A., Mtetwa, S., Bukhari, M., Fiorentino, G., & Passaro, R. (2023). Circular economy and green chemistry: The need for radical innovative approaches in the design for new products. *Energies*, *16*, 1752.

Nguyen, C., Hoang, K., & Gan, C. E. (2024). *Fintech and green investment: Transforming challenges into opportunities.* World Scientific.

Nikookaran, N., Karakostas, G., & Todd, T. D. (2017). Combining capital and operating expenditure costs in vehicular roadside unit placement. *IEEE Transactions on Vehicular Technology*, *66*, 7317–7331.

Ogbuagu, O. O., Mbata, A. O., Balogun, O., Oladapo, O., Ojo, O. O., & Muonde, M. (2024). Sustainable pharmaceutical supply chains: Green chemistry approaches to drug production and distribution. *IRE Journals*, *8*, 761–767.

Padulles, R., Kountouris, I., Kanbur, B. B., Jensen, J. K., & Elmegaard, B. (2025). Break-even coefficient of performance: Scenario-based analysis of industrial heat electrification feasibility in Europe. *Energy Conversion and Management*, *326*, 119488.

Patel, A., Patel, S., Mehta, M., Patel, Y., Patel, R., Shah, D., Patel, D., Shah, U., Patel, M., & Patel, S. (2022). A review on synthetic investigation for quinoline—Recent green approaches. *Green Chemistry Letters and Reviews*, *15*, 337–372.

Plass, C., Adebar, N., Hiessl, R., Kleber, J., Grimm, A., Langsch, A., Otter, R., Liese, A., & Gröger, H. (2021). Structure-performance guided design of sustainable plasticizers from biorenewable feedstocks. *European Journal of Organic Chemistry*, *2021*, 6086–6096.

Prajapati, P., Ahir, H., Prajapati, B., & Shah, S. (2023). Chemometric-based AQbD and green chemistry approaches to chromatographic analysis of remogliflozin etabonate and vildagliptin. *Journal of AOAC International, 106*, 239–249.

Pratama, P. D. A., Padang, R. R., Pratiwi, N. M. F., & Narartha, K. S. (2025). Brand image as a mediator in the relationship between green marketing and purchase decision at IKEA company. In *Proceedings of TEAMS: The International Conference on Tourism, Economic, Accounting, Management and Social Science* (pp. 224–236).

Prokhorenko, O., & Kokoshynska, Y. (2024). Integrated CSR management model for the pharmaceutical industry: A framework for sustainable impact. Труди, 27–30.

Raman, R., Sreenivasan, A., Ma, S., Patwardhan, A., & Nedungadi, P. (2023). Green supply chain management research trends and linkages to UN sustainable development goals. *Sustainability, 15*, 15848.

Ratti, R. (2020). Industrial applications of green chemistry: Status, challenges and prospects. *SN Applied Sciences, 2*, 263.

Ren, L., Wang, J., Zhang, L., Hu, X., Ning, Y., Cong, J., Li, Y., Zhang, W., Xu, T., & Shi, X. (2025). Quantitative assessment of the carbon border adjustment mechanism: Impacts on China–EU trade and provincial-level vulnerabilities. *Sustainability, 17*, 1699.

Renné, D. S. (2022). Progress, opportunities and challenges of achieving net-zero emissions and 100% renewables. *Solar Compass, 1*, 100007.

Robinson, J., & Dow, S. (2021). *Economic philosophy*. Routledge.

Roschangar, F., Sheldon, R. A., & Senanayake, C. H. (2015). Overcoming barriers to green chemistry in the pharmaceutical industry – The green aspiration level™ concept. *Green Chemistry, 17*, 752–768.

Ruiz-Mercado, G. J., Carvalho, A., & Cabezas, H. (2016). Using green chemistry and engineering principles to design, assess, and retrofit chemical processes for sustainability. *ACS Sustainable Chemistry & Engineering, 4*, 6208–6221.

Sandoval, D. P. (2024). *Environmental impact extrapolation of environmental processes at low technology readiness level (TRL): Application to microalgae-based processes* (Doctoral dissertation). Université Côte d'Azur.

Schwager, P., Decker, N., & Kaltenegger, I. (2016). Exploring green chemistry, sustainable chemistry and innovative business models such as chemical leasing in the context of international policy discussions. *Current Opinion in Green and Sustainable Chemistry, 1*, 18–21.

Sgroi, F., Foderà, M., Di Trapani, A. M., Tudisca, S., & Testa, R. (2015). Cost-benefit analysis: A comparison between conventional and organic olive growing in the Mediterranean area. *Ecological Engineering, 82*, 542–546.

Sharma, S., Dutta, S., & Sharma, R. K. (2018). Recyclability of reagents. In *Hazardous reagent substitution* (Chapter 2). Royal Society of Chemistry. https://doi.org/10.1039/9781782623847-00018

Sharma, S., Gallou, F., & Handa, S. (2024). Towards a sustainable tomorrow: Advancing green practices in organic chemistry. *Green Chemistry, 26*, 6289–6317.

Sheldon, R. A. (2014). Green and sustainable manufacture of chemicals from biomass: State of the art. *Green Chemistry, 16*, 950–963.

Sheldon, R. A. (2016a). Engineering a more sustainable world through catalysis and green chemistry. *Journal of the Royal Society Interface, 13*, 20160087.

Sheldon, R. A. (2016b). Green chemistry and resource efficiency: Towards a green economy. *Green Chemistry, 18*, 3180–3183.

Sheldon, R. A., & Brady, D. (2022). Green chemistry, biocatalysis, and the chemical industry of the future. *ChemSusChem, 15*, e202102628.

Singh, A., & Trivedi, A. (2016). Sustainable green supply chain management: Trends and current practices. *Competitiveness Review, 26*, 265–288.

Souza, A. S. D., Ferreira, P. G., Jesus, I. S. D., Carvalho, A. S. D., Futuro, D. O., & Ferreira, V. F. (2024). Sustainability, sustainable development, sustainable chemistry, and green chemistry: Parallels and interconnections. Química Nova, *47*, e-20240049.

Sovacool, B. K., Newell, P., Carley, S., & Fanzo, J. (2022). Equity, technological innovation and sustainable behaviour in a low-carbon future. *Nature Human Behaviour, 6*, 326–337.

Sundqvist, S. (2022). *EU taxonomy and sustainable finance: Key levers for climate change mitigation*(Master's thesis). Turku.

Tahir, M. M., Abbas, A., & Dickson, R. (2025). Green hydrogen and chemical production from solar energy in Pakistan: A geospatial, techno-economic, and environmental assessment. *International Journal of Hydrogen Energy*, *116*, 613–626.

Tan, R., Aviso, K. B., & Ng, D. (2019). Optimization models for financing innovations in green energy technologies. *Renewable and Sustainable Energy Reviews*, *113*, 109258.

Thormann, L., Neuling, U., & Kaltschmitt, M. (2021). Opportunities and challenges of the European Green Deal for the chemical industry: An approach measuring innovations in bioeconomy. *Resources*, *10*, 91.

Tiwari, A. K., Kumar, S., & Abakah, E. J. A. (2025). Correlation and price spillover effects among green assets. *Annals of Operations Research*, *347*, 419–444.

Van Berkel, R. (2006). Regional resource synergies for sustainable development in heavy industrial areas: An overview of opportunities and experiences. *Curtin University of Technology*, 135–139.

Van Fan, Y., Klemeš, J. J., Tan, R. R., & Varbanov, P. S. (2019). Graphical break-even based decision-making tool (BBDM) to minimise GHG footprint of biomass utilisation: Biochar by pyrolysis. *Chemical Engineering Transactions*, *76*.

Vartiainen, E., Masson, G., Breyer, C., Moser, D., & Román Medina, E. (2020). Impact of weighted average cost of capital, capital expenditure, and other parameters on future utility-scale PV levelised cost of electricity. *Progress in Photovoltaics: Research and Applications*, *28*, 439–453.

Veleva, V. R., & Cue, B. W.Jr. (2019). The role of drivers, barriers, and opportunities of green chemistry adoption in the major world markets. *Current Opinion in Green and Sustainable Chemistry*, *19*, 30–36.

Venkatesan, K., Sundarababu, J., & Anandan, S. S. (2024). The recent developments of green and sustainable chemistry in multidimensional way: Current trends and challenges. *Green Chemistry Letters and Reviews*, *17*, 2312848.

Walker, A. (2020). Future regulatory, market and technology trends in the global passenger car and commercial vehicle sectors: Key challenges to achieving net zero emissions for road vehicles. *Johnson Matthey Technology Review*, *64*, 320–337.

Wang, X., & Wang, Q. (2021). Research on the impact of green finance on the upgrading of China's regional industrial structure from the perspective of sustainable development. *Resources Policy*, *74*, 102436.

Wang, Y., Zhao, N., Lei, X., & Long, R. (2021). Green finance innovation and regional green development. *Sustainability*, *13*, 8230.

White, R., Marzano, M., Fesenko, E., Inman, A., Jones, G., Agstner, B., & Mumford, R. (2022). Technology development for the early detection of plant pests: A framework for assessing technology readiness levels (TRLs) in environmental science. *Journal of Plant Diseases and Protection*, *129*, 1249–1261.

Wijaya, D. (2025). *Comparative policy frameworks for promoting organic farming and biodiversity conservation: Case studies from emerging and developed economies. Studies in Knowledge Discovery, Intelligent Systems, and Distributed Analytics*, *15*, 13–20.

William, R., Goodwell, A., Richardson, M., Le, P. V., Kumar, P., & Stillwell, A. S. (2016). An environmental cost-benefit analysis of alternative green roofing strategies. *Ecological Engineering*, *95*, 1–9.

Winterton, N. (2021). The green solvent: A critical perspective. *Clean Technologies and Environmental Policy, 23*, 2499–2522.

Xia, Z., Jia, G., Ma, Z., Wang, J., Zhang, Y., & Jin, L. (2021). Analysis of economy, thermal efficiency and environmental impact of geothermal heating system based on life cycle assessments. *Applied Energy, 303*, 117671.

Xie, Z., Wang, J., & Zhao, G. (2022). Impact of green innovation on firm value: Evidence from listed companies in China's heavy pollution industries. *Frontiers in Energy Research, 9*, 806926.

Yadav, A., & Dubey, A. (2024). A review on green chemistry and its applications. *Asian Journal of Pharmaceutical Research and Development, 12*, 35–41.

Yin, H.T., Chang, C.P., & Wang, H. (2022). The impact of monetary policy on green innovation: Global evidence. *Technological and Economic Development of Economy, 28*, 1933–1953.

Zhang, Y. (2024). A review on the economic impact of green manufacturing practices. *Research Journal in Civil, Industrial and Mechanical Engineering, 1*, 01–09.

Zhang, Y., Liu, F., Zhang, J., Gao, W., Yu, J., & Wang, L. (2024). Life cycle thinking-based greenness framework for advancing green chemistry: Case study with typical ionic liquids for cellulose dissolution and regeneration. *ACS Sustainable Chemistry & Engineering, 12*, 6400–6411.

Zhao, Y., Pohl, O., Bhatt, A. I., Collis, G. E., Mahon, P. J., Rüther, T., & Hollenkamp, A. F. (2021). A review on battery market trends, second-life reuse, and recycling. *Sustainable Chemistry, 2*, 167–205.

Zuin, V. G., Eilks, I., Elschami, M., & Kümmerer, K. (2021). Education in green chemistry and in sustainable chemistry: Perspectives towards sustainability. *Green Chemistry, 23*, 1594–1608.

Index

For Product Safety Concerns and Information please contact our EU representative GPSR@taylorandfrancis.com
Taylor & Francis Verlag GmbH, Kaufingerstraße 24, 80331 München, Germany

www.ingramcontent.com/pod-product-compliance
Ingram Content Group UK Ltd.
Pitfield, Milton Keynes, MK11 3LW, UK
UKHW022218190926
475702UK00003B/432

* 9 7 8 1 0 4 1 0 7 8 5 7 9 *